Nucleic Acid-Based Nanomaterials

Nucleic Acid-Based Nanomaterials

Stabilities and Applications

Edited by Yunfeng Lin and Shaojingya Gao

WILEY-VCH

Editors

Prof. Yunfeng Lin
Sichuan University
West China College of Stomatology
State Key Laboratory of Oral Diseases
No. 14, 3rd Section of Ren Min Nan Road
Chengdu 610041
PR China

Prof. Shaojingya Gao
Sichuan University
West China College of Stomatology
State Key Laboratory of Oral Diseases
No. 14, 3rd Section of Ren Min Nan Road
Chengdu 610041
PR China

Cover Image: © CHRISTOPH BURGSTEDT/SCIENCE PHOTO LIBRARY/Getty Images

All books published by **WILEY-VCH** are carefully produced. Nevertheless, authors, editors, and publisher do not warrant the information contained in these books, including this book, to be free of errors. Readers are advised to keep in mind that statements, data, illustrations, procedural details or other items may inadvertently be inaccurate.

Library of Congress Card No.: applied for

British Library Cataloguing-in-Publication Data
A catalogue record for this book is available from the British Library.

Bibliographic information published by the Deutsche Nationalbibliothek
The Deutsche Nationalbibliothek lists this publication in the Deutsche Nationalbibliografie; detailed bibliographic data are available on the Internet at <http://dnb.d-nb.de>.

© 2024 WILEY-VCH GmbH, Boschstraße 12, 69469 Weinheim, Germany

All rights reserved (including those of translation into other languages). No part of this book may be reproduced in any form – by photoprinting, microfilm, or any other means – nor transmitted or translated into a machine language without written permission from the publishers. Registered names, trademarks, etc. used in this book, even when not specifically marked as such, are not to be considered unprotected by law.

Print ISBN: 978-3-527-35205-0
ePDF ISBN: 978-3-527-84189-9
ePub ISBN: 978-3-527-84190-5
oBook ISBN: 978-3-527-84191-2

Typesetting Straive, Chennai, India
Printing and Binding CPI Group (UK) Ltd, Croydon, CR0 4YY

Contents

Preface *xi*

1 Introductions of Nucleic Acid-Based Nanomaterials *1*
Shaojingya Gao and Yunfeng Lin
1.1 History of DNA-Based Nanomaterials – Design and Construction *3*
1.1.1 DNAzymes *4*
1.1.2 Aptamers *5*
1.1.3 Triplex DNA *5*
1.1.4 DNA Origami and DNA Tiles *6*
References *9*

2 The Methods to Improve the Stability of Nucleic Acid-Based Nanomaterials *15*
Xueping Xie
2.1 Introduction *15*
2.2 Methods to Improve Stability *16*
2.2.1 Artificial Nucleic Acids *17*
2.2.2 Backbone Modification of Nucleic Acids *18*
2.2.2.1 Phosphate Group Modifications *18*
2.2.2.2 Nucleobase or Ribose Modifications *19*
2.2.3 Coating with Protective Structures *20*
2.2.4 Covalent Crosslinking *22*
2.2.5 Tuning Buffer Conditions *23*
2.2.6 Construction of Novel NAN *26*
2.3 Conclusion and Recommendations *26*
References *27*

3 Framework Nucleic Acid-Based Nanomaterials: A Promising Vehicle for Small Molecular Cargos *37*
Yanjing Li
3.1 Basis of FNAs as Potential Drug Carriers *38*
3.1.1 Classification and Construction of FNAs *38*
3.1.2 Physical and Chemical Properties *40*

3.1.3	Biological Properties	*40*
3.2	Small-molecule Cargos	*41*
3.2.1	Antitumor Agents	*42*
3.2.1.1	Chemotherapeutic Drugs	*42*
3.2.1.2	Phototherapeutic Agents	*43*
3.2.2	Antibiotic Agents	*44*
3.2.3	Phytochemicals	*45*
3.3	Merits of FNA Delivery Systems in Biomedical Application	*46*
3.3.1	Efficient Drug Delivery	*46*
3.3.2	Targeted Drug Delivery	*46*
3.3.3	Controlled Drug Release	*49*
3.3.4	Overcoming Drug Resistance	*51*
3.4	Conclusions and Prospects	*52*
	References	*54*

4 The Application of Framework Nucleic Acid-Based Nanomaterials in the Treatment of Mitochondrial Dysfunction *61*
Lan Yao and Tao Zhang

4.1	Introduction	*61*
4.2	Treatment Mechanisms in Mitochondrial Dysfunction	*61*
4.2.1	Treating in mtDNA	*62*
4.2.1.1	Clearing Mutations	*62*
4.2.1.2	Inhibiting Replication	*64*
4.2.2	Treating in mRNA, tRNA, and rRNA	*64*
4.2.2.1	Increase Normal RNA	*64*
4.2.2.2	Silencing Abnormal RNA	*64*
4.2.2.3	Treating in Noncoding RNA	*65*
4.3	Nucleic Acid Nanomaterial-Based Delivery System in Mitochondrial Treatment	*65*
4.3.1	Cell and Mitochondria Targeting	*66*
4.3.1.1	Cell Targeting	*66*
4.3.1.2	Mitochondria Targeting	*66*
4.3.2	Framework Nucleic Acid-Based Delivery System in Mitochondria Treatment	*68*
4.4	Challenges and Prospectives	*71*
	Funding	*72*
	References	*72*

5 Regeneration of Bone-Related Diseases by Nucleic Acid-Based Nanomaterials: Perspectives from Tissue Regeneration and Molecular Medicine *81*
Xiaoru Shao

5.1	Introduction	*81*
5.2	The Development Process of Functional Nucleic Acid	*82*
5.2.1	DNA Tile	*83*

5.2.2	DNA Origami	*83*
5.2.3	Three-dimensional DNA Self-assembly	*83*
5.2.4	DNA Nanobots and DNA Microchips	*84*
5.3	Nucleic Acid-Based Functional Nanomaterials	*84*
5.3.1	Nanomaterials That Can Bind to Functional Nucleic Acids	*84*
5.3.1.1	Metal-Based Nanomaterials	*84*
5.3.1.2	Carbon-Based Nanomaterials	*85*
5.3.1.3	Bionanomaterials	*86*
5.3.1.4	Quantum Dots	*86*
5.3.1.5	Magnetic Nanomaterials	*86*
5.3.1.6	Composite Nanomaterials	*87*
5.3.2	Combination of Functional Nucleic Acids and Nanomaterials	*87*
5.4	Multiple Roles of Nucleic Acid-Based Functional Nanomaterials in Bone Tissue Repair and Regeneration	*89*
5.4.1	Sustained Release	*89*
5.4.2	Bone Targeting	*91*
5.4.3	Scaffold Material for Bone Regeneration	*92*
5.4.4	Bioimaging of Bone Tissue Regeneration	*93*
5.5	Conclusion and Perspectives	*94*
	References	*94*

6	***In Situ* Fluorescence Imaging and Biotherapy of Tumor Based on Hybridization Chain Reaction**	**101**
	Ye Chen, Songhang Li, and Taoran Tian	
6.1	Hybridization Chain Reaction	*102*
6.2	Nucleic Acid Detection	*102*
6.2.1	miRNA Detection	*103*
6.2.1.1	Autocatalytic HCR Biocircuit	*103*
6.2.1.2	Nonlinear HCR System	*104*
6.2.2	Single-Nucleotide Variants Detection	*105*
6.3	Protein Detection	*107*
6.3.1	Antibody-Based HCR System	*107*
6.3.2	Aptamer-Based HCR System	*108*
6.4	Multiple Target Detection	*109*
6.4.1	Combined HCR-Based Probe	*109*
6.4.2	HCR-Based Logic Gate	*110*
6.5	HCR-Based Assembly Nanoplatforms	*113*
6.6	HCR-Based Tumor Biotherapy	*115*
6.6.1	Chemotherapy	*115*
6.6.2	Photodynamic Therapy	*115*
6.6.3	RNA Interfering Therapy	*116*
6.7	Conclusion	*116*
	References	*116*

7	**Application and Prospects of Framework Nucleic Acid-Based Nanomaterials in Tumor Therapy** *123*	
	Tianyu Chen and Xiaoxiao Cai	
7.1	Development of Nucleic Acid Nanomaterials *124*	
7.2	Properties and Applications of Nucleic Acid Nanomaterials *125*	
7.2.1	tFNAs *125*	
7.2.2	DNA Origami *127*	
7.2.3	Dynamic DNA Nanostructure *130*	
7.3	Conclusion *133*	
	References *133*	

8	**Application of Framework Nucleic Acid-Based Nanomaterials in the Treatment of Endocrine and Metabolic Diseases** *139*	
	Jingang Xiao	
8.1	Endocrine and Metabolic Diseases *139*	
8.2	Nucleic Acid Nanomaterials *141*	
8.3	Nucleic Acid and Drugs *141*	
8.4	Nucleic Acid Nanomaterials for Endocrine and Metabolic Diseases *144*	
8.4.1	Diabetes Mellitus *144*	
8.4.2	Osteoporosis *146*	
8.4.3	Obesity *147*	
8.4.4	Nonalcoholic Fatty Liver Disease *148*	
8.5	Conclusion and Outlook *149*	
	References *151*	

9	**The Antibacterial Applications of Framework Nucleic Acid-Based Nanomaterials: Current Progress and Further Perspectives** *161*	
	Zhiqiang Liu and Yue Sun	
9.1	Some Advantages of DNA Nanostructures in the Antibacterial Field *163*	
9.1.1	Compatibility of DNA Nanostructures *163*	
9.1.2	Stability of DNA Nanostructures *163*	
9.1.3	Editability of DNA Nanostructures *163*	
9.1.4	Drug-loading Performance of DNA Nanostructures *164*	
9.2	Application of 2D Nanostructures in the Antibacterial Field *164*	
9.2.1	Five "Holes" DNA Nanostructure *164*	
9.2.2	Super Silver Nanoclusters Based on Branched DNA *164*	
9.2.3	Melamine-DNA-AgNC Complex *165*	
9.2.4	NET-like Nanogel Based on 2D DNA Networks *166*	
9.2.5	ε-poly-L-lysine-DNA Nanocomplex *166*	
9.3	Application of 3D DNA Nanostructures in the Antibacterial Field *166*	
9.3.1	Tetrahedral Framework DNA *166*	
9.3.1.1	Delivery of Traditional Antibiotics Based on Tetrahedral Framework DNA *168*	
9.3.1.2	Delivery of Nucleic Acid Antibiotics Based on Tetrahedral Framework DNA *168*	

9.3.1.3	Delivery of Polypeptide Antibiotics Based on Tetrahedral Framework DNA *169*	
9.3.2	DNA Six-Helix Bundle *169*	
9.3.3	DNA Nanoribbon *170*	
9.3.4	DNA Pom-Pom Nanostructure *170*	
9.4	Application of DNA Hydrogel Nanostructures in the Antibacterial Field *170*	
9.5	Challenges and Further Perspectives *172*	
	References *174*	

10 Framework Nucleic Acid Nanomaterial-Based Therapy for Osteoarthritis: Progress and Prospects *181*
Yangxue Yao, Hongxiao Huang, and Sirong Shi

10.1	Introduction *181*	
10.2	Pathology of OA *181*	
10.3	Risk Factors of OA *183*	
10.4	Challenges for OA Therapy *183*	
10.5	Nucleic Acid Nanomaterial-Based Therapy for OA *184*	
10.5.1	Vector-Independent Nucleic Acid Nanomaterials for OA Therapy *184*	
10.5.1.1	Tetrahedral Framework Nucleic Acids (tFNAs) *184*	
10.5.1.2	Antisense Oligonucleotides (ASOs) *187*	
10.5.1.3	Aptamers *187*	
10.5.2	Vector-Dependent Nucleic Acid Nanomaterials for OA Therapy *188*	
10.5.2.1	MicroRNA (miRNA) Mimics *188*	
10.5.2.2	Small Interfering RNA (siRNA) *188*	
10.5.2.3	cDNA *192*	
10.5.2.4	mRNA *192*	
10.5.2.5	Circular RNA (CircRNA) *192*	
10.5.3	Nucleic Acid Nanomaterials as Carriers for OA Therapy *194*	
10.6	Conclusion and Prospects *194*	
	References *195*	

Index *205*

Preface

Here, the thematic issue is dedicated to further introducing *Nucleic Acid-Based Nanomaterials: Stabilities and Applications*. Several review chapters provided by researchers focus on biostability and the applications of nucleic acid nanomaterials in tissue engineering, antimicrobial therapy, disease treatment, and medicine delivery. We truly expect the book will attract experts and students in the field.

Nucleic acid nanomaterials, with the goal of building next-generation biomaterials, combine the advantages of nucleic acid and nanomaterials. The interactions between nucleic acid and nanomaterials have established themselves as hot research areas, where target recognition, response, and self-assembly ability, combined with stability, stimuli response, and delivery potentials, give rise to a variety of novel and fascinating applications.

The self-assembled nucleic acid nanomaterials are programmable, intelligent, biocompatible, non-immunogenic, and noncytotoxic. The stability of nucleic acid nanomaterials is often affected by cation concentrations, enzymatic degradation, and organic solvents. To deal with this problem, a lot of methods have been attempted to improve the stability of nucleic acid nanomaterials, including artificial nucleic acids, modification with specific groups, encapsulation with protective structures.

Mitochondrial dysfunction is considered highly related to the development and progression of diseases. We discuss the challenges and opportunities of nucleic acid nanomaterial delivery systems in mitochondrial dysfunction. Nucleic acid nanomaterials provide a promising approach for small-molecule delivery. We review and discuss the advantages, applications, and current challenges of tFNAs for the delivery of small molecular cargo. The good designability, biocompatibility, designable responsiveness, biodegradability, and mechanical strength provided by DNA building blocks facilitate the application of DNA hydrogels in cytoscaffolds, drug delivery systems, immunotherapeutic carriers, biosensors, and nanozyme-protected scaffolds. We provide an overview of the main classification and synthesis methods of DNA hydrogels and highlight the application of DNA hydrogels in biomedical fields. Because of their high biocompatibility and editability, nucleic acid nanomaterials are frequently employed in disease diagnosis and therapy, including the treatment of metabolic diseases, severe bacterial infections, osteoarthritis, and autoimmune diseases.

1

Introductions of Nucleic Acid-Based Nanomaterials

Shaojingya Gao and Yunfeng Lin

Sichuan University, West China College of Stomatology, State Key Laboratory of Oral Diseases, No. 14, 3rd Sec, Ren Min Nan Road, Chengdu 610041, PR China

Nanotechnology is a science and technology that produces substances from a single atom or molecule in a size range of 1–100 nm. As early as 1986, American scientists put forward nanotechnology in the creation of machines, but due to the low level of science and technology at that time, the technology did not achieve obvious results. Researchers believe that nanotechnology is to make the combination of molecules in a machine practical, so as to arbitrarily combine all kinds of molecules to produce different molecular structures. Nanotechnology has been extensively applied in various fields and has had a profound impact on our lives. The concept of nanotechnology was first introduced to the public in the speech "There's Plenty of Room at the Bottom" in 1960 by Nobel Prize laureate Richard P. Feynman. With the continuous development of science and technology, people's research on nanotechnology is also in-depth, and the corresponding branch of the subject has also developed. Nanotechnology integrates quantum mechanics, molecular biology, nanobiology, nanochemistry, and other disciplines with the ultimate goal of directly constructing products with specific functions from atoms or molecules. Nanotechnology transformed the drug delivery system dramatically by delivering microtherapeutic drugs to parts of the body that are difficult to reach otherwise. With an expanding array of strategies that allow nanomaterials to tailor their properties to specific indications, nanomaterials are entering the clinic at an unprecedented rate [1]. In nanotechnology, nanomaterials are often defined as the creation of materials with new properties and functions at the nanoscale. There are two main approaches of constructing nanomaterials so far: the "top-down" approach is to reduce the size of large structures to nanoscale, while the "bottom-up" approach, which is also called "molecular nanotechnology," is to engineer materials from molecular or atom components through assembly or self-assembly. In 1953, the discovery of deoxyribonucleic acid (DNA)'s complementary base pairing principle and double helix structure ushered in the era of molecular biology. DNA became a potential material for nanofabrication due to its unique properties and high controllability [2–4]. Seeman and his coworkers first reported the synthesis rules of DNA-based nanomaterials in the 1980s, bringing DNA nanotechnology into the limelight as a research hotspot. The most common approach to building DNA-based nanomaterials is the "bottom-up"

Nucleic Acid-Based Nanomaterials: Stabilities and Applications, First Edition.
Edited by Yunfeng Lin and Shaojingya Gao.
© 2024 WILEY-VCH GmbH. Published 2024 by WILEY-VCH GmbH.

approach. In recent years, DNA-based nanomaterials have been widely used in bioimaging, biosensing, gene transfer, drug delivery, disease diagnosis, and treatment due to their inherent biodegradability and biocompatibility. In addition, DNA-based nanomaterials are easy to customize in size and shape and have good structural stability.

DNA-based nanomaterials come in a variety of sizes and shapes and are designed in a variety of structures, including two-dimensional and three-dimensional structures. According to different molecular construction methods of functional DNA-based materials, synthetic DNA-based nanomaterials include monolayer and multilayer nanomaterials. These DNA-based nanomaterials can also be divided into circular, linear, and branching forms, and they have been extensively constructed and studied [5]. DNA-based nanomaterials were commonly treated as drug delivery systems. In immunotherapy, traditional materials, including liposomes and adenoviruses, have been used as drug delivery systems in the past, but their defects limit their clinical application. For example, they share the same disadvantage of low targeting ability. Separately, adenoviruses are difficult to build and are usually toxic, while liposomes are easy to build but have low portability and low toxicity. Compared with traditional materials, DNA-based nanomaterials have many advantages, including structural stability, unparalleled programmability, natural biocompatibility, and negligible immunogenicity. These advantages may make DNA-based nanomaterials more favorable in immunotherapy.

DNA-based nanomaterials with different structures are made for different biomedical purposes. With the development of DNA nanomaterials advancing, various new DNA-based nanomaterials were constructed and widely used in immune engineering, drug delivery, molecular biology, tissue engineering, disease diagnosis or biosensing, etc. [6–18]. In recent years, successful attempts in immunotherapy have been reported, suggesting that DNA-based nanomaterials may possess therapeutic potential. Other hyper-polymeric compounds and nanomaterials such as avidin [19], polyethylenimine (PEI) [20–22], chitosan [23–25], and gold nanoparticles (AuNPs) [26–28] were loaded on DNA-based nanomaterials to enhance their therapeutic effect. It has been reported in previous research that, together with other materials or specific structures, polymeric DNA-based nanomaterials could influence the biological behavior of cells, such as proliferation [8, 9, 11, 29], autophagy [10], differentiation [30, 31], cell viability [29, 32], morphology [33], and migration [34]. Thanks to these special properties, polymeric DNA-based nanomaterials can be potential treatments for certain diseases and applied for tissue regeneration engineering [35–37]. When combined with other materials, such as proteins and some chemical drugs, polymeric DNA-based nanomaterials could treat some autoimmune diseases [38–41]. When combined with aptamers, the DNA-based nanomaterials could promote the antitumor effects [29, 42, 43], the inhibition of malignant cells [42], and the ability to target [29, 32, 42]. In the past few decades, DNA-based nanomaterials have made great progress and development, providing new options for the effective treatment of a variety of diseases and making significant contributions to the public health of society. At present, scientists have developed a variety of delivery systems, such as DNA origami, DNA tiles, and tetrahedral DNA-based nanomaterials (TDNs). In this chapter, we will make a summary of the self-assembly and structural design of DNA-based nanomaterials and highlight their therapeutic potential in immunotherapy.

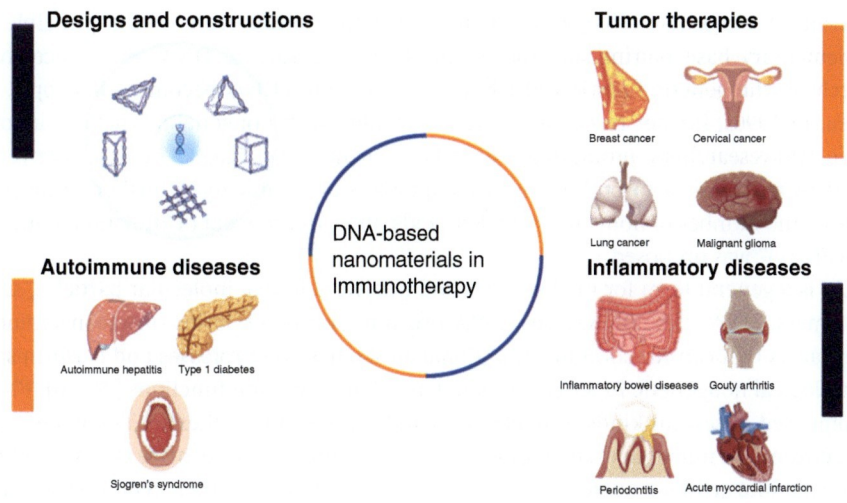

1.1 History of DNA-Based Nanomaterials – Design and Construction

Nucleic acid is a biological macromolecule used by living organisms to store genetic information [4, 44]. Being the basic genetic material in nature, it is not only closely related to normal life activities such as growth and reproduction, genetic variation, and cell differentiation but also closely related to abnormal life activities such as the occurrence of tumor, radiation damage, genetic disease, metabolic disease, viral infection, and so on. Moreover, nucleic acids have many unique properties besides their biological function; their molecular recognition ability, biocompatibility, and controllability at the nanoscale contribute to the construction of a variety of complex inorganic and organic nanostructures. Therefore, the study of nucleic acids is an important field in the development of modern biochemistry, molecular biology, and medicine. Nucleic acids are usually found in cells in the form of nucleoproteins that bind to proteins. Natural nucleic acids are divided into two main groups, namely ribonucleic acid (RNA) and DNA. The high controllability and high precision of Watson–Crick base pairing made DNA-based nanomaterials a potential substance for nanofabrication [2–4, 45]. Through the process of self-assembly, a large number of DNA-based nanomaterials of different shapes and sizes have been designed and constructed based on the classical Watson–Crick base pairing principle. Some DNA-based nanomaterials can change the biological behavior of cells functionally, such as cell migration, cell proliferation, cell differentiation, autophagy, and anti-inflammatory effects; hence, DNA nanotechnology has been greatly developed. DNA-based nanomaterials are employed in different scientific directions for various biological applications such as tissue regeneration, disease prevention, inflammation inhibition, bioimaging, biosensing, diagnosis, antitumor drug delivery, and therapeutics. In this section, we hope to introduce you to a comprehensive history of DNA-based nanomaterials.

The era of molecular biology began in 1953 with the discovery of the principle of complementary base pairing and the double helix structure of DNA. Ever since then research on the genetic function of DNA at the microlevel has become a hot topic. In the year of 1990s, further research sparked new interest in nongenetic functions among nucleic acid researchers, prompting scientists to further investigate. The first discovered were DNA aptamers of thrombin and RNA aptamers of organic dyes, further discoveries increased the number of nongenetic nucleic acids; thus, the concept of "functional nucleic acids (FNAs)" was proposed.

FNAs is a general term for nucleic acids and nucleic acid-like molecular particles, such as aptamers, DNA tiles, DNAzymes, DNA origami, and other forms of unconventional nucleic acids that can function like traditional antibodies and proteases and perform specific biological nongenetic functions with independent structural functions [46–50]. FNAs are composed of several kinds of nucleotides and are easy to synthesize. There are more than a dozen base nucleotides, including A, G, C, T, X, and Y [51]. As more and more effort is being put into artificial nucleic acid synthesis, the technology of their synthesis becomes more sophisticated, and the cost has decreased. The structure of FNAs is diverse and can be expressed as single-stranded, double-stranded, three-stranded, and four-stranded DNA. Due to the high compatibility of FNAs, a variety of targets can be attached to them. Recently, FNAs have been widely used in molecular imaging, nucleic acid self-assembly, biomolecular detection, and other biological fields, showing great advantages [52]. In the following section, four different kinds of well-known FNAs will be introduced.

1.1.1 DNAzymes

DNAzymes are a class of DNA molecules with catalytic functions. Like protein and RNA-catalyzing enzymes, DNAzymes are capable of catalyzing many types of biochemical reactions and are extensively applied in asymmetric catalysis, biosensors, DNA nanotechnology, and clinical diagnostics [53]. Proteins are thought to be the only biological molecules that can function as catalysts; however, in the 1980s and 1990s, scientists discovered some RNA with catalytic properties, which prompted widespread research for DNA enzymes. DNAzymes are generally obtained through the SELEX (Systematic Evolution of Ligands by EXponential enrichment) *in vitro* screening technique. (The basic idea of SELEX technology is to chemically synthesize a large, randomized DNA library *in vitro* and mix it with the target substance. The complex of the target substance and nucleic acid is mixed in a solution, and after the nucleic acids that are not bound to the target substance are washed away, the bound molecules are separated. The nucleic acid molecules are then used as the template for PCR amplification and the next round of the screening process. By repeating the screening and amplification process, some DNA or RNA molecules that do not bind to or have low or medium affinity with the target substance are washed away. Adaptor proteins, namely DNA with high affinity with the target substance, are isolated from very large random libraries, and their purity increases with the SELEX process.) According to their different catalytic functions, DNAzymes can be divided into different categories. The discovery of DNAzymes is mostly exciting because of the realization of enzyme-free catalytic reactions, which overcame the dependence on natural enzymes. Therefore, DNAzymes have been widely used as switches in the fields of biosensors and bioimaging.

1.1.2 Aptamers

Aptamers are a series of single-chain nucleic acid molecules that bind to specific target molecules. Their specificity is similar to that of antibodies, and they have strict recognition ability and a high affinity for binding ligands. They first appeared in 1990, when two research groups reported using aptamers to target small ligands and proteins with high affinity [54]. Aptamers are constructed similarly to DNAzymes by using the SELEX method of random RNA or DNA sequence libraries. In the past decades, some progress has been made in the optimization of aptamers. The *in vitro* screening process was automated in the year of 2001, and then in the year of 2005, fluorescence magnetic bead-SELEX (FluMag-SELEX) technology was developed and applied for DNA quantitative analysis and aptamer selection. In 2010, cell-SELEX was introduced to generate aptamers that bind to specific cell types, which further improved the synthesis method of aptamers. Compared with antibodies, aptamers have significant advantages such as (i) target cells can be screened at low toxicity and low immunogenicity; (ii) better specificity and affinity than antibodies; (iii) easy to be chemically modified; (iv) can be easily and economically obtained; (v) good thermal stability; (vi) can be used in combination with other drugs for combined treatment; and (vii) have easily customizable properties, such as deletion, splitting, fusion, extension, and substitution, to improve performance, particularly as a result of the development of splitting aptamer technology, which enables the substitution of protein-based antibodies and the direct detection of small molecules. Due to their inherent thermal stability, aptamers can undergo multiple denaturations and regenerations. This makes them easy to regenerate and reuse when they are applied to a variety of biosensors. In the absence of a target, aptamers are indistinguishable from ordinary nucleic acids. The presence of the target induces a conformational change of the aptamer. When an aptamer is bound to some nanomaterials, conformational changes in the aptamer may separate some groups of nanomaterials and lead to charge changes and subsequent recorded potential changes. This feature has been used for electrochemical detection of various target types.

1.1.3 Triplex DNA

Triplex DNA was first introduced to the world in 1957 by Felsenfeld and Rich [55]. In the year of 1995, the structure of triplex DNA was described in detail [56], that is, based on the double helix structure, a third oligodeoxynucleotide can be combined in the large furrow region of double-stranded DNA by Hoogsteen hydrogen bond pairing to form triplex DNA. Each strand segment of the triplex DNA must be an all-purine sequence or an all-pyrimidine sequence, which is bound to the target double-stranded DNA to form triplex DNA oligodeoxynucleotides, called TFOs (triplex-forming oligonucleotides). At the beginning of the discovery of triplex DNA, due to insufficient evidence that triplex DNA could exist in the body, its actual biological significance did not draw much attention. It was not until Helene and coworkers [57, 58] confirmed that triplex DNA could be used for gene expression regulation that researchers began to show strong interest in triplex DNA. According to the composition and orientation of triplex DNA, intermolecular triplex DNA can be divided into three types: the first type, TC triplex DNA (C+·GC and T·AT triplets, parallel; A is adenine, G is guanine, T thymine, C is cytosine); the second type, GT triple-stranded DNA (G·GC and T·AT triplets, parallel or anti-parallel); and the third type, GA-type triplex DNA (G·GC and A·AT triplets, antiparallel) [56]. In the last decade, triplex DNA has been widely used in sensing technology. These methods utilize triplex DNA not only as recognition elements but also as functional structure conversion units, allowing output signals to be generated during target recognition. Therefore, detection targets involving triplex DNA are not limited to specific nucleic acid sequences but cover a wide range of molecular targets, including antibodies, proteins, heavy metal ions, and small molecules.

1.1.4 DNA Origami and DNA Tiles

Among FNAs, DNA origami and DNA tiles are two components commonly used in nucleic acid self-assembly techniques to construct high-order nucleic acid nanomaterials. DNA tiles are completely dependent on the assembly of short DNA single strands, which are usually first assembled into unique or identical blocks and then further assembled into highly ordered finite structures. The sequence of each DNA block is related to the spatial position it occupies and is individually addressable [59]. In the 1990s, to construct interesting 2-dimension (2D) nanomaterials, scientists began designing DNA tiles with branches that resembled natural Holliday junctions [60, 61]. DNA-based structure nanotechnology was first proposed by Ned Seeman [62] and his coworkers in 1982. They creatively used specific sequences of DNA molecules to build out stable four-arm nanostructures (Holliday juncture/junction), which opened the structure of DNA nanotechnology to this new field of science. Since then, various nanostructures based on DNA self-assembly have been designed and constructed. They introduced the idea of branching DNA junctions and combining sticky end cohesion to manufacture geometric objects and periodic 2D lattices. By creating a structural motif with two four-way junctions, a set of branched complexes called double-crossed (DX) molecules is constructed. The proper adhesive end design enables DX molecules to perfectly self-assemble into periodic, two-dimensional lattices [63]. After further study, intersecting DNA tiles, triangles, and three-point star patterns were constructed into two-dimensional or even three-dimensional lattices, such as triple-crossovers (TX) and paramedic crossovers (PX) [64]. The development of DNA nanotechnology has benefited from the invention of DNA tiles. However, the synthesis of tile-based nanomaterials contains many interactions between short oligonucleotides. The synthesis output of nanomaterials requiring multiple reaction steps and purification is limited [51]. First reported by Rothemund in 2006, DNA origami, a DNA-based nanomaterial that uses a long strand of DNA as a microscaffold with many small fixed strands, has emerged to help solve the problems associated with the DNA tile method [51]. DNA origami is the folding together of scaffold strands (long DNA single strands) and hundreds of designed short DNA single strands. Each short DNA single strand has multiple binding domains with the scaffold strand, which are bound together by complementary base pairing and folded into arbitrary shapes in a manner similar to knitting. The DNA origami assembly scheme generally shows higher yield, stability, and ability to construct complex geometric shapes compared to the DNA tile-based assembly scheme. Researchers have used this technique to produce various 1D (1-dimension), 2D (2-dimension), and 3D (3-dimension) DNA nanostructures. The technology was first used to synthesize a number of monolayer and flat structures, such as triangles, simple rectangles, five-point stars, and some complicated graphics. All the structures have their own unique sizes, about 100 nm in diameter, and the processes associated with the self-assembly as well as the design of such DNA-based nanomaterials were thoroughly described by Rothemund [51]. From then on, more related research has been inspired, and more two-dimensional graphics have been successfully synthesized, including a map of China. Currently, DNA origami self-assembled structures typically have a surface area of 8000–10 000 nm^2 and contain approximately 200 addressable points within this regional range, thus allowing researchers to design arbitrary structures and apply them to multiple fields.

After binding with various nanomaterials such as metal-based nanomaterials, carbon-based nanomaterials, silicon-based nanomaterials, bionanomaterials, magnetic nanomaterials, and other composite nanomaterials, FNA-nanomaterial composites are formed. Thanks to the exceeding advantages like targeting, signal conversion, and amplification capabilities of FNA, as well as the stability and versatility of nanomaterials, FNA-nanomaterial composites are mainly applied in several categories, which will be illustrated in the following section.

The biological imaging function of FNA nanomaterials has greatly promoted the development of early disease diagnosis and lesion imaging. Targeting and fluorescence recovery rate are the keys to FNA nanomaterial bioimaging. The basic principle of biological imaging is to focus on the specific affinity between the target (including tumor cells, proteins, metal ions, or specific DNA sequences) and the probe, as well as the effective delivery and recovery of fluorescent substances. Therefore, FNA nanomaterials fully meet the requirements of biological imaging. The nanomaterials prevent the degradation of FNA through the action of various nucleases in the body, providing the nanomaterials with the ability of targeting, fluorescence signal generation, and amplification. Some nanomaterials, like MNPs, also have photothermal effects that further activate the release of fluorescent substances.

Among the many applications of FNA nanomaterials, the most common one is biosensing. Biosensing can be divided into four steps: signal perception, signal conversion, signal amplification, and signal output, each of which could involve FNA nanomaterials. In the process of signal perception, the ability of targeting and decoding ensures the accurate acquisition of the target, while in the signal conversion step, various chemical groups could be labeled using a number of fluorescence signatures of FNAs. Various types of biosensors have been constructed due to the fluorescence and luminescence properties of various nanomaterials and FNA nanomaterials' ability to convert abstract molecular signals into realistic visual, fluorescent, or electrochemical signals. Also, the cleavage and extension of FNA strands are of great help in enabling biosensors to realize the signal amplification process. In the signal output step, according to the different detection principles, sensing process, and sensitivity, FNA nanomaterial biosensors could be divided into visual biosensors, fluorescent biosensors, and electrochemical biosensors.

FNA nanomaterials can be widely applied in the biomedical field thanks to their ability of drug delivery and molecular recognition. They are mainly applied in disease diagnosis and treatment [65]. Nanomaterials like AuNRs, AuNPs, and magnetic NPs are well-known for their advantages and optical properties in biocoupling, synthesis, focal imaging, and photothermal therapy that could prove of great benefit in biomedical applications. FNAs combined with such substances have been extensively explored for the diagnosis of diseases by scientists [66–68]. Self-assembled 3D FNA nanostructures, such as DNA origami and DNA hydrogels, have good biocompatibility, stability, flexibility, and precise programmability, as well as switching characteristics and ease of synthesis and modification [6]. Thus, FNA nanomaterials are expected to have potential in areas such as disease analysis and drug delivery [69, 70].

Today's world has completely entered the era of big data, and all life-related activities involve data storage and processing [71, 72]. The exponential growth of modern data has outpaced the capacity growth of existing memory devices. However, existing storage media, such as magnetic storage (magnetic tape or hard disk drive), optical storage (such as Blu-ray), and solid-state storage (such as flash memory), have been unable to meet the growing demand for storage capacity and have become a problem that human beings have to face. Molecular data storage is a novel data storage method with high stability and high storage density showing great potential. It is expected to address the growing gap between the amount of information available today and the capacity to store it. As a typical molecular data storage method, DNA data storage can be used as an alternative and transformative storage medium to break through the physical limit of existing storage methods and meet the ever-increasing demand for data storage. In recent days, a series of proof-of-principle experiments have demonstrated the feasibility and value of DNA as a storage medium, showing great potential for changing the way we store data [73, 74]. In a review reported by Panda et al., they critically analyze the emergence of the concept of DNA as a storage medium and its historical perspective, feasibility, recent breakthroughs, and the challenges that need to be overcome in order to make it a marketable data storage medium. They conclude that storing astronomical amounts of data in nucleic acids is no longer the stuff of science fiction [74].

After discovering that each DNA unit could bind with four adjacent DNA units or more, varieties of 3D DNA-based nanomaterial frameworks were manufactured. Ned Seeman's pioneering use of DNA as a building block to assemble high-dimensional materials has led to various methods of making DNA nanostructures of different sizes and morphs [75, 76]. One of the research focuses on making arbitrary structures in high dimensions, such as DNA paint fabricated using 2D tiles or micron-sized 3D DNA-based nanomaterials. Some three-dimensional DNA nanostructures with complex curvature have been prepared and characterized using DNA origami folding techniques [77]. For example, the double-helix DNA is bent to follow the circular outline of the target object, and potential chain crossings are subsequently identified. Concentric circles of DNA are used to generate in-plane curvature, constrained to two dimensions by appropriately designed geometry and crossover networks, resulting in a series of DNA nanostructures with high curvature and ellipsoidal shells. Interestingly, in 2009, Anderson et al. created an addressable DNA box that could be opened in the presence of externally provided DNA "keys" derived from the principle of complementary base pairing [78]. Controlling access to the inner compartments of such DNA nanocapsules could yield some interesting applications. At the same time, the closing and opening mechanisms of the DNA box have inspired researchers to build complex and multifunctional iterations of 3D nanorobots and nanocargo. As an example, the nucleolin-targeting aptamer serves as the targeting domain and as a molecular trigger that the DNA nanobots mechanically open. As a result, the internal thrombin is exposed and activates clotting at the tumor site. In later times, different kinds of 3D DNA-based nanomaterials were constructed. These DNA nanostructures can not only facilitate the study of molecular interactions in chemical and biological systems by building spatially organized molecular networks that can be used as molecular devices with more complex information processing capabilities than before but also lead to the construction of more complex structural components in DNA robots and localized DNA circuits [79]. Douglas et al. demonstrated the design and self-assembly process of six different shapes of DNA nanostructures, including boulders, balustrade bridges, square nuts, stacked crosses, and monster bottles, with precisely controlled sizes ranging from 10 to 100 nm; he also reported on the effectiveness of the design approach by using honeycomb alignment to combine an integral, square mesh, and slotted cross [80]. In another study, Ke et al. reported a novel approach to the design of multilayered DNA structures based on quadruple-helix bundles. In this design, despite the high density of the DNA helix, the square lattice can be folded into nanostructures of the design size by a one-step annealing process [81]. With further study, Ke et al. reported the successful folding of a multilayered DNA structure, helically arranged on a tightly packed hexagonal lattice. The study also showed that hybrid DNA structures can contain three different shapes in a single design, including a square lattice, a honeycomb lattice, and a hexagonal spiral structure.

Although complex DNA origami technology provides a research platform and a versatile building block, its complex manufacturing process, high cost, relatively low yield, and technical sensitivity may limit its application unless breakthroughs are made in the production, folding, and purification of scaffolds. First introduced in 2004, Turberfield and his coworkers reported a convenient, one-step synthesis method of fabricating 3D DNA-based nanomaterials [82], which achieved progress in simple design, simple structure, low cost, and high yield. Of all the 3D DNA-based nanomaterials constructed, the DNA tetrahedron is supposed to be the simplest 3D DNA-based nanomaterial and one of the most classical 3D frameworks [83–85]. This ideal nanomaterial can be synthesized by a simple procedure in which four single strands of DNA are mixed in equal mole quantities in a saline buffer solution and denatured at 95 °C, then annealed by cooling to 4 °C. As the most typical 3D DNA-based nanomaterials, TDNs are composed of four isometric single-stranded DNAs and have strong mechanical strength as well as unique advantages over other types of DNA-based nanomaterials [60, 83, 86–90]. DNA, as a biological macromolecule, cannot enter the cell directly through the plasma membrane for its polyanionic nature, but because of its structural stability, it can be endocytosed by caveolin-mediated cells and then transported to the lysosome via a microtubule-dependent manner, meaning it can be maintained longer in cells [91]. In addition, TDNs can be specifically directed to targets that are capable of lysosomal escape when connected to nuclear loci aptamers, which may be key to gene delivery [92]. Zagorovsky et al. found that DNA-based nanomaterials possess better serum stability with more condensed spatial structure and higher DNA density, meaning that TDNs, due to their spatial simplicity, may be more stable than other structural 3D DNA-based nanomaterials [90]. Blessed with these specialties and with further discoveries, TDNs have become one of the most advanced nanomaterials in various fields.

References

1 Anselmo, A.C. and Mitragotri, S. (2016). Nanoparticles in the clinic. *Bioeng. Transl. Med.* 1 (1): 10–29.
2 Seeman, N.C. (2018). DNA nanotechnology: from the pub to information-based chemistry. *Methods Mol. Biol.* 1811: 1–9.

3 Seeman, N.C. (1998). DNA nanotechnology: novel DNA constructions. *Annu. Rev. Biophys. Biomol. Struct.* 27: 225–248.

4 Dong, Y., Yao, C., Zhu, Y. et al. (2020). DNA functional materials assembled from branched DNA: design, synthesis, and applications. *Chem. Rev.* 120 (17): 9420–9481.

5 Roh, Y.H., Ruiz, R.C., Peng, S. et al. (2011). Engineering DNA-based functional materials. *Chem. Soc. Rev.* 40 (12): 5730–5744.

6 Li, J., Mo, L., Lu, C.H. et al. (2016). Functional nucleic acid-based hydrogels for bioanalytical and biomedical applications. *Chem. Soc. Rev.* 45 (5): 1410–1431.

7 Samanta, A. and Medintz, I.L. (2016). Nanoparticles and DNA – a powerful and growing functional combination in bionanotechnology. *Nanoscale* 8 (17): 9037–9095.

8 Shi, S., Peng, Q., Shao, X. et al. (2016). Self-assembled tetrahedral DNA nanostructures promote adipose-derived stem cell migration via lncRNA XLOC 010623 and RHOA/ROCK2 signal pathway. *ACS Appl. Mater. Interfaces* 8 (30): 19353–19363.

9 Shao, X., Lin, S., Peng, Q. et al. (2017). Tetrahedral DNA nanostructure: a potential promoter for cartilage tissue regeneration via regulating chondrocyte phenotype and proliferation. *Small* 13 (12): 1602770.

10 Shi, S., Lin, S., Li, Y. et al. (2018). Effects of tetrahedral DNA nanostructures on autophagy in chondrocytes. *Chem. Commun.* 54 (11): 1327–1330.

11 Ma, W., Shao, X., Zhao, D. et al. (2018). Self-assembled tetrahedral DNA nanostructures promote neural stem cell proliferation and neuronal differentiation. *ACS Appl. Mater. Interfaces* 10 (9): 7892–7900.

12 Ma, W., Xie, X., Shao, X. et al. (2018). Tetrahedral DNA nanostructures facilitate neural stem cell migration via activating RHOA/ROCK2 signalling pathway. *Cell Proliferation* 51 (6): e12503.

13 Li, M., Wang, C., Di, Z. et al. (2019). Engineering multifunctional DNA hybrid nanospheres through coordination-driven self-assembly. *Angew. Chem. Int. Ed. Engl.* 58 (5): 1350–1354.

14 Maeda, M., Kojima, T., Song, Y., and Takayama, S. (2019). DNA-based biomaterials for immunoengineering. *Adv. Healthcare Mater.* 8 (4): e1801243.

15 Liu, S., Jiang, Q., Wang, Y., and Ding, B. (2019). Biomedical applications of DNA-based molecular devices. *Adv. Healthcare Mater.* 8 (10): e1801658.

16 Liu, N., Zhang, X., Li, N. et al. (2019). Tetrahedral framework nucleic acids promote corneal epithelial wound healing in vitro and in vivo. *Small* 15 (31): e1901907.

17 Chandrasekaran, A.R., Punnoose, J.A., Zhou, L. et al. (2019). DNA nanotechnology approaches for microRNA detection and diagnosis. *Nucleic Acids Res.* 47 (20): 10489–10505.

18 Khajouei, S., Ravan, H., and Ebrahimi, A. (2020). DNA hydrogel-empowered biosensing. *Adv. Colloid Interface Sci.* 275: 102060.

19 Morpurgo, M., Radu, A., Bayer, E.A., and Wilchek, M. (2004). DNA condensation by high-affinity interaction with avidin. *J. Mol. Recognit.* 17 (6): 558–566.

20 Utsuno, K. and Uludağ, H. (2010). Thermodynamics of polyethylenimine-DNA binding and DNA condensation. *Biophys. J.* 99 (1): 201–207.

21 Tian, T., Zhang, T., Zhou, T. et al. (2017). Synthesis of an ethyleneimine/tetrahedral DNA nanostructure complex and its potential application as a multi-functional delivery vehicle. *Nanoscale* 9 (46): 18402–18412.

References

22 Zhang, Y., Lin, L., Liu, L. et al. (2018). Ionic-crosslinked polysaccharide/PEI/DNA nanoparticles for stabilized gene delivery. *Carbohydr. Polym.* 201: 246–256.

23 Liu, W., Sun, S., Cao, Z. et al. (2005). An investigation on the physicochemical properties of chitosan/DNA polyelectrolyte complexes. *Biomaterials* 26 (15): 2705–2711.

24 Gu, T., Wang, J., Xia, H. et al. (2014). Direct electrochemistry and electrocatalysis of horseradish peroxidase immobilized in a DNA/chitosan-Fe_3O_4 magnetic nanoparticle bio-complex film. *Materials* 7 (2): 1069–1083.

25 Kumar, S., Garg, P., Pandey, S. et al. (2015). Enhanced chitosan-DNA interaction by 2-acrylamido-2-methylpropane coupling for an efficient transfection in cancer cells. *J. Mater. Chem. B* 3 (17): 3465–3475.

26 Lee, H., Dam, D.H., Ha, J.W. et al. (2015). Enhanced human epidermal growth factor receptor 2 degradation in breast cancer cells by lysosome-targeting gold nanoconstructs. *ACS Nano* 9 (10): 9859–9867.

27 Edwardson, T.G., Lau, K.L., Bousmail, D. et al. (2016). Transfer of molecular recognition information from DNA nanostructures to gold nanoparticles. *Nat. Chem.* 8 (2): 162–170.

28 Liu, B., Song, C., Zhu, D. et al. (2017). DNA-origami-based assembly of anisotropic plasmonic gold nanostructures. *Small* 13 (23).

29 Zhan, Y., Ma, W., Zhang, Y. et al. (2019). DNA-based nanomedicine with targeting and enhancement of therapeutic efficacy of breast cancer cells. *ACS Appl. Mater. Interfaces* 11 (17): 15354–15365.

30 Hendrikson, W.J., Zeng, X., Rouwkema, J. et al. (2016). Biological and tribological assessment of poly(ethylene oxide terephthalate)/poly(butylene terephthalate), polycaprolactone, and poly(L'DL) lactic acid plotted scaffolds for skeletal tissue regeneration. *Adv. Healthcare Mater.* 5 (2): 232–243.

31 Taniguchi, J., Pandian, G.N., Hidaka, T. et al. (2017). A synthetic DNA-binding inhibitor of SOX2 guides human induced pluripotent stem cells to differentiate into mesoderm. *Nucleic Acids Res.* 45 (16): 9219–9228.

32 Ma, W., Zhan, Y., Zhang, Y. et al. (2019). An intelligent DNA nanorobot with in vitro enhanced protein lysosomal degradation of HER2. *Nano Lett.* 19 (7): 4505–4517.

33 Jayme, C.C., de Paula, L.B., Rezende, N. et al. (2017). DNA polymeric films as a support for cell growth as a new material for regenerative medicine: compatibility and applicability. *Exp. Cell. Res.* 360 (2): 404–412.

34 Wang, M., He, F., Li, H. et al. (2019). Near-infrared light-activated DNA-agonist nanodevice for nongenetically and remotely controlled cellular signaling and behaviors in live animals. *Nano Lett.* 19 (4): 2603–2613.

35 Feng, G., Zhang, Z., Dang, M. et al. (2017). Injectable nanofibrous spongy microspheres for NR4A1 plasmid DNA transfection to reverse fibrotic degeneration and support disc regeneration. *Biomaterials* 131: 86–97.

36 Basu, S., Pacelli, S., Feng, Y. et al. (2018). Harnessing the noncovalent interactions of DNA backbone with 2D silicate nanodisks to fabricate injectable therapeutic hydrogels. *ACS Nano* 12 (10): 9866–9880.

37 Zhang, Y., Ma, W., Zhan, Y. et al. (2018). Nucleic acids and analogs for bone regeneration. *Bone Res.* 6: 37.

38 Zhang, M., Zhang, X., Tian, T. et al. (2022). Anti-inflammatory activity of curcumin-loaded tetrahedral framework nucleic acids on acute gouty arthritis. *Bioact. Mater.* 8: 368–380.

39 Zhou, M., Gao, S., Zhang, X. et al. (2021). The protective effect of tetrahedral framework nucleic acids on periodontium under inflammatory conditions. *Bioact. Mater.* 6 (6): 1676–1688.

40 Gao, S., Li, Y., Xiao, D. et al. (2021). Tetrahedral framework nucleic acids induce immune tolerance and prevent the onset of type 1 diabetes. *Nano Lett.* 21 (10): 4437–4446.

41 Gao, S., Wang, Y., Li, Y. et al. (2021). Tetrahedral framework nucleic acids reestablish immune tolerance and restore saliva secretion in a Sjögren's syndrome mouse model. *ACS Appl. Mater. Interfaces* 13 (36): 42543–42553.

42 Li, Q., Zhao, D., Shao, X. et al. (2017). Aptamer-modified tetrahedral DNA nanostructure for tumor-targeted drug delivery. *ACS Appl. Mater. Interfaces* 9 (42): 36695–36701.

43 Meng, L., Ma, W., Lin, S. et al. (2019). Tetrahedral DNA nanostructure-delivered DNAzyme for gene silencing to suppress cell growth. *ACS Appl. Mater. Interfaces* 11 (7): 6850–6857.

44 Broker, T.R. and Lehman, I.R. (1971). Branched DNA molecules: intermediates in T4 recombination. *J. Mol. Biol.* 60 (1): 131–149.

45 Tian, T., Li, Y., and Lin, Y. (2022). Prospects and challenges of dynamic DNA nanostructures in biomedical applications. *Bone Res.* 10 (1): 40.

46 Ahmed, S., Kintanar, A., and Henderson, E. (1994). Human telomeric C-strand tetraplexes. *Nat. Struct. Biol.* 1 (2): 83–88.

47 Cho, E.J., Yang, L., Levy, M., and Ellington, A.D. (2005). Using a deoxyribozyme ligase and rolling circle amplification to detect a non-nucleic acid analyte, ATP. *JACS* 127 (7): 2022–2023.

48 Ellington, A.D. and Szostak, J.W. (1990). In vitro selection of RNA molecules that bind specific ligands. *Nature* 346 (6287): 818–822.

49 Joyce, G.F. (2001). RNA cleavage by the 10–23 DNA enzyme. *Methods Enzymol.* 341: 503–517.

50 Travascio, P., Bennet, A.J., Wang, D.Y., and Sen, D. (1999). A ribozyme and a catalytic DNA with peroxidase activity: active sites versus cofactor-binding sites. *Chem. Biol.* 6 (11): 779–787.

51 Zhang, Y., Ptacin, J.L., Fischer, E.C. et al. (2017). A semi-synthetic organism that stores and retrieves increased genetic information. *Nature* 551 (7682): 644–647.

52 Rothemund, P.W. (2006). Folding DNA to create nanoscale shapes and patterns. *Nature* 440 (7082): 297–302.

53 Zhou, W., Saran, R., and Liu, J. (2017). Metal sensing by DNA. *Chem. Rev.* 117 (12): 8272–8325.

54 Tuerk, C. and Gold, L. (1990). Systematic evolution of ligands by exponential enrichment: RNA ligands to bacteriophage T4 DNA polymerase. *Science* 249 (4968): 505–510.

55 Felsenfeld, G. and Rich, A. (1957). Studies on the formation of two- and three-stranded polyribonucleotides. *Biochim. Biophys. Acta* 26 (3): 457–468.

56 Frank-Kamenetskii, M.D. and Mirkin, S.M. (1995). Triplex DNA structures. *Annu. Rev. Biochem.* 64: 65–95.

57 Hélène, C., Montenay-Garestier, T., Saison, T. et al. (1985). Oligodeoxynucleotides covalently linked to intercalating agents: a new class of gene regulatory substances. *Biochimie* 67 (7–8): 777–783.
 58 Zerial, A., Thuong, N.T., and Hélène, C. (1987). Selective inhibition of the cytopathic effect of type A influenza viruses by oligodeoxynucleotides covalently linked to an intercalating agent. *Nucleic Acids Res.* 15 (23): 9909–9919.
 59 Hung, A.M., Micheel, C.M., Bozano, L.D. et al. (2010). Large-area spatially ordered arrays of gold nanoparticles directed by lithographically confined DNA origami. *Nat. Nanotechnol.* 5 (2): 121–126.
 60 Chen, J.H. and Seeman, N.C. (1991). Synthesis from DNA of a molecule with the connectivity of a cube. *Nature* 350 (6319): 631–633.
 61 Kochoyan, M., Havel, T.F., Nguyen, D.T. et al. (1991). Alternating zinc fingers in the human male associated protein ZFY: 2D NMR structure of an even finger and implications for "jumping-linker" DNA recognition. *Biochemistry* 30 (14): 3371–3386.
 62 Seeman, N.C. (1982). Nucleic acid junctions and lattices. *J. Theor. Biol.* 99 (2): 237–247.
 63 Seeman, N.C. (2005). Structural DNA nanotechnology: an overview. *Methods Mol. Biol.* 303: 143–166.
 64 Zheng, J., Birktoft, J.J., Chen, Y. et al. (2009). From molecular to macroscopic via the rational design of a self-assembled 3D DNA crystal. *Nature* 461 (7260): 74–77.
 65 Chen, T., Ren, L., Liu, X. et al. (2018). DNA nanotechnology for cancer diagnosis and therapy. *Int. J. Mol. Sci.* 19 (6): 1671.
 66 Qing, T., He, X., He, D. et al. (2017). Dumbbell DNA-templated CuNPs as a nano-fluorescent probe for detection of enzymes involved in ligase-mediated DNA repair. *Biosens. Bioelectron.* 94: 456–463.
 67 Xu, H., Liao, C., Zuo, P. et al. (2018). Magnetic-based microfluidic device for on-chip isolation and detection of tumor-derived exosomes. *Anal. Chem.* 90 (22): 13451–13458.
 68 Wang, Z., Ye, S., Zhang, N. et al. (2019). Triggerable mutually amplified signal probe based SERS-microfluidics platform for the efficient enrichment and quantitative detection of miRNA. *Anal. Chem.* 91 (8): 5043–5050.
 69 Wang, Y., Shang, X., Liu, J., and Guo, Y. (2018). ATP mediated rolling circle amplification and opening DNA-gate for drug delivery to cell. *Talanta* 176: 652–658.
 70 Jiang, Q., Song, C., Nangreave, J. et al. (2012). DNA origami as a carrier for circumvention of drug resistance. *JACS* 134 (32): 13396–13403.
 71 Extance, A. (2016). How DNA could store all the world's data. *Nature* 537 (7618): 22–24.
 72 Zhirnov, V., Zadegan, R.M., Sandhu, G.S. et al. (2016). Nucleic acid memory. *Nat. Mater.* 15 (4): 366–370.
 73 Erlich, Y. and Zielinski, D. (2017). DNA fountain enables a robust and efficient storage architecture. *Science* 355 (6328): 950–954.
 74 Panda, D., Molla, K.A., Baig, M.J. et al. (2018). DNA as a digital information storage device: hope or hype? *3 Biotech* 8 (5): 239.
 75 Winfree, E., Liu, F., Wenzler, L.A., and Seeman, N.C. (1998). Design and self-assembly of two-dimensional DNA crystals. *Nature* 394 (6693): 539–544.
 76 Seeman, N.C. (2003). DNA in a material world. *Nature* 421 (6921): 427–431.
 77 Han, D., Pal, S., Nangreave, J. et al. (2011). DNA origami with complex curvatures in three-dimensional space. *Science* 332 (6027): 342–346.

78 Andersen, E.S., Dong, M., Nielsen, M.M. et al. (2009). Self-assembly of a nanoscale DNA box with a controllable lid. *Nature* 459 (7243): 73–76.
79 Chao, J., Wang, J., Wang, F. et al. (2019). Solving mazes with single-molecule DNA navigators. *Nat. Mater.* 18 (3): 273–279.
80 Douglas, S.M., Dietz, H., Liedl, T. et al. (2009). Self-assembly of DNA into nanoscale three-dimensional shapes. *Nature* 459 (7245): 414–418.
81 Ke, Y., Douglas, S.M., Liu, M. et al. (2009). Multilayer DNA origami packed on a square lattice. *JACS* 131 (43): 15903–15908.
82 Goodman, R.P., Berry, R.M., and Turberfield, A.J. (2004). The single-step synthesis of a DNA tetrahedron. *Chem. Commun.* 12: 1372–1373.
83 Goodman, R.P., Schaap, I.A., Tardin, C.F. et al. (2005). Rapid chiral assembly of rigid DNA building blocks for molecular nanofabrication. *Science* 310 (5754): 1661–1665.
84 Gao, Y., Chen, X., Tian, T. et al. (2022). A lysosome-activated tetrahedral nanobox for encapsulated siRNA delivery. *Adv. Mater.* 34: e2201731.
85 Li, S., Liu, Y., Tian, T. et al. (2021). Bioswitchable delivery of microRNA by framework nucleic acids: application to bone regeneration. *Small* 17 (47): e2104359.
86 Shih, W.M., Quispe, J.D., and Joyce, G.F. (2004). A 1.7-kilobase single-stranded DNA that folds into a nanoscale octahedron. *Nature* 427 (6975): 618–621.
87 Simmel, S.S., Nickels, P.C., and Liedl, T. (2014). Wireframe and tensegrity DNA nanostructures. *Acc. Chem. Res.* 47 (6): 1691–1699.
88 Vindigni, G., Raniolo, S., Ottaviani, A. et al. (2016). Receptor-mediated entry of pristine octahedral DNA nanocages in mammalian cells. *ACS Nano* 10 (6): 5971–5979.
89 Walsh, A.S., Yin, H., Erben, C.M. et al. (2011). DNA cage delivery to mammalian cells. *ACS Nano* 5 (7): 5427–5432.
90 Zagorovsky, K., Chou, L.Y., and Chan, W.C. (2016). Controlling DNA-nanoparticle serum interactions. *PNAS* 113 (48): 13600–13605.
91 Zhang, W. and Li, L.J. (2011). Observation of phonon anomaly at the armchair edge of single-layer graphene in air. *ACS Nano* 5 (4): 3347–3353.
92 Cortez, M.A., Godbey, W.T., Fang, Y. et al. (2015). The synthesis of cyclic poly(ethylene imine) and exact linear analogues: an evaluation of gene delivery comparing polymer architectures. *JACS* 137 (20): 6541–6549.

2

The Methods to Improve the Stability of Nucleic Acid-Based Nanomaterials

Xueping Xie

Zhejiang University School of Medicine, The First Affiliated Hospital, Department of Oral and Maxillofacial Surgery, Hangzhou 310003, PR China

2.1 Introduction

Nucleic acid-based nanomaterials (NANs) are two- or three-dimensional nanostructures composed of nucleic acid sequences. Because of the high specificity of base binding, NANs with different shapes can be obtained by precise design aided by computer software, including tetrahedral nanostructure, octahedral nanostructure, nanoribbon, nanoflower, and so on. The excellent properties of NANs make them have a wide application prospect in the field of biomedicine, including clear size and shape, programmable and predictable structure, good biological compatibility, and good biological stability [1-4]. In addition, the robust scaffold structure of NAN allows different groups, such as targeting groups or drug molecules, to be connected accurately. On the other hand, different types of NANs exhibit different properties [5]. All of these make NAN a promising candidate in the field of biomedicine, including tumor detection and targeted drug delivery. In recent years, tetrahedral framework nucleic acids (tFNAs), a specially designed NAN, have been studied extensively. It plays an important role in tissue regeneration. Researchers found that tFNAs have outstanding performance in the treatment of neurogenic disease [6-8], acute liver failure [9], severe acute pancreatitis [10], and diffuse microvascular endothelial cell injury after subarachnoid hemorrhage [11]. Moreover, tFNAs might be an excellent drug carrier. Li et al. found that clindamycin-tFNAs complex could repair infected bone defect [12]. Zhang et al. suggested that curcumin-loaded tFNAs had the function of resisting acute gouty arthritis [13]. Zhao's research demonstrated that angiogenic peptide-loaded tFNAs could prevent bisphosphonate-related osteonecrosis of the jaw by promoting angiogenesis [14]. What's more, NANs, such as tFNAs, have significant advantages in carrying nucleic acid drugs because of their homogeneity and compatibility. MicroRNA-155-loaded tFNAs could inhibit choroidal neovascularization by regulating the polarization of macrophages [15]. tFNAs could carry siRNA into cells to produce stronger gene silencing [16]. So, NAN is an important research hotspot with extensive biomedical applications.

However, the application of NANs is restricted by many factors. The stability of NAN is one of the most important limiting factors. The stability of NANs is dependent on high

Nucleic Acid-Based Nanomaterials: Stabilities and Applications, First Edition.
Edited by Yunfeng Lin and Shaojingya Gao.
© 2024 WILEY-VCH GmbH. Published 2024 by WILEY-VCH GmbH.

cation concentrations; NANs might be destroyed by enzymes and denatured in organic solvents. NANs are traditionally assembled in TBE or TAE buffer containing 5–20 mM magnesium (Mg^{2+}) to neutralize the negative charges between the phosphate backbone and ensure the helices are correctly folded [17]. However, the physiological salt concentration in cell medium and blood is much lower than that needed to keep NAN stable. It was found that the concentration of Mg^{2+}, potassium (K^+), and sodium (Na^+) in cell medium is 0.4–0.8, ~5.5, and ~150 mM, respectively, and 0.6–1, ~4, and ~140 mM, respectively, in blood [18–21]. Because of the low cation concentrations, NANs might be significantly unstable under physiological conditions. Nuclease degradation in the biological system is another big obstacle to the stability of NANs. Different endonuclease and exonuclease enzymes in cell culture and in cells exhibit different destructive powers to NANs. Exo VIII and T7 Exo appear to be harmless to NAN, while Exo Ie, Exo T, T7 Endo, and Exo III have been shown to be destructive. Among all the nucleases, the endonuclease DNase I, which exists both inside and outside the cell, is the most destructive to NANs [22]. The negative effects of nucleases on the stability of NANs seriously affect their clinical application. Since the half-life of NANs should be greater than 30 hours [23]. Moreover, organic solvents also undergo NAN denaturation, which limits their wide use [24, 25]. For example, polar aprotic solvents such as dimethylformamide (DMF) and dimethyl sulfoxide (DMSO) are often used in polymer synthesis and peptide synthesis, which leads to denaturation, reduced thermal stability, and conformational changes in the DNA double strand [26, 27]. Improving their stability in organic solvents could be one way to expand their properties and applications.

Furthermore, NAN is only stable below the melting temperature, too low or high temperatures could cause the nanostructures to disintegrate [28]. DNA nanostructures are versatile templates for low-cost and high-resolution nanofabrication. The synthesis of inorganic materials usually requires a high temperature of more than 500 °C. However, DNA breaks down when heated to 250 °C, making DNA nanostructures seem impossible to achieve mode transfer under these conditions [29]. Their limited chemical stability has limited their application in nanofabrication to low-temperature processes [30–32]. Therefore, it is important to improve the thermal stability of DNA nanostructures and then broaden their application in high-temperature conditions.

In recent decades, the field of NAN design and application has recognized and focused on careful quantification of stability issues. A great deal of methods for protecting NAN from low cation concentrations, nucleases, temperature, and organic solvents have been developed. In this review, we summarize the stabilization strategies recently developed.

2.2 Methods to Improve Stability

Since the stability of NAN plays an important role in its widespread application, the field has started exploring methods to prevent fatal destruction. The methods could be categorized into six themes: (i) changing the composition of nucleic acids; (ii) backbone modification of nucleic acids; (iii) coating with protective structures; (iv) covalent crosslinking; (v) tuning buffer conditions; (vi) construction of novel NAN. We will discuss the present examples per class in the following section.

2.2.1 Artificial Nucleic Acids

In order to improve the stability of NAN, many studies have focused on replacing natural nucleic chains with unnatural analogs. At present, many artificial nucleic acids (XNAs) that mimic natural DNA and RNA have been synthesized, such as L-DNA, peptide nucleic acids (PNA) and acyclic L-threoninol nucleic acid (L-aTNA). In addition, a special stable structure, G4, was synthesized (Table 2.1).

It has been reported that nucleic acid cargo interferes with NAN construction. L-DNA is a mirror form of natural D-DNA and has the same thermodynamic properties. L-DNA was used as the skeleton for the assembly of nanocarriers to avoid the hybridization interference between the loaded sequence and the carrier sequence. Kim et al. prepared the L-DNA tetrahedron (L-Td). They concluded that L-Td exhibited higher nuclease resistance in serum and was more readily absorbed by cells than D-Td [33]. In addition, L-Td showed stronger antitumor effect as a tumor therapy vector [39, 40].

PNA is a special type of functional nucleic acid, through the 2-aminoethyl-glycine link to replace the phosphodiester backbone, synthetic neutral peptide-like backbone [41]. Due to the lack of charge repulsion, PNA has many excellent properties, including high nuclease and protease resistance, high chemical stability, and extremely high binding affinity for DNA and RNA [42, 43]. Therefore, PNA is suitable for synthetic and biomedical applications of NAN [34]. More importantly, Kumar et al. demonstrated that gamma-modified PNA (γPNA) could be used to form complex self-assembled nanostructures in polar organic solvent mixtures [35].

Table 2.1 Artificial nucleic acids (XNAs) improved the stability of NAN.

XNAs	Characteristics	Significant results	References
L-DNA	The mirror form of nature D-DNA	Higher nuclease resistance in serum	[33]
PNA	A neutral peptide-like backbone by replacing the phosphodiester backbone with 2-aminoethyl-glycine linkage	High nuclease and protease resistance, high chemical stability	[34]
γPNA	Gamma-modified PNA	Keep stable in polar organic solvent mixtures	[35]
L-aTNA	The acyclic XNAs modified with L-threoninol	High stability in serum	[36]
FANA	The sugar moiety of DNA contained an arabinose ring, and the 2'-hydroxyl group was replaced by a fluorine atom	Higher thermal stability and better nuclease resistance	[37]
G4	Four-stranded structures comprising stacked G-tetrads that are formed by the self-assembly of G-rich nucleic acid sequences containing the tandem nucleobase, guanine	High stability and biocompatibility under physiological conditions	[38]

XNAs with flexible acyclic scaffolds have attracted much attention due to nuclease resistance in cells. However, most of them could not form stable duplexes with DNA and RNA. Thus, Murayama et al. modified acyclic XNAs with L-threoninol to synthesize the acyclic L-aTNA and demonstrated that L-aTNA formed more stable duplexes with DNA and RNA than D-aTNA [44]. Later, Märcher et al. form a four-way junction (4WJ) using L-aTNA. The L-aTNA 4WJ is serum stable, has no cytotoxicity, no immune response, and no non-targeted uptake [36]. In addition, Taylor et al. prepared four tetrahedral nanostructures using different XNAs. Their results showed that tetrahedral nanostructures constructed from XNAs had a wider range of structural and physicochemical properties, including enhanced biostabilities [45]. It was reported that when the sugar moiety of DNA contained an arabinose ring and the 2′-hydroxyl group was replaced by a fluorine atom, 2′-fluoroarabinonucleic acid (FANA) was produced, which was another type of XNA [46]. The FANA could hybridize with DNA and RNA through Watson–Crick base pairing rule. The hybrid duplexes were thermally stable and nuclease resistant [47, 48]. Thus, FANA could be assembled into 3D nanostructures using the same rules of DNA. Taylor has utilized FANA to synthesize tetrahedron and octahedron nanostructures [45]. Recently, Wang et al. synthesized FANA nanostructures and confirmed their higher thermal stability and better nuclease resistance than their DNA counterparts [37].

G4 is a four-stranded structure consisting of stacked G-tetrads formed by self-assembly of G-rich nucleic acid sequences containing the tandem base, guanine [49]. According to the orientation of the four strands, G4 can form parallel or antiparallel stable structures under certain conditions. In addition, G4 has high stability and biocompatibility under physiological conditions. It has been widely used as functional DNA motifs in DNA nanomaterials [38]. Huang et al. prepared functional DNA hydrogel system with G4 cross-strand. Their results showed that the hydrogels containing G4 had excellent stability and sensitivity [50].

To summarize, the synthesis of NAN in combination with the stable nucleic acids and structures mentioned above can enhance its stability and benefit its biological applications.

2.2.2 Backbone Modification of Nucleic Acids

There are three important sites for modification in natural nucleosides; they could be on nucleobase, ribose, or phosphate group (Figure 2.1). A large number of different modification methods have been reported to improve the stability of nucleic acids [51].

2.2.2.1 Phosphate Group Modifications

For a long time, modification of phosphate groups was limited to phosphorothioates (PS) and methyl phosphonates [52, 53]. PS oligonucleotides were designed by replacing non-bridged phosphate oxygen with sulfur. This method is one of the most popular oligonucleotide modifications, providing oligonucleotide-enhanced pharmacokinetic properties, bioavailability, and *in vivo* nuclease resistance [54, 55].

With the development of research, various new phosphate substituents have been formed on phosphorus atoms. Depending on the modification and the chemical method used, the net charge of the chain between nucleotides can be negative, neutral, or even positively [56].

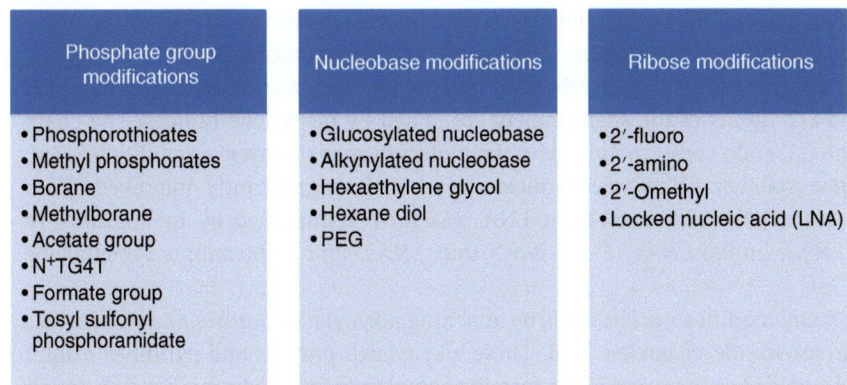

Figure 2.1 Backbone modification of nucleic acids.

Kumar et al. worked on the chemical modification of oligodeoxynucleotides. The phosphoric acid group was modified with borane, methyl borane, acetic acid group, and formic acid group. All of these modified oligodeoxynucleotides were metabolically stable and, in most cases, were effectively internalized by cells [57]. More recently, charge-neutral modification of DNA and RNA phosphate main strands has received a lot of attention in recent years because of its ability to improve both nuclease resistance and binding affinity with complementary DNA/RNA/dsDNA. Su et al. designed a G-rich oligodeoxynucleotide (TG4T) that replaces all phosphates with 4-(trimethylammonio)butylsulfonyl phosphoramidate group (N^+), resulting in a formal charge-neutral zwitterionic N^+TG4T. The N^+-modified G-quadruplex had enhanced nuclease resistance and thermal stability [58]. Encouraged by this finding, the team investigated the properties of tosyl sulfonyl phosphoramidate (Ts) modified oligodeoxynucleotide. They demonstrated that both N^+ and Ts modifications could improve the aqueous solubility and chemical stability of oligodeoxynucleotide [59]. All these methods could be used to synthesize NAN to improve stability.

2.2.2.2 Nucleobase or Ribose Modifications

In addition to the phosphate groups, there are many sites on the nucleobase and ribose that can be modified. 5-Methylcytosine (5mC) and its oxidized congeners, including 5-hydroxymethylcytosine (5hmC), 5-formylcytosine (5fC), and 5-carboxycytosine (5caC), are the most important epigenetic DNA bases [60]. However, 5-hydroxymethyluracil (5hmU) is not yet fully understood. Some bacteriophage genomes containing highly modified pyrimidine bases, i.e. O-glucosylated 5hmC, O-phosphorylated 5hmU, or O-glucosylated 5-hmU, have shown resistance to degradation by certain type II restriction endonucleases (REs) [61–63]. Inspired by these phenomena, Chakrapani et al. synthesized DNA containing glucosylated pyrimidines and found that this DNA was resistant to the cleavage of type II REs [64].

Previous studies have reported that modifications to the 2′-position of ribose, especially 2′-fluoro, 2′-amino, and 2′-Omethyl, can improve the stability of nucleic acid [65]. Over the past two decades, the fluorination of nucleosides has emerged as the most promising tool for obtaining bioactive compounds. The very strong C—F bond gives fluorine-containing compounds unusual metabolic stability, which is very useful in drug design [66]. In addition to increasing metabolic stability, functional groups in fluoro-substituted nucleosides increase lipophilicity and binding interactions with enzymes and macromolecules [67].

Locked nucleic acid (LNA) is synthesized by another modification of ribose, in which the C4' and O2' atoms of the sugar moiety are joined by methylene bridges. The chain ribose in the C3' endo conformation gives the molecule an extra layer of stability [68, 69]. The dynamic stability of oligodeoxynucleotides can be significantly improved when LNA is inserted [70]. The stability of LNA was further improved by modification of 5'-methyl-LNAs, amido-LNAs, 5'-Me-β-D-2'-thio LNAs, and N-Meaminooxymethylene LNAs [71–73].

Another study modifies nucleic acids by attaching alkynylated purines and pyridazines to the main grove side of nucleic acid. These alkynylated purines and pyridines exhibit unique and stable pairing properties by forming complementary hydrogen bonds between the pseudo-nucleobases of the DNA backbone. Therefore, the modified DNA duplex had higher stability and sequence-specificity [74].

The stable nucleic acids obtained by the above methods can be used to synthesize NANs and improve their stability. However, synthesizing these oligonucleotides is often too complex and expensive. Simple modifications can be urged to enhance stability. Single modifications may be commercially available and simple ways to improve the stability of DNA strands. Conway et al. modified oligonucleotide ends with hexadecylglycerol and hexandiol to significantly increase nuclease resistance in serum. DNA prismatic cages constructed with modified oligonucleotides showed increased resistance to nuclease degradation and had a lifetime of 62 hours in serum [75]. In addition, Hiroshi Abe and coworkers reported a chemical modification to obtain DNA-PEG conjugates by simply covalently binding a polyethylene glycol unit (PEG) to the 5' DNA terminal. The designed PEG-DNA structure is soluble in organic solvents and stable. Another method of obtaining PEG-DNA structure relies on the electrostatic grafting of amino-terminal PEG polymer to the backbone of the DNA to neutralize the negative charge of DNA strand. This strategy improves the stability of DNA in salt-free or low-ionic-strength aqueous medium [76]. These modification strategies are low-cost and easy to insert into DNA strands.

2.2.3 Coating with Protective Structures

Negatively charged DNA nanostructures are inevitably degraded in low-salt conditions or by nucleases in physiological fluids. Coating DNA nanostructures with protective structures could help these objects overcome difficulties and improve their stability in organic solvents. A number of structures have been reported to be coated with DNA nanostructures, including cowpea chlorotic mottle virus capsid proteins, cationic poly(2-dimethylaminoethylmethacrylate) (PDMAEMA)-based polymers, liposomes, and other proteins such as bovine serum albumin (BSA) [77–81]. Mikkilä et al. wrapped DNA origami with purified cowpea chlorotic mottle virus capsid proteins through electrostatic interaction. Protein-coated DNA origami had higher cell adhesion and drug delivery efficiency, as well as greater stability than naked DNA origami [77]. Kiviaho's research has shown that controllable polymer coatings can improve the transfection rate of DNA origami and enhance the stability of origami in a biological environment [78]. In 2015, Peng et al. coated PHBHHx nanoparticles with BSA and proposed that the presence of

BSA might be a promising method for enhancing the biological stability of nanoparticles [82]. Subsequently, Auvinen et al. synthesized BSA-coated DNA origami. Their result showed that the origami stability of endonucleases (DNase I) was significantly improved and that the BSA coating attenuated the activation of the immune response [83]. Xu et al. used cationic human serum albumin-coated DNA origami to resist the digestion of endonuclease, maintain stability under low salt concentration, and improve cellular uptake efficiency [84]. However, the coating strategies described above usually induce aggregation of DNA nanostructures, and the synthesis methods are very complex and expensive. In recent years, there have been more and more reports on low-cost and efficient methods of coating DNA nanostructures. Ahmadi et al. designed several DNA origami nanostructures coated with chitosan and synthetic linear polyethyleneimine (LPEI), both of which are low-cost natural cationic polysaccharide [85]. Chitosan and LPEI coatings protect DNA nanostructures from Mg^{2+}-depleted media and enzymatic degradation such as DNase I and 10% FBS. In addition, LPEI-coated DNA nanostructures are more efficient than chitosan-coated DNA nanostructures due to their higher charge density and stronger polyvalent binding to DNA. Another exciting finding was that the antinuclease ability of DNA nanostructures coated with oligolysines to 0.5 : 1 N:P was 10 times higher than that of uncoated DNA nanostructures. For further optimization, in media containing 10% FBS, PEG-coupled oligolysine coating showed a stronger slowing of nuclease degradation by up to ~400 times. PEGylated oligolysine coating is a low-cost and efficient method for DNA nanostructure protection [86, 87]. However, this remarkable improvement *in vitro* still falls short of the criteria for some biomedical applications. Anastassacos et al. prepared glutaraldehyde-mediated chemical crosslinking of oligolysine-PEG5K-coated DNA nanostructures. Compared with naked DNA nanostructures, the nuclease resistance was improved by ~1 000 000 times. Compared with oligolysine-PEG5K-coated DNA nanostructures, its resistance increased by ~250-fold [88]. Recently, it has been reported that 1,2-dioleoly-3-trimethylammonium-propane cationic lipid coated by DNA origami is significantly resistant to nuclease degradation [89].

The coating strategy creates additional problems that need to be addressed. The entire outer surface of DNA nanostructures is often covered, making it necessary for functional molecules to pass through a thick layer of covering material to bind to DNA strands or nanostructures. Kim and Yin found that nanostructures containing certain DNA brick motifs can maintain structural stability under low salt conditions, and wrapping the DNA brick nanostructures with dendritic oligonucleotides can protect the nanostructures from nuclease degradation. In addition, dendritic oligonucleotide-coated DNA brick nanostructures still show DNA strand accessibility on the surface [90]. More recently, Ma et al. modified DNA tetrahedron with HER2 target DNA aptamer and then coated the modified nanostructure with biomimetic camouflage (specially prepared erythrocyte membrane). Mixed eperytosomal nanoparticles exhibit longer circulation times and better HER2-positive cancer inhibition due to avoidance of the reticuloendothelial system and blood clearance [91]. This inspired us that the strategy of modification before coating could avoid the problem of unreachable modification sites. Palazzol et al. synthesized a compact short tube DNA origami (STDO) precisely fitting inside a stealthy liposome-loaded doxorubicin. This strategy provided a 6-hours *in vivo* circulation half-time as compared to

20 minutes for unprotected origami and a lesser inflammatory response in mice [23, 92]. Nevertheless, further research is needed to address this issue.

The application of DNA nanostructures in high-temperature conditions, such as templates for nanofabrication, is also the research focus. Their low thermal stability limited their broad application. Some studies make efforts to improve their thermal stability at high temperatures. With a thin Al_2O_3 film coating, DNA nanostructures could be transferred to carbon nanostructures at high temperatures (c. 800 °C), and their nanoscale topography was preserved [93]. Due to its good thermal stability and high Yang' modulus, graphene was reported to encapsulate water, DNA nanostructures, and nanoparticles [94, 95]. Graphene-encapsulated DNA nanostructures could avoid damage by water and have increased thermal stability deposited on mica at <150 °C [96]. Ricardo et al. synthesized chemical vapor deposition (CVD) graphene-encapsulated triangle DNA nanostructures and investigated their thermal stability by atomic force microscopy (AFM). They found that the topographical features of the DNA nanostructure disappeared after annealing at 300 °C for 5 hours but reappeared after 23 hours. However, without the graphene coating, the topographical features disappeared after heating at 300 °C for 45 minutes [97]. These results might be helpful in developing new applications of DNA nanostructure at high temperatures.

2.2.4 Covalent Crosslinking

The self-assembly of DNA strands into nanostructures is a simplified process. Molecules are bonded by non-covalent forces. In addition, DNA strands self-assemble leaving multiple nick points. All these can cause the nanostructures to wear out or break down [28, 98]. Covalent crosslinking can reduce the instability of DNA nanostructure. These cross-locks determine the arrangement of individual chains and prevent the dissociation of components.

Enzymatic ligation of the nicks is often used in molecular biology [99]. Thermal stability is another obstacle that DNA nanostructures must overcome. DNA nanostructures can only be used in solution below ~55 °C, which limits their applications and synthesis of more structures. Enzymatic ligation can improve the thermal stability of DNA nanostructures. With the presence of T4 ligase, a covalent phosphodiester bond can be formed between 5' phosphate and 3' hydroxl of the DNA molecules. In 2006, the Fygenson lab reported successful ligation of DNA nanotubes with T4 DNA ligase. The ligated nanotubes can withstand high temperatures of more than 70 °C and can be stably preserved in pure water for more than one month [99]. However, the tightly packed DNA origami can be a challenge for enzymatic ligation, as it blocks the passage of enzymes into the interior. It is necessary to optimize the ligation conditions and perform a good characterization of ligation. Recently, Rajendran et al. conducted a detailed analysis and conditional optimization of the enzymatic ligation of four two-dimensional DNA origami. They demonstrated that the ligation performed efficiently overnight at 37 °C instead of the usual 16 °C or room temperature and required a higher concentration of T4 DNA ligase than is usually used. DNA origami is enzymatically ligated by T4 DNA ligase, and the thermal stability of origami can be improved from 5 to 20 °C depending on the structure. In addition, origami is more compact after ligated [100]. Other studies have reported that enzymatic ligation enables DNA nanostructures to be super-stable at up to 37 °C and remain intact in the presence of 6 M urea without affecting their shape [101]. To solve the problem of limited enzymatic ligation sites, chemical ligation of 5′ phosphorylated strands and 3′ amino termini of oligonucleotide can seal the nick present between these two ends, increasing the thermal stability of DNA origami [102, 103].

Photochemical crosslinking is another strategy for stabilizing NANs [24]. Crosslinking agents are known to stabilize nucleic acids and are expected to improve the heat resistance of DNA nanostructures [104]. 8-Methoxypsoralen (8-MOP) is a natural drug found in many vegetables. It can form covalent adducts with pyrimidine bases in nucleic acids under ultraviolet (UV) light. The photo-cross-linking of 8-MOP can increase the thermal stability of DNA nanostructures by about 30 °C [105]. Gerling et al. induced the formation of covalent cyclobutene pyrimidine dimer (CPD) bonds by placing hymidine near DNA nanostructures under 310 nm irradiation. Then, by joining the free strand ends, the chain breaks at the crossing sites are bridged, and additional interhelix connections are created to design multilayered DNA origami objects by attaching covalently. Multilayer DNA origami objects were designed to maintain structural integrity at temperatures up to ∼90 °C and in double-distilled water without additional cations and showed enhanced nuclease resistance [22]. Engelhardt et al. are currently using CPD to stabilize NANs and open opportunities for risk. They suggest that this crosslinking provides unmet thermal stability and complete DNA origami stability, but requires optimization and validation of the crosslinking time for each origami structure [106]. Cassinelli et al. constructed a 24-oligonucletide 6-helix DNA origami tube with the 5′ end of each oligonucleotide modified with azide and the 3′ end modified with acetylene. Aided by a slight clicking reaction, the ends of the oligonucleotides are covalently linked to form a DNA strand alkane. At lower cationic concentrations, the structure remains intact even in pure water, and is resistant to exo-nuclease degradation, showing high thermal stability [107].

Even though enzymatic ligation and photochemical crosslinking could efficiently improve the stability of DNA nanostructures, the ligase cannot reach internal positions of the larger folding motifs, and crosslinking sites are not easy to adjust and locate independently [108]. Disulfide crosslinking has recently been applied to small nanostructures with up to 100 base pairs, and the thermal stability was up to 60 °C [109]. However, the size of nanostructures was limited, and the increase in thermal stability was modest. Recently, Wolfrum et al. reported disulfide crosslinking between 2′-deoxyuridine or 2′-deoxycytidine residues. The crosslinking sites could be any position of oligonucleotides. The cross-linked nanostructures ranging in size from 56 to 1456 nucleotides were assembled from protected oligonucleotides and had increased thermal stability, with UV-melting points 9–50 °C above those of the control structures (Figure 2.2) [110].

2.2.5 Tuning Buffer Conditions

The difference in cation concentration between NAN buffer and cell culture media or physiological conditions is another significant factor that threatens the stability of NAN. The cation concentration in typical cell culture media and physiological conditions is much lower than that in NAN buffer. So, the negative charges between the phosphate backbones are not enough to be neutralized. Electrostatic repulsion causes the strands to disassemble,

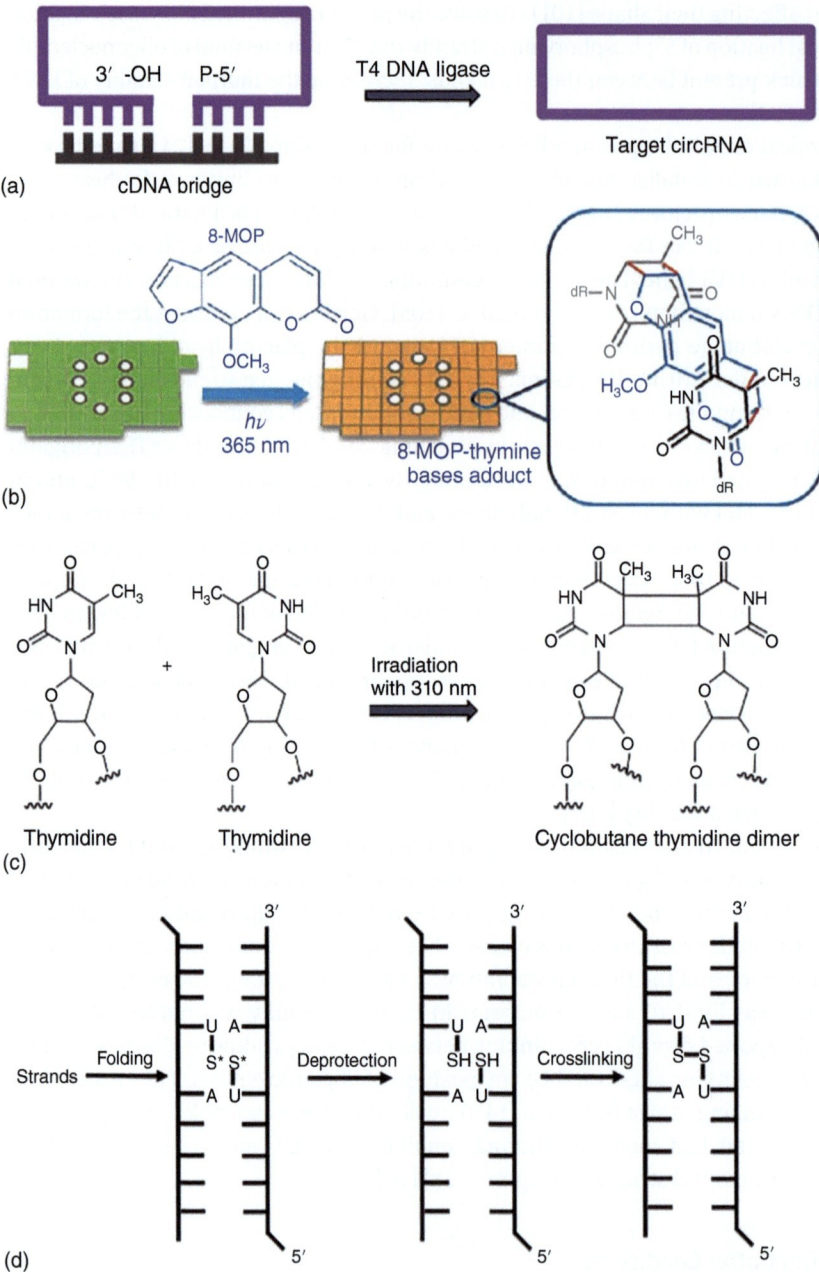

Figure 2.2 Covalent crosslinking of staple strands. (a) T4 DNA ligase could help ligate double-stranded duplexes (from Ref. [84]); (b) DNA origami structures and 8-MOP were exposed to 365 nm light to crosslink the strands (from Ref. [88]); (c) Adjacent thymidine forms cyclobutene thymidine dimer exposure to 310 nm light (from Ref. [21]). (d) Disulfide crosslinking using synthetic oligonucleotides with 5-mercaptoalkinyl-2′-deoxyuridine residues (from Ref. [110]).

leading to NAN's rapid disintegration. Designing NAN with a large interhelical spacing greatly reduces electrostatic repulsion, is the simplest way to avoid this effect. As previously reported, three-dimensional wireframe DNA origami containing only one to two helices per edge were indeed stable in low cation concentration buffer (e.g. ~140 mM NaCl) [111, 112]. Even so, DNA nanostructures with dense helices were required in many applications. Additional efforts are needed to focus on adjusting buffer conditions to stabilize these objects, even without high Mg^{2+} concentrations. Shih and Perrault reported in 2014 that the use of 6 mM Mg^{2+} and 200 nM actin protein could stabilize DNA nanostructures obviously, but only *in vitro* [28]. In 2018, Kroener et al. tested the short- and long-term stability of DNA origamis with four and six helices under conditions relevant to most biophysical experiments. The results showed that both origamis were stable over a 10 minutes period in 20 mM Tris buffer with 0, 1, 5, and 20 mM of Mg^{2+} and long-term stable for about two hours under low single-digit Mg^{2+} concentrations, resembling physiological conditions in cells [113]. Furthermore, addition of 1xTAE buffer containing 1 mM of Mg^{2+} chelator EDTA would lead DNA nanostructures rapidly falling apart by displacing Mg^{2+} ions from the DNA backbone, while removing the EDTA would restore structure stability. The addition of phosphate buffer caused damage to several structures by reducing the strength of the Mg^{2+}-DNA interaction, while the addition of 100–200 mM NaCl to screen the DNA backbone charges restored their stability [20]. Chopra et al. in 2016 synthesized biocompatible DNA origami structures stable in cell lysate at approximately physiological spermidine (Spd^{3+}) concentrations ([Spd^{3+}] < 1 mM) which considerably reduced ionic strength of the fold buffer. Spd-folded origami structures appear to be stable in cell lysate [114]. Although previous studies reported that an increasing cation concentration could prevent NANs from denaturation, an increasing cation concentration promoted denaturation of NANs in 4 M GdmCl. This cation-induced DNA origami denaturation effect can be used for selective denaturation for purification purposes [115]. In addition, ethylenediamine could neutralize the negative repulsive charges of DNA and enhance the nuclease stability of DNA origami in metal-ion-free conditions [116]. Thus, the judicious use of buffer conditions allows DNA nanostructures to remain stable in conditions compatible with cell culture (Figure 2.3).

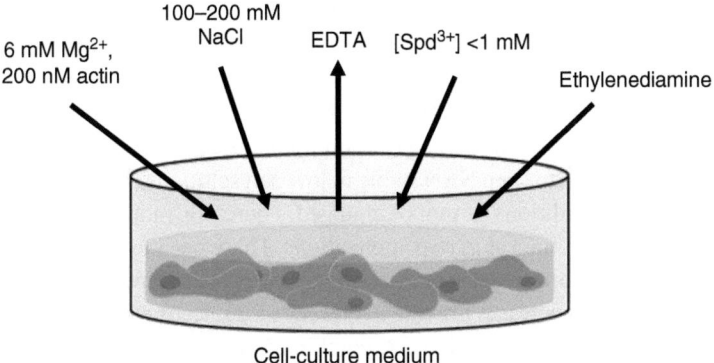

Figure 2.3 Tuning buffer conditions.

2.2.6 Construction of Novel NAN

DNA nanostructures assembled by the Watson–Crick base-pairing were denatured under heating or lack of Mg^{2+} condition. The stability of DNA nanostructures could be affected by their shape and internal structure. Wang and coworkers synthesized the DNA nanomesh through a novelty strategy. Briefly, three ssDNAs were mixed in 1 : 1 : 1 ratio. The mixture was assembled into net by gradient cooling from 95 to 37 °C. Then, the nanonet was transferred to a stable nanomesh under the presence of topoisomerase. The DNA nanomesh was thermally stable, Mg^{2+} free, and storable in a nonsolution state [117]. Star-shaped nanostructures such as DNA dendrimers and *n*-arm junctions have attracted much attention in biomedical applications, including delivery systems and scaffolds with immunoactivity and antibacterial activity [118–120]. However, the size of these star-shaped nanostructures was several to tens of nanometers, it is challenging to reach larger sizes (>100 nm) because of the use of short oligonucleotides [121, 122]. Researchers found that long double-stranded DNA (dsDNA) instead of ssDNA could be used to build large star-shaped nanostructures [123]. Sanchez-Rueda et al. built DNA nanostars with a higher number of arms through the use of dsDNA and coated them with DNA-binding proteins. They found that the hybrid protein-dsDNA complexes could resist enzymatic attack and elevated temperatures [124].

2.3 Conclusion and Recommendations

Self-assembled NANs exhibit excellent properties due to shape and size control, non-cytotoxicity, high efficiency in cells, and degradation to nontoxic products. Nevertheless, DNA nanostructures have been widely studied due to more mature synthesis techniques, and RNA nanotechnology is also advancing rapidly. NAN has attracted much more attention in the field of biomedicine. However, the stability of NAN is one of the biggest obstacles restricting its widespread applications. These catalyzed a design wave of protection strategies.

Salt concentration, organic solvent, nuclease degradation, and temperatures are the main factors affecting the stability of NANs. In this review, we summarized many methods to improve the stability of NANs under low salt concentrations, organic solvents, nuclease present, and high temperature conditions. However, many problems remain. A lot of strategies have been adopted to improve the stability of NAN in organic solvents, which facilitates the synthesis of a wide variety of nanostructures. However, how nanostructures synthesized in organic solvents remain stable under physiological conditions is another question. In addition, adjusting buffer conditions to keep NAN stable in low-salt solutions is also a great challenge because physiological conditions cannot be changed. Therefore, more research is needed because *in vitro* studies are aimed at *in vivo* applications. These need to be carefully handled in the context of cell types and substances associated with *in vivo* experiments.

The strategies described in this review can provide a reference for selecting the best protection methods for cells and *in vivo* applications. A combination of strategies may further stabilize nanostructures and influence cellular responses. Many protection strategies depend on structures, so the effectiveness of the protection methods chosen for different structures always needs to be verified. In addition, the protection method should maintain accessibility to active sites on NANs and show more functional groups in addition to protecting the shell. Moreover, persisting in constructing novel NANs is another way to improve the stability of NANs. The stability of NANs in cell and *in vivo* environments is of great significance for their biological applications. Due to the wide selection of protection methods currently available, it is recommended that all future reports on cellular and *in vivo* effects of NANs provide sufficient stability analysis data and select appropriate protection methods.

References

1 Loescher, S., Groeer, S., and Walther, A. (2018). 3D DNA origami nanoparticles: from basic design principles to emerging applications in soft matter and (bio-)nanosciences. *Angew. Chem. Int. Ed. Engl.* 57: 10436–10448. https://doi.org/10.1002/anie.201801700.

2 Chidchob, P. and Sleiman, H.F. (2018). Recent advances in DNA nanotechnology. *Curr. Opin. Chem. Biol.* 46: 63–70. https://doi.org/10.1016/j.cbpa.2018.04.012.

3 Mei, Q., Wei, X., Su, F. et al. (2011). Stability of DNA origami nanoarrays in cell lysate. *Nano Lett.* 11: 1477–1482. https://doi.org/10.1021/nl1040836.

4 Keum, J.W. and Bermudez, H. (2009). Enhanced resistance of DNA nanostructures to enzymatic digestion. *Chem. Commun.* 7036–7038. https://doi.org/10.1039/b917661f.

5 Ramakrishnan, S., Shen, B., Kostiainen, M.A. et al. (2019). Real-time observation of superstructure-dependent DNA origami digestion by DNase I using high-speed atomic force microscopy. *ChemBioChem* 20: 2818–2823. https://doi.org/10.1002/cbic.201900369.

6 Zhou, M., Zhang, T., Zhang, B. et al. (2022). A DNA nanostructure-based neuroprotectant against neuronal apoptosis *via* inhibiting toll-like receptor 2 signaling pathway in acute ischemic stroke. *ACS Nano* 16: 1456–1470. https://doi.org/10.1021/acsnano.1c09626.

7 Zhu, J., Yang, Y., Ma, W. et al. (2022). Antiepilepticus effects of tetrahedral framework nucleic acid via inhibition of gliosis-induced downregulation of glutamine synthetase and increased AMPAR internalization in the postsynaptic membrane. *Nano Lett.* 22: 2381–2390. https://doi.org/10.1021/acs.nanolett.2c00025.

8 Li, J., Yao, Y., Wang, Y. et al. (2022). Modulation of the crosstalk between Schwann cells and macrophages for nerve regeneration: a therapeutic strategy based on a multifunctional tetrahedral framework nucleic acids system. *Adv. Mater.* 34: e2202513. https://doi.org/10.1002/adma.202202513.

9 Chen, Y., Shi, S., Li, B. et al. (2022). Therapeutic effects of self-assembled tetrahedral framework nucleic acids on liver regeneration in acute liver failure. *ACS Appl. Mater. Interfaces* 14: 13136–13146. https://doi.org/10.1021/acsami.2c02523.

10 Wang, Y., Li, Y., Gao, S. et al. (2022). Tetrahedral framework nucleic acids can alleviate taurocholate-induced severe acute pancreatitis and its subsequent multiorgan injury in mice. *Nano Lett.* 22: 1759–1768. https://doi.org/10.1021/acs.nanolett.1c05003.

11 Chen, R., Wen, D., Fu, W. et al. (2022). Treatment effect of DNA framework nucleic acids on diffuse microvascular endothelial cell injury after subarachnoid hemorrhage. *Cell Prolif.* 55: e13206. https://doi.org/10.1111/cpr.13206.

12 Li, J., Li, M., Chen, X. et al. *Repair of Infected Bone Defect with Clindamycin-Tetrahedral DNA Nanostructure Complex-Loaded 3D Bioprinted Hybrid Scaffold*. Social Science Electronic Publishing.

13 Zhang, M., Zhang, X., Tian, T. et al. (2022). Anti-inflammatory activity of curcumin-loaded tetrahedral framework nucleic acids on acute gouty arthritis. *Bioact. Mater.* 8: 368–380. https://doi.org/10.1016/j.bioactmat.2021.06.003.

14 Zhao, D., Xiao, D., Liu, M. et al. (2022). Tetrahedral framework nucleic acid carrying angiogenic peptide prevents bisphosphonate-related osteonecrosis of the jaw by promoting angiogenesis. *Int. J. Oral Sci.* 14: 23. https://doi.org/10.1038/s41368-022-00171-7.

15 Qin, X., Xiao, L., Li, N. et al. (2022). Tetrahedral framework nucleic acids-based delivery of microRNA-155 inhibits choroidal neovascularization by regulating the polarization of macrophages. *Bioact. Mater.* 14: 134–144. https://doi.org/10.1016/j.bioactmat.2021.11.031.

16 Gao, Y., Chen, X., Tian, T. et al. (2022). A lysosome-activated tetrahedral nanobox for encapsulated siRNA delivery. *Adv. Mater.* 34: e2201731. https://doi.org/10.1002/adma.202201731.

17 Andersen, E.S., Dong, M., Nielsen, M.M. et al. (2009). Self-assembly of a nanoscale DNA box with a controllable lid. *Nature* 459: 73–76. https://doi.org/10.1038/nature07971.

18 McKee, T.J. and Komarova, S.V. (2017). Is it time to reinvent basic cell culture medium? *Am. J. Physiol. Cell Physiol.* 312: C624–C626. https://doi.org/10.1152/ajpcell.00336.2016.

19 Doye, J.P., Ouldridge, T.E., Louis, A.A. et al. (2013). Coarse-graining DNA for simulations of DNA nanotechnology. *Phys. Chem. Chem. Phys.* 15: 20395–20414. https://doi.org/10.1039/c3cp53545b.

20 Kielar, C., Xin, Y., Shen, B. et al. (2018). On the stability of DNA origami nanostructures in low-magnesium buffers. *Angew. Chem. Int. Ed. Engl.* 57: 9470–9474. https://doi.org/10.1002/anie.201802890.

21 Jahnen-Dechent, W. and Ketteler, M. (2012). Magnesium basics. *Clin. Kidney J.* 5: i3–i14. https://doi.org/10.1093/ndtplus/sfr163.

22 Gerling, T., Kube, M., Kick, B., and Dietz, H. (2018). Sequence-programmable covalent bonding of designed DNA assemblies. *Sci. Adv.* 4: eaau1157. https://doi.org/10.1126/sciadv.aau1157.

23 Palazzolo, S., Hadla, M., Russo Spena, C. et al. (1997). An effective multi-stage liposomal DNA origami nanosystem for in vivo cancer therapy. *Cancers* 11. https://doi.org/10.3390/cancers11121997.

24 Stephanopoulos, N. (2019). Strategies for stabilizing DNA nanostructures to biological conditions. *ChemBioChem* 20: 2191–2197. https://doi.org/10.1002/cbic.201900075.

25 Yang, H. and Xi, W. (2017). Nucleobase-containing polymers: structure, synthesis, and applications. *Polymers* 9. https://doi.org/10.3390/polym9120666.

26 Ke, F., Luu, Y.K., Hadjiargyrou, M., and Liang, D. (2010). Characterizing DNA condensation and conformational changes in organic solvents. *PLoS One* 5: e13308. https://doi.org/10.1371/journal.pone.0013308.

27 Sen, A. and Nielsen, P.E. (2007). On the stability of peptide nucleic acid duplexes in the presence of organic solvents. *Nucleic Acids Res.* 35: 3367–3374. https://doi.org/10.1093/nar/gkm210.

28 Hahn, J., Wickham, S.F., Shih, W.M., and Perrault, S.D. (2014). Addressing the instability of DNA nanostructures in tissue culture. *ACS Nano* 8: 8765–8775. https://doi.org/10.1021/nn503513p.

29 Kim, H., Surwade, S.P., Powell, A. et al. (2014). Stability of DNA origami nanostructure under diverse chemical environments. *Chem. Mater.* 26: 5265–5273.

30 He, Y., Tao, Y., Ribbe, A.E., and Mao, C. (2011). DNA-templated fabrication of two-dimensional metallic nanostructures by thermal evaporation coating. *JACS* 133: 1742.

31 Pearson, A.C., Liu, J., Pound, E. et al. (2012). DNA origami metallized site specifically to form electrically conductive nanowires. *J. Phys. Chem. B* 116: 10551.

32 Zhou, F., Michael, B., Surwade, S.P. et al. (2015). Mechanistic study of the nanoscale negative-tone pattern transfer from DNA nanostructures to SiO_2. *Chem. Mater.* 27: 1692–1698.

33 Kim, K.R., Lee, T., Kim, B.S., and Ahn, D.R. (2014). Utilizing the bioorthogonal base-pairing system of L-DNA to design ideal DNA nanocarriers for enhanced delivery of nucleic acid cargos. *Chem. Sci.* 5: 1533–1537.

34 Zhang, Y., Zhu, L., Tian, J. et al. (2021). Smart and functionalized development of nucleic acid-based hydrogels: assembly strategies, recent advances, and challenges. *Adv. Sci.* 8: 2100216. https://doi.org/10.1002/advs.202100216.

35 Kumar, S., Pearse, A., Liu, Y., and Taylor, R.E. (2020). Modular self-assembly of gamma-modified peptide nucleic acids in organic solvent mixtures. *Nat. Commun.* 11: 2960. https://doi.org/10.1038/s41467-020-16759-8.

36 Märcher, A., Kumar, V., Andersen, V.L. et al. (2022). Functionalized acyclic (L)-threoninol nucleic acid four way junction with high stability in vitro and in vivo. *Angew. Chem. Int. Ed. Engl.* https://doi.org/10.1002/anie.202115275.

37 Wang, Q., Chen, X., Li, X. et al. (2020). 2′-Fluoroarabinonucleic acid nanostructures as stable carriers for cellular delivery in the strongly acidic environment. *ACS Appl. Mater. Interfaces* https://doi.org/10.1021/acsami.0c11684.

38 Zhou, L., Sun, N., Xu, L. et al. (2016). Dual signal amplification by an "on-command" pure DNA hydrogel encapsulating HRP for colorimetric detection of ochratoxin A. *RSC Adv.* 6: 114500–114504.

39 Kim, S., Ahn, D.-R., Bang, D. et al. (2016). Self-assembled mirror DNA nanostructures for tumor-specific delivery of anticancer drugs. *J. Controlled Release* 243: 121–131.

40 Kim, K.R., Hwang, D., Kim, J. et al. (2018). Streptavidin-mirror DNA tetrahedron hybrid as a platform for intracellular and tumor delivery of enzymes. *J. Controlled Release* 280: 1–10.

41 Xu, Y., Wei, Y., Cheng, N. et al. (2018). Nucleic acid biosensor synthesis of an all-in-one universal blocking linker recombinase polymerase amplification with a peptide nucleic acid-based lateral flow device for ultrasensitive detection of food pathogens. *Anal. Chem.* 90: 708–715. https://doi.org/10.1021/acs.analchem.7b01912.

42 Chu, T.W., Feng, J., Yang, J., and Kopeček, J. (2015). Hybrid polymeric hydrogels via peptide nucleic acid (PNA)/DNA complexation. *J. Controlled Release* 220: 608–616. https://doi.org/10.1016/j.jconrel.2015.09.035.

43 Nielsen, P.E. and Egholm, M. (1999). An introduction to peptide nucleic acid. *Curr. Issues Mol. Biol.* 1: 89–104.

44 Murayama, K., Kashida, H., and Asanuma, H. (2015). Acyclic L-threoninol nucleic acid (L-aTNA) with suitable structural rigidity cross-pairs with DNA and RNA. *Chem. Commun.* 51: 6500–6503. https://doi.org/10.1039/c4cc09244a.

45 Taylor, A.I., Beuron, F., Peak-Chew, S.Y. et al. (2016). Nanostructures from synthetic genetic polymers. *ChemBioChem* 17: 1107–1110. https://doi.org/10.1002/cbic.201600136.

46 Wilds, C.J. and Damha, M.J. (2000). 2′-Deoxy-2′-fluoro-β-D-arabinonucleosides and oligonucleotides (2′F-ANA): synthesis and physicochemical studies. *Nucleic Acids Res.* 28: 3625–3635. https://doi.org/10.1093/nar/28.18.3625.

47 Watts, J.K. and Damha, M.J. (2008). 2′F-Arabinonucleic acids (2′F-ANA) — history, properties, and new frontiers. *Can. J. Chem.* 86: 641–656.

48 Watts, J.K., Katolik, A., Viladoms, J., and Damha, M.J. (2009). Studies on the hydrolytic stability of 2′-fluoroarabinonucleic acid (2′F-ANA). *Org. Biomol. Chem.* 7: 1904–1910. https://doi.org/10.1039/b900443b.

49 Huppert, J.L. and Balasubramanian, S. (2007). G-quadruplexes in promoters throughout the human genome. *Nucleic Acids Res.* 35: 406–413.

50 Huang, Y., Xu, W., Liu, G., and Tian, L. (2017). A pure DNA hydrogel with stable catalytic ability produced by one-step rolling circle amplification. *Chem. Commun.* 53: 3038.

51 Meanwell, M., Silverman, S.M., Lehmann, J. et al. (2020). A short de novo synthesis of nucleoside analogs. *Science* 369: 725–730. https://doi.org/10.1126/science.abb3231.

52 Saccà, B., Lacroix, L., and Mergny, J.L. (2005). The effect of chemical modifications on the thermal stability of different G-quadruplex-forming oligonucleotides. *Nucleic Acids Res.* 33: 1182–1192. https://doi.org/10.1093/nar/gki257.

53 Pozmogova, G.E., Zaitseva, M.A., Smirnov, I.P. et al. (2010). Anticoagulant effects of thioanalogs of thrombin-binding DNA-aptamer and their stability in the plasma. *Bull. Exp. Biol. Med.* 150: 180–184. https://doi.org/10.1007/s10517-010-1099-5.

54 Jie, J., Xia, Y., Huang, C.H. et al. (2019). Sulfur-centered hemi-bond radicals as active intermediates in S-DNA phosphorothioate oxidation. *Nucleic Acids Res.* 47: 11514–11526. https://doi.org/10.1093/nar/gkz987.

55 Zhou, J., Li, T., Geng, X. et al. (2021). Antisense oligonucleotide repress telomerase activity via manipulating alternative splicing or translation. *Biochem. Biophys. Res. Commun.* 582: 118–124. https://doi.org/10.1016/j.bbrc.2021.10.034.

56 Su, Y., Fujii, H., Burakova, E.A. et al. (2019). Neutral and negatively charged phosphate modifications altering thermal stability, kinetics of formation and monovalent ion dependence of DNA G-quadruplexes. *Chem. Asian J.* 14: 1212–1220. https://doi.org/10.1002/asia.201801757.

57 Kumar, P. and Caruthers, M.H. (2020). DNA analogues modified at the nonlinking positions of phosphorus. *Acc. Chem. Res.* 53: 2152–2166. https://doi.org/10.1021/acs.accounts.0c00078.

58 Su, Y., Edwards, P.J.B., Stetsenko, D.A., and Filichev, V.V. (2020). The importance of phosphates for DNA G-quadruplex formation: evaluation of zwitterionic G-rich oligodeoxynucleotides. *ChemBioChem* 21: 2455–2466. https://doi.org/10.1002/cbic.202000110.

59 Su, Y., Bayarjargal, M., Hale, T.K., and Filichev, V.V. (2021). DNA with zwitterionic and negatively charged phosphate modifications: formation of DNA triplexes, duplexes and cell uptake studies. *Beilstein J. Org. Chem.* 17: 749–761. https://doi.org/10.3762/bjoc.17.65.

60 Schön, A., Kaminska, E., Schelter, F. et al. (2020). Analysis of an active deformylation mechanism of 5-formyl-deoxycytidine (fdC) in stem cells. *Angew. Chem. Int. Ed. Engl.* 59: 5591–5594. https://doi.org/10.1002/anie.202000414.

61 Flodman, K., Corrêa, I.R. Jr., Dai, N. et al. (2020). In vitro type II restriction of bacteriophage DNA with modified pyrimidines. *Front. Microbiol.* 11: 604618. https://doi.org/10.3389/fmicb.2020.604618.

62 Hutinet, G., Lee, Y.J., de Crécy-Lagard, V., and Weigele, P.R. (2021). Hypermodified DNA in viruses of *E. coli* and *Salmonella*. *EcoSal Plus* 9: eESP00282019. https://doi.org/10.1128/ecosalplus.ESP-0028-2019.

63 Kieft, R., Zhang, Y., Marand, A.P. et al. (2020). Identification of a novel base J binding protein complex involved in RNA polymerase II transcription termination in trypanosomes. *PLoS Genet.* 16: e1008390. https://doi.org/10.1371/journal.pgen.1008390.

64 Chakrapani, A., Ruiz-Larrabeiti, O., Pohl, R. et al. (2022). Glucosylated 5-hydroxymethylpyrimidines as epigenetic DNA bases regulating transcription and restriction cleavage. *Chemistry* https://doi.org/10.1002/chem.202200911.

65 Shigdar, S., Macdonald, J., O'Connor, M. et al. (2013). Aptamers as theranostic agents: modifications, serum stability and functionalisation. *Sensors* 13: 13624–13637. https://doi.org/10.3390/s131013624.

66 Johnson, B.M., Shu, Y.Z., Zhuo, X., and Meanwell, N.A. (2020). Metabolic and pharmaceutical aspects of fluorinated compounds. *J. Med. Chem.* 63: 6315–6386. https://doi.org/10.1021/acs.jmedchem.9b01877.

67 Pal, S., Chandra, G., Patel, S., and Singh, S. (2022). Fluorinated nucleosides: synthesis, modulation in conformation and therapeutic application. *Chem. Rec.* e202100335. https://doi.org/10.1002/tcr.202100335.

68 Nielsen, J.T., Arar, K., and Petersen, M. (2006). NMR solution structures of LNA (locked nucleic acid) modified quadruplexes. *Nucleic Acids Res.* 34: 2006–2014. https://doi.org/10.1093/nar/gkl144.

69 Nielsen, J.T., Arar, K., and Petersen, M. (2009). Solution structure of a locked nucleic acid modified quadruplex: introducing the V4 folding topology. *Angew. Chem. Int. Ed. Engl.* 48: 3099–3103. https://doi.org/10.1002/anie.200806244.

70 Pal, R., Deb, I., Sarzynska, J., and Lahiri, A. (2022). LNA-induced dynamic stability in a therapeutic aptamer: insights from molecular dynamics simulations. *J. Biomol. Struct. Dyn.* 41: 1–10. https://doi.org/10.1080/07391102.2022.2029567.

71 Seth, P.P., Allerson, C.R., Siwkowski, A. et al. (2010). Configuration of the 5′-methyl group modulates the biophysical and biological properties of locked nucleic acid (LNA) oligonucleotides. *J. Med. Chem.* 53: 8309–8318. https://doi.org/10.1021/jm101207e.

72 Yahara, A., Shrestha, A.R., Yamamoto, T. et al. (2012). Amido-bridged nucleic acids (AmNAs): synthesis, duplex stability, nuclease resistance, and in vitro antisense potency. *ChemBioChem* 13: 2513–2516. https://doi.org/10.1002/cbic.201200506.

73 Goswami, A., Prasad, A.K., Maity, J., and Khaneja, N. (2022). Synthesis and applications of bicyclic sugar modified locked nucleic acids: a review. *Nucleosides Nucleotides Nucleic Acids* 41: 1–27. https://doi.org/10.1080/15257770.2022.2052316.

74 Okamura, H., Trinh, G.H., Dong, Z. et al. (2022). Selective and stable base pairing by alkynylated nucleosides featuring a spatially-separated recognition interface. *Nucleic Acids Res.* https://doi.org/10.1093/nar/gkac140.

75 Conway, J.W., McLaughlin, C.K., Castor, K.J., and Sleiman, H. (2013). DNA nanostructure serum stability: greater than the sum of its parts. *Chem. Commun.* 49: 1172–1174. https://doi.org/10.1039/c2cc37556g.

76 Chakraborty, G., Balinin, K., Portale, G. et al. (2019). Electrostatically PEGylated DNA enables salt-free hybridization in water. *Chem. Sci.* 10: 10097–10105. https://doi.org/10.1039/c9sc02598g.

77 Mikkilä, J., Eskelinen, A.P., Niemelä, E.H. et al. (2014). Virus-encapsulated DNA origami nanostructures for cellular delivery. *Nano Lett.* 14: 2196–2200. https://doi.org/10.1021/nl500677j.

78 Kiviaho, J.K., Linko, V., Ora, A. et al. (2016). Cationic polymers for DNA origami coating – examining their binding efficiency and tuning the enzymatic reaction rates. *Nanoscale* 8: 11674–11680. https://doi.org/10.1039/c5nr08355a.

79 Perrault, S.D. and Shih, W.M. (2014). Virus-inspired membrane encapsulation of DNA nanostructures to achieve in vivo stability. *ACS Nano* 8: 5132–5140. https://doi.org/10.1021/nn5011914.

80 Agarwal, N.P., Matthies, M., Gür, F.N. et al. (2017). Block copolymer micellization as a protection strategy for DNA origami. *Angew. Chem. Int. Ed. Engl.* 56: 5460–5464. https://doi.org/10.1002/anie.201608873.

81 Ramakrishnan, S., Ijäs, H., Linko, V., and Keller, A. (2018). Structural stability of DNA origami nanostructures under application-specific conditions. *Comput. Struct. Biotechnol. J.* 16: 342–349. https://doi.org/10.1016/j.csbj.2018.09.002.

82 Peng, Q., Wei, X.Q., Yang, Q. et al. (2015). Enhanced biostability of nanoparticle-based drug delivery systems by albumin corona. *Nanomedicine* 10: 205–214. https://doi.org/10.2217/nnm.14.86.

83 Auvinen, H., Zhang, H., Nonappa et al. (2017). Protein coating of DNA nanostructures for enhanced stability and immunocompatibility. *Adv. Healthcare Mater.* 6: 1700692. https://doi.org/10.1002/adhm.201700692.

84 Xu, X., Fang, S., Zhuang, Y. et al. (2019). Cationic albumin encapsulated DNA origami for enhanced cellular transfection and stability. *Materials* 12: 949.

85 Ahmadi, Y., De Llano, E., and Barišić, I. (2018). (Poly)cation-induced protection of conventional and wireframe DNA origami nanostructures. *Nanoscale* 10: 7494–7504. https://doi.org/10.1039/c7nr09461b.

86 Ponnuswamy, N., Bastings, M.M.C., Nathwani, B. et al. (2017). Oligolysine-based coating protects DNA nanostructures from low-salt denaturation and nuclease degradation. *Nat. Commun.* 8: 15654. https://doi.org/10.1038/ncomms15654.

87 Bertosin, E., Stömmer, P., Feigl, E. et al. (2021). Cryo-electron microscopy and mass analysis of oligolysine-coated DNA nanostructures. *ACS Nano* 15: 9391–9403. https://doi.org/10.1021/acsnano.0c10137.

88 Anastassacos, F.M., Zhao, Z., Zeng, Y., and Shih, W.M. (2020). Glutaraldehyde cross-linking of oligolysines coating DNA origami greatly reduces susceptibility to nuclease degradation. *JACS* 142: 3311–3315. https://doi.org/10.1021/jacs.9b11698.

89 Julin, S., Nonappa, Shen, B. et al. (2021). DNA-origami-templated growth of multilamellar lipid assemblies. *Angew. Chem. Int. Ed. Engl.* 60: 827–833. https://doi.org/10.1002/anie.202006044.

90 Kim, Y. and Yin, P. (2020). Enhancing biocompatible stability of DNA nanostructures using dendritic oligonucleotides and brick motifs. *Angew. Chem. Int. Ed. Engl.* 59: 700–703. https://doi.org/10.1002/anie.201911664.

91 Ma, W., Yang, Y., Zhu, J. et al. (2022). Biomimetic nanoerythrosome-coated aptamer-DNA tetrahedron/maytansine conjugates: pH-responsive and targeted cytotoxicity for HER2-positive breast cancer. *Adv. Mater.* 34: e2109609. https://doi.org/10.1002/adma.202109609.

92 Palazzolo, S., Hadla, M., Spena, C.R. et al. (2019). Proof-of-concept multistage biomimetic liposomal DNA origami nanosystem for the remote loading of doxorubicin. *ACS Med. Chem. Lett.* 10: 517–521. https://doi.org/10.1021/acsmedchemlett.8b00557.

93 Zhou, F., Sun, W., Ricardo, K.B. et al. (2016). Programmably shaped carbon nanostructure from shape-conserving carbonization of DNA. *ACS Nano* 10: 3069–3077. https://doi.org/10.1021/acsnano.5b05159.

94 Berman, D., Erdemir, A., and Sumant, A.V. (2014). Graphene: a new emerging lubricant. *Mater. Today* 17: 31–42.

95 Zhou, F., Li, Z., Shenoy, G.J. et al. (2013). Enhanced room-temperature corrosion of copper in the presence of graphene. *ACS Nano* 7: 6939–6947. https://doi.org/10.1021/nn402150t.

96 Matkovic, A., Vasić, B., Pesic, J. et al. (2016). Enhanced structural stability of DNA origami nanostructures by graphene encapsulation. *New J. Phys.* 18: 025016.

97 Ricardo, K.B. and Liu, H. (2018). Graphene-encapsulated DNA nanostructure: preservation of topographic features at high temperature and site-specific oxidation of graphene. *Langmuir* 34: 15045–15054. https://doi.org/10.1021/acs.langmuir.8b02129.

98 Wei, X., Nangreave, J., Jiang, S. et al. (2013). Mapping the thermal behavior of DNA origami nanostructures. *JACS* 135: 6165–6176. https://doi.org/10.1021/ja4000728.

99 O'Neill, P., Rothemund, P.W., Kumar, A., and Fygenson, D.K. (2006). Sturdier DNA nanotubes via ligation. *Nano Lett.* 6: 1379–1383. https://doi.org/10.1021/nl0603505.

100 Rajendran, A., Krishnamurthy, K., Giridasappa, A. et al. (2021). Stabilization and structural changes of 2D DNA origami by enzymatic ligation. *Nucleic Acids Res.* 49: 7884–7900. https://doi.org/10.1093/nar/gkab611.

101 Ramakrishnan, S., Schärfen, L., Hunold, K. et al. (2019). Enhancing the stability of DNA origami nanostructures: staple strand redesign versus enzymatic ligation. *Nanoscale* 11: 16270–16276. https://doi.org/10.1039/c9nr04460d.

102 Kalinowski, M., Haug, R., Said, H. et al. (2016). Phosphoramidate ligation of oligonucleotides in nanoscale structures. *ChemBioChem* 17: 1150–1155. https://doi.org/10.1002/cbic.201600061.

103 Weizenmann, N., Scheidgen-Kleyboldt, G., Ye, J. et al. (2021). Chemical ligation of an entire DNA origami nanostructure. *Nanoscale* 13: 17556–17565. https://doi.org/10.1039/d1nr04225d.

104 Rajendran, A., Magesh, C.J., and Perumal, P.T. (2008). DNA-DNA cross-linking mediated by bifunctional [SalenAlIII]+ complex. *Biochim. Biophys. Acta* 1780: 282–288. https://doi.org/10.1016/j.bbagen.2007.11.012.

105 Rajendran, A., Endo, M., Katsuda, Y. et al. (2011). Photo-cross-linking-assisted thermal stability of DNA origami structures and its application for higher-temperature self-assembly. *JACS* 133: 14488–14491. https://doi.org/10.1021/ja204546h.

106 Engelhardt, F.A.S., Praetorius, F., Wachauf, C.H. et al. (2019). Custom-size, functional, and durable DNA origami with design-specific scaffolds. *ACS Nano* 13: 5015–5027. https://doi.org/10.1021/acsnano.9b01025.

107 Cassinelli, V., Oberleitner, B., Sobotta, J. et al. (2015). One-step formation of "chain-armor"-stabilized DNA nanostructures. *Angew. Chem. Int. Ed. Engl.* 54: 7795–7798. https://doi.org/10.1002/anie.201500561.

108 Kashida, H., Doi, T., Sakakibara, T. et al. (2013). p-Stilbazole moieties as artificial base pairs for photo-cross-linking of DNA duplex. *JACS* 135: 7960–7966. https://doi.org/10.1021/ja401835j.

109 De Stefano, M. and Vesterager Gothelf, K. (2016). Dynamic chemistry of disulfide terminated oligonucleotides in duplexes and double-crossover tiles. *ChemBioChem* 17: 1122–1126. https://doi.org/10.1002/cbic.201600076.

110 Wolfrum, M., Schwarz, R.J., Schwarz, M. et al. (2019). Stabilizing DNA nanostructures through reversible disulfide crosslinking. *Nanoscale* 11: 14921–14928. https://doi.org/10.1039/c9nr05143k.

111 Benson, E., Mohammed, A., Gardell, J. et al. (2015). DNA rendering of polyhedral meshes at the nanoscale. *Nature* 523: 441–444. https://doi.org/10.1038/nature14586.

112 Veneziano, R., Ratanalert, S., Zhang, K. et al. (2016). Designer nanoscale DNA assemblies programmed from the top down. *Science* 352: 1534. https://doi.org/10.1126/science.aaf4388.

113 Kroener, F., Traxler, L., Heerwig, A. et al. (2019). Magnesium-dependent electrical actuation and stability of DNA origami rods. *ACS Appl. Mater. Interfaces* 11: 2295–2301. https://doi.org/10.1021/acsami.8b18611.

114 Chopra, A., Krishnan, S., and Simmel, F.C. (2016). Electrotransfection of polyamine folded DNA origami structures. *Nano Lett.* 16: 6683–6690. https://doi.org/10.1021/acs.nanolett.6b03586.

115 Ramakrishnan, S., Krainer, G., Grundmeier, G. et al. (2017). Cation-induced stabilization and denaturation of DNA origami nanostructures in urea and guanidinium chloride. *Small* 13. https://doi.org/10.1002/smll.201702100.

116 Li, S., Jiang, Q., Liu, S. et al. (2018). A DNA nanorobot functions as a cancer therapeutic in response to a molecular trigger in vivo. *Nat. Biotechnol.* 36: 258–264. https://doi.org/10.1038/nbt.4071.

117 Wang, X., Yu, J., Lan, W. et al. (2020). Novel stable DNA nanoscale material and its application on specific enrichment of DNA. *ACS Appl. Mater. Interfaces* 12: 19834–19839. https://doi.org/10.1021/acsami.0c02242.

118 Qu, Y., Yang, J., Zhan, P. et al. (2017). Self-assembled DNA dendrimer nanoparticle for efficient delivery of immunostimulatory CpG motifs. *ACS Appl. Mater. Interfaces* 9: 20324–20329. https://doi.org/10.1021/acsami.7b05890.

119 Mohri, K., Kusuki, E., Ohtsuki, S. et al. (2015). Self-assembling DNA dendrimer for effective delivery of immunostimulatory CpG DNA to immune cells. *Biomacromolecules* 16: 1095–1101. https://doi.org/10.1021/bm501731f.

120 Du, S.M., Zhang, S., and Seeman, N.C. (1992). DNA junctions, antijunctions, and mesojunctions. *Biochemistry* 31: 10955–10963. https://doi.org/10.1021/bi00160a003.

121 Huang, K., Yang, D., Tan, Z. et al. (2019). Self-assembly of wireframe DNA nanostructures from junction motifs. *Angew. Chem. Int. Ed. Engl.* 58: 12123–12127. https://doi.org/10.1002/anie.201906408.

122 Wah, J.L., David, C., Rudiuk, S. et al. (2016). Observing and controlling the folding pathway of DNA origami at the nanoscale. *ACS Nano* 10: 1978–1987. https://doi.org/10.1021/acsnano.5b05972.

123 Tian, Y., He, Y., Ribbe, A.E., and Mao, C. (2006). Preparation of branched structures with long DNA duplex arms. *Org. Biomol. Chem.* 4: 3404–3405. https://doi.org/10.1039/b605464a.

124 Sanchez-Rueda, E.G., Rodriguez-Cristobal, E., Moctezuma González, C.L., and Hernandez-Garcia, A. (2019). Protein-coated dsDNA nanostars with high structural rigidity and high enzymatic and thermal stability. *Nanoscale* 11: 18604–18611. https://doi.org/10.1039/c9nr05225a.

3

Framework Nucleic Acid-Based Nanomaterials: A Promising Vehicle for Small Molecular Cargos

Yanjing Li

Tianjin Medical University School and Hospital of Stomatology, Department of Prosthodontics, 12 Qixiangtai Road, Tianjin 300070, PR China

The biological polymer DNA carries information about life and stores genetic codes in living organisms. DNA is constructed from four basic blocks, including guanine (G), cytosine (C), adenine (A), and thymine (T), and is assembled through Watson–Crick base-pairing. In 1982, Prof. Seeman pioneered specific DNA architectures through rigid junctions [1], thus leading to the development of DNA nanotechnology. Thereafter, great progress has been made in the construction, modification, and application of nucleic acid nanomaterials. A variety of nucleic acid nanomaterials have been constructed, from DNA Holliday junctions to two-dimensional crystalline structures [2], two-dimensional nanogrids [3], advanced DNA origami [4], and DNA tetrahedra [5].

Among these architectures, framework nucleic acids (FNAs), which include DNA nanostructures such as DNA origami, tetrahedra, and polyhedra, have been widely researched and applied in the biomedical field because of their controllable structure, rapid assembly, and high yield. In addition, FNAs demonstrate excellent physical and chemical properties and biological characteristics and display a number of advantages in disease treatment, drug delivery, and biosensing. First, based on the principle of complementary base pairing, FNAs exhibit controlled self-assembly and excellent programmability [6]. According to the desired requirements, FNAs with different sizes, shapes, and chiral nanostructures can be constructed via specific designs. It is to be noted that the small size of these nanostructures leads to passive targeting properties and wide distribution in organisms, which facilitates certain delivery strategies. Furthermore, functional nucleic acid fragments can be connected to the ends of single-stranded DNA (ssDNA) through sequence lengthening or sticky ends, such as targeted aptamers. Second, the spatial arrangement between each successive base pair provides abundant sites for the insertion of small molecules, which is beneficial for the efficient loading and delivery of small molecular cargo. Third, FNAs possess excellent biocompatibility and biodegradability, low immunogenicity [7], and relative structural stability [8], which are essential for ideal drug carriers. Fourth, unlike linear nucleic acids, FNAs have unique interactions with cells, such as remarkable cell membrane penetrability and regulation of cell behavior [6]. It is difficult for ssDNA to enter cells directly, whereas FNAs can enter cells without the help of transfection agents, which may

Nucleic Acid-Based Nanomaterials: Stabilities and Applications, First Edition.
Edited by Yunfeng Lin and Shaojingya Gao.
© 2024 WILEY-VCH GmbH. Published 2024 by WILEY-VCH GmbH.

be because of the unique spatial structure of the latter. Spontaneous cellular internalization can ensure the efficient entry of the loaded cargo into the cell, and the biological effect of FNAs and cargos may act synergistically through a pleiotropic mechanism.

Based on these merits, FNAs are regarded as ideal carriers of small molecular agents that can overcome the typical limitations of using these agents in clinical applications, such as poor stability in the physiological environment, low targetability, low cell uptake efficiency, and undesired high off-target toxicity. Therefore, FNAs have gradually become a research hotspot in the field of small-molecule transport. Here, we summarize the current status of FNAs for the delivery of small molecules. This review focuses on the classification, synthesis, and properties of FNAs and emphasizes their potential in various strategies for the delivery of small molecules, their biomedical applications, and the underlying mechanisms. Finally, the prospects and challenges in the use of FNAs for drug delivery are discussed.

3.1 Basis of FNAs as Potential Drug Carriers

3.1.1 Classification and Construction of FNAs

FNAs include various architectures such as compact folded DNA origami, tetrahedral or polyhedral DNA frameworks, and spherical DNA structures. DNA origami, one of the widely researched FNAs, is a kind of self-assembled DNA nanostructure with a precisely designed shape and size. DNA origami is constructed by folding a long ssDNA, known as a "DNA scaffold strand," and introducing a set of short ssDNA, which binds to the DNA scaffold and acts as "staples" to hold nonarbitrary structures (Figure 3.1). The DNA origami technique dates back to the pioneering work of Rothemund in 2006 [9]. After that seminal publication of Rothemund, hundreds of DNA origami structures have been fabricated using this method, including two-dimensional structures such as triangles, rectangles, five-pointed stars, smiley-face circles [9], and crystals, as well as three-dimensional structures such as nanotubes [10], nanocubes [11], honeycombs, gridirons, wireframe nanopolyhedra, and polyhedral meshes [12]. Recently, scaffold-free structures [13], shape-complementary DNA components [13], and dynamic architectures that can respond to stimuli have been developed. In Rothemund's research, the ssDNA scaffold and staples were designed using a computer program that assists with manual correction, and the origami structures were fabricated through DNA self-assembly in a one-pot reaction. Using this approach, a series of nanoscale two-dimensional shapes have been fabricated, including triangles, rectangles, stars, and combinations, as well as disks with smiley-face holes, patterns representing letters, maps, and even Mona Lisa [14]. Based on the "bottom-up" self-assembly route, Douglas et al. improved the design program and developed a graphical interface-based design program named caDNAno to save time and reduce errors in origami design [15]. They subsequently extended the one-spot DNA origami method to build custom three-dimensional structures [16], such as a 10-layer monolith, a square nut with a hexagonal tunnel, and a stacked cross consisting of two domains. Moreover, a larger structure, icosahedral wireframe DNA origami, has also been constructed using a two-step hierarchical assembly method [16]. Gradually, this technique has been generalized to DNA origami fabrication, where these architectures have played a prominent role in the biomedical field because of their precisely controllable dimensions, shapes, and sizes.

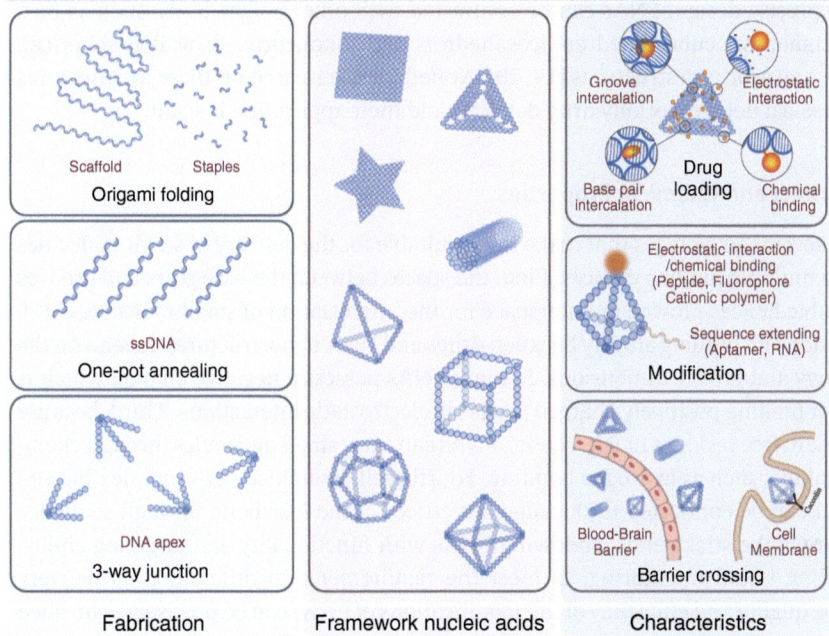

Figure 3.1 The fabrication and properties of FNAs that are suitable for the delivery of small molecules.

Another widely researched FNA is the DNA tetrahedron, which is the simplest polyhedron. Tetrahedrons, including DNA and RNA tetrahedrons, are a family of wireframe structures with lengths of approximately 10 nm on a side and are assembled using a few ssDNAs. The fabrication of DNA tetrahedrons can be traced back to 2005, when Goodman's team designed a family of DNA tetrahedrons by using four ssDNAs through one-pot annealing [5]. The synthesized tetrahedron possesses a high assembly yield (~95%), flexible junctions in the vertices, and a rigid structure that is capable of resisting deformation. A single tetrahedron can act as a block for assembling larger constructs via programmable DNA linkers and is assembled via a rapid process. Hence, DNA tetrahedra have been widely studied in the biomedical field, paving the way for the development and application of tetrahedral FNAs (tFNAs). DNA tetrahedrons can also be fabricated using other methods. In contrast to the method used by Ge et al., DNA tetrahedrons have been formed by four DNA triplexes. The DNA triplex has a sticky end such that four triplexes can interact with each other to form a tetrahedron [17]. Three-way-junction hybridization is generally used for the fabrication of DNA polyhedra, enabling the formation of large structures. Nevertheless, this process is hampered by cost and yield [18]. Veneziano et al. introduced a DNA wireframe design procedure, DAEDALUS, and constructed a DNA tetrahedron using a scaffold and a set of staples via the origami approach [19]. Compared with one-pot annealing, the scaffold folding method can be used to construct larger and more rigid DNA tetrahedrons. However, this approach is expensive and is only applicable to large structures. Therefore, owing to their structural simplicity, simple synthesis procedure, and high yield, DNA tetrahedra fabricated via one-pot annealing have been widely researched and applied.

Based on precise design, FNAs can be contrasted with other polyhedrons, such as hexahedron, octahedron, cuboctahedron, icosahedron, and zonohedron, as well as spherical, bioinspired, and arbitrary structures [19, 20]. Nonetheless, research on these architectures in the biomedical field, especially drug delivery, and their application is scant.

3.1.2 Physical and Chemical Properties

FNAs offer several advantages that make them suitable for the delivery of small molecules and provide multiple binding choices. First, the spaces between the base pairs and grooves in DNA double helices provide an interspace for the intercalation of small molecules [21]. The interaction and binding affinity between drugs and DNA nanostructures depend on the DNA topology and groove dimensions. Second, FNAs possess a negative charge, which is beneficial for binding positively charged drugs via electrostatic interactions. Third, because of the abundant free residues in DNA bases, FNAs can carry small molecules through chemical cross-linking, such as hydrogen bonding. Fourth, self-assembled FNAs are flexible and editable and can be connected to the edge or vertices of the backbone through sequence extension or via the sticky ends, endowing FNAs with functionality and targeting ability, while enabling visual monitoring, to meet the requirements of different drug delivery systems. The quality and efficiency of the modifications of FNAs can be precisely controlled owing to the exact localization of nucleic acids. Consequently, FNAs exhibit relative structural stability both *in vitro* and *in vivo* [22]. Under storage conditions, that is, in a buffer containing magnesium ions, at constant pH, and at low temperature, FNAs can remain stable for several days or even weeks without polymerization or degradation. In biological environments, FNAs are exposed to low cation concentrations, variable pH, and active DNA-degrading enzymes, and their structural integrity can persist for hours or several days. However, compared to double-stranded DNA, FNAs are more resistant to DNase, and the degradation rate is dependent on the structure of the FNAs. For instance, closely packed structures can be degraded more slowly [23]. In addition, modifications with lipid bilayers, proteins, or cationic polymers can protect FNAs from DNase digestion, enhancing their stability and function under physiological conditions [24–26]. Finally, with the advantages of precise design and sophisticated manufacturing, the dimension and shape of FNAs can be controlled, affording specific vectors for the delivery and application of different drugs.

3.1.3 Biological Properties

Owing to their excellent biocompatibility and biodegradability, rapid and autonomous endocytosis, and extensive intracellular distribution, FNAs have been widely used in the field of drug delivery. Moreover, DNA is an essential molecule for life and can be degraded by active nucleases *in vivo*. Therefore, the nature of DNA endows FNAs with tremendous biosafety. Biocompatibility is an essential characteristic of carriers. Although many currently researched nanocarriers can efficiently improve the bioavailability and biological effects of cargos, their poor biocompatibility and biodegradability hamper their in-depth research and application. Therefore, FNAs provide valuable advantages for drug delivery.

FNAs are more readily endocytosed compared to ssDNA. FNAs can rapidly enter cells in large quantities without the help of transfection agents and can retain their structural integrity. The efficiency of cellular uptake depends on the shape and size of the DNA structure. By using a time-lapse live-cell imaging system, Zeng et al. demonstrated that three-dimensional rigid DNA structures are better uptaken by the cells than two-dimensional DNA origami structures [27]. Three-dimensional FNAs, especially tFNAs with an edge length of 21 bp, are more effectively internalized than other nucleic acid materials of various sizes and shapes, which may be due to the unique spatial structure and dimensions of the former [28]. A proteomic identification method has been used to screen proteins that interact with tFNAs during endocytosis. The results revealed that both caveolin- and micropinocytosis-related proteins are involved in this process, indicating caveolin-mediated endocytosis and micropinocytosis facilitate the internalization of tFNAs [8, 29]. Once they enter the cell, they reach the lysosome via a microtubule-dependent pathway and become localized there [30], precluding their use in drug delivery. To achieve lysosomal escape, a cationic polymer named ethylene imine (PEI) is utilized to modify tFNAs [31]. The PEI-modified nanocomplex can avoid lysosomal degradation and displays enhanced structural stability and cell-entry ability, making it a prospective tool for drug delivery. Furthermore, the intracellular distribution of FNAs can be altered by modification with functional monomers, thereby expanding their applications as delivery agents. For instance, aptamer AS1411-modified tFNAs can reach the nucleus because of the high affinity of AS1411 for the nuclear membrane protein nucleolin [32]. In addition to notable cellular internalization, FNAs have excellent tissue penetration characteristics, enabling transdermal drug delivery by various approaches. FNAs penetrate tissue in a size-dependent manner. tFNAs that are ~20 nm in size have been demonstrated to be effective carriers with the best skin penetration and drug-loading ability among different FNAs with a wide size range [33]. The excellent biocompatibility and cell-entry ability enable FNAs to enhance the bioavailability, targeting ability, and biological effects of small molecules.

3.2 Small-molecule Cargos

Small-molecule drugs are widely used in the treatment of diseases, such as tumors, inflammation, and degenerative diseases, as well as in antibiotic therapy. However, many small molecules face challenges, such as low bioavailability, inability to target specific tissues/cells, and serious side effects, which greatly limit their clinical application. Therefore, there is an urgent need to develop an ideal drug carrier to overcome these deficiencies. FNAs have been widely studied in the field of small-molecule delivery because of their excellent biocompatibility, high drug-loading efficiency, unique editability, and excellent cell endocytosis. The interaction between DNA and small-molecule drugs can be divided into the following four modalities [34]: (i) the molecule sits in the grooves of DNA duplexes, (ii) the molecule is sandwiched between base pairs, (iii) the molecule chemically reacts with DNA, and (iv) the molecule cleaves DNA. FNAs can effectively overcome the shortcomings of the delivery modes of drugs and enhance the biological effects of the drug molecules. Below, we summarize current research on the use of FNAs for the delivery of small-molecule drugs, identifying opportunities for the in-depth development and clinical transformation of FNAs.

3.2.1 Antitumor Agents

3.2.1.1 Chemotherapeutic Drugs

Chemotherapeutic drugs such as doxorubicin (DOX), paclitaxel (PTX), and cisplatin are classic antitumor agents. However, these traditional drugs possess a variety of drawbacks, including poor selectivity and low accumulation in tumors. The therapeutic requirements are not met by small doses of these drugs, whereas large doses can produce adverse side effects, toxicity, and resistance. Therefore, FNAs have been exploited to build intelligent drug delivery systems (Figure 3.2a), enabling antitumor agents to directly reach and accumulate in tumors. These systems can release their cargo at a predefined time or location, realize anti-resistance, improve therapeutic potency, and reduce drug toxicity and side effects.

For instance, to address the challenges in the application of DOX, extensive studies of the interaction between DOX and DNA base pairs have been conducted [35]. Zhang et al. first adopted three types of two-dimensional DNA nanostructures, triangle, square, and tubular DNA origami, as DOX delivery vehicles. They demonstrated that the triangular-shaped DNA origami exhibited the highest accumulation in tumors, whereas the other forms were also distributed in the liver and kidney [36]. Hence, triangle DNA origami loaded with DOX shows long-lasting properties in orthotopic breast cancer and prominent therapeutic efficacy in a mouse model. In attempts to further improve the efficiency of DOX loading, researchers determined that tightly folded DNA origami possesses higher loading efficiency [37]. Straight nanotubes with 18-helix bundles, and a twisted version, have been designed

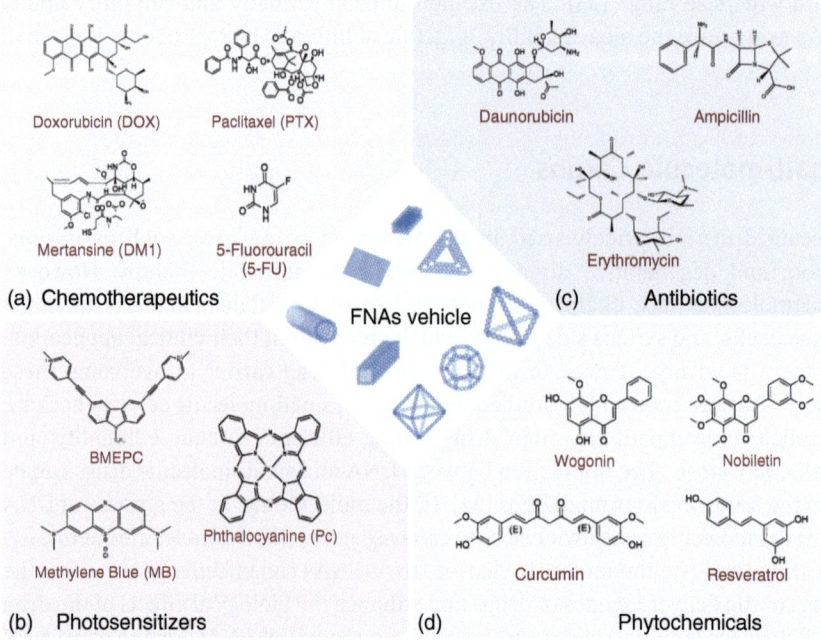

Figure 3.2 Small molecular cargo delivered by FNAs. (a) Chemotherapeutics. (b) Photosensitizers. (c) Antibiotics. (d) Phytochemicals.

for DOX delivery, where the twisted nanotube shows better drug loading capacity and controlled release behavior, enabling a release of the loaded drugs into tumor cells rather than in the environment, using a low drug concentration. The twisted DNA nanotubes can be localized in the nucleus, and they induce more cellular apoptosis than free DOX. This indicates the higher potential of the former for targeted antitumor drug delivery. In addition, a DNA origami of six helical bundles has been reported, which can be loaded with DOX at a high stoichiometric ratio of 1 : 3200 between DNA origami and DOX molecules [38]. These DOX-loaded FNAs have been modified with targeting ligands or tumor-penetrating peptides to achieve targeted internalization and accumulation in tumor cells [39–41]. This can lead to an effective therapy with less collateral damage.

FNAs have been widely utilized for the transport of chemotherapeutic agents, including DOX, PTX, cisplatin, 5-fluorouracil (5-FU), and mertansine (DM1). Shi et al. designed an effective PTX delivery strategy using tFNAs [42]. To the best of our knowledge, free PTX cannot cross the blood–brain barrier (B-BB), resulting in restricted effects on glioblastoma multiforme. In that study, two aptamers were appended to tFNAs to guide the delivery system to penetrate the B-BB and selectively enter glioblastoma multiforme cells. Consequently, the prepared nanocomplex could notably inhibit the proliferation, migration, and invasion of tumor cells. The team also utilized tFNAs to overcome the resistance of non-small cell lung cancer cells to PTX [43]. For the delivery of cisplatin, a tubular structure with dimensions of $6 \times 5.4 \times 24$ nm, a small cubic structure with dimensions of $8 \times 8 \times 22$ nm, and a large cubic structure with dimensions of $12 \times 12 \times 22$ nm were constructed [44]. These structures showed similar drug release behavior, whereas the cubic structures displayed better stability in the presence of nuclease, which may be due to the compact arrangement of the DNA duplexes. The large cubic vehicle exhibited the highest accumulation and long-lasting properties in tumors, thereby efficiently inducing cell apoptosis and overcoming drug resistance. Furthermore, rectangular DNA origami was developed for the delivery of 5-FU via covalent grafting on a DNA backbone [45], and nanoerythrosome-coated tFNAs were designed for the transmission of maytansine (DM1) [46]. These strategies enabled remarkable improvements to address the limitations of small-molecule chemotherapeutic agents, such as short half-life, low efficacy, and poor targeting ability.

3.2.1.2 Phototherapeutic Agents

Photodynamic therapy (PDT) and photothermal therapy (PTT) mediated by photosensitizers represent promising strategies for tumor therapy, as they provide a noninvasive approach. Phototherapeutic agents can be activated by light at an appropriate wavelength, and then they generate transient but superfluous reactive oxygen species (ROS) or heat, resulting in tumor cell death and solid tumor elimination. However, free photosensitizers face transport, target, and resistance problems; hence, achieving accurate delivery, controlled release, and superable drug resistance are currently the main challenges. As excellent drug carriers, FNAs can be used to construct various multifunctional delivery systems (Figure 3.2b), thus overcoming the drawbacks of free photosensitizers and enhancing their therapeutic effect. FNAs have provided an advanced strategy for delivering small-molecule phototherapeutic agents with the following advantages: (i) actively targeted photosensitizer delivery, (ii) stimulus-responsive and accurate photosensitizer release, (iii) overcoming resistance and hypoxia, and (iv) maintaining photophysical stability.

For instance, the carbazole derivative BMEPC is a traditional near-infrared light-excited photosensitizer that possesses a high ROS yield, deep tumor penetrability, and hypoxic applicability. However, it is restricted by poor solubility and stability. Liang and coworkers constructed triangular DNA origami to load BMEPC through intercalation [47]. Owing to the excellent biocompatibility, low immunogenicity, and excellent cell-entry ability of the triangular-shaped DNA origami, the drawbacks of BMEPC can be significantly reduced. Thus, the delivery system displays enhanced solubility, fluorescence emission, and stability in an aqueous solution, which enhances the photodynamic efficiency of BMEPC. Furthermore, Shaukat et al. developed a 6-helix bundle DNA origami for the delivery of phthalocyanine [48], another classic photosensitizer for PDT. Phthalocyanine can be loaded onto DNA origami through electrostatic adsorption, where the DNA origami acts as a vehicle to enhance the stability and optical properties of phthalocyanine, which protects the DNA origami against enzymatic digestion. Wireframe FNAs have also been utilized for the delivery of photosensitizers [49]. Methylene blue was loaded into the easily constructed DNA tetrahedron to address the poor cell penetration and stability of methylene blue in biological environments. The nanocomplex effectively suppressed tumor growth both *in vitro* and *in vivo*. By virtue of their excellent solubility, stability, modifiability, and penetrability, FNAs can ameliorate the shortcomings of free photosensitizers, such as low solubility, poor tumor penetrability, and inevitable side effects caused by the lack of selectivity. Hence, FNAs are promising candidates for the accurate, specific, and enhanced delivery of phototherapeutic agents, which may greatly expand the application of these agents.

3.2.2 Antibiotic Agents

Antibiotics, one of the greatest accomplishments of the twentieth century, are viewed as "wonder drugs" because they provide an ideal way to cure serious bacterial infections. A series of small-molecule antibiotics, including penicillin, cephalosporin, erythromycin, and tetracycline, have been widely applied in clinical therapy. However, the resistance induced by the widespread use and toxicity caused by the lack of targeted action clearly demands an alternative approach [50]. It has been reported that antibacterial drugs such as fluoroquinolones, cefoxitin, and gatifloxacin can bind to DNA through an intercalation mechanism [34, 51]. Hence, DNA structure-based FNAs have provided a promising strategy for the efficient loading of these antibiotics, achieving targeted delivery, and overcoming resistance (Figure 3.2c).

Halley et al. constructed a rod-like DNA origami with four open cavities for daunorubicin transport [52]. The delivery system significantly increases the entry and retention of drugs in cells, circumvents efflux pump-mediated drug resistance, and consequently enhances the therapeutic efficacy. Wireframe DNA architectures, tFNAs, have also been employed for the delivery of broad-spectrum antibiotics [53]. Ampicillin-loaded tFNAs exhibit a sufficient encapsulation rate, combined with good structural stability. The nanoplatforms evidently boosted the effects of ampicillin. Specifically, they enhanced drug uptake in methicillin-resistant *Staphylococcus aureus* (MRSA) and evidently reduced the resistance level. The same group also developed an erythromycin-loaded tFNA cargo tank through intercalation between DNA and erythromycin [54]. As expected, the vehicle was investigated as a means of enhancing the antimicrobial effect by facilitating membrane destabilization in *Escherichia coli* and reducing drug resistance by improving the internalization of erythromycin. Aptamer-modified DNA origami has been utilized for the targeted delivery of lysozyme [55], where the prepared nanostructures can accurately target Gram-positive and Gram-negative bacteria and suppress their growth. Hence, in addition to providing alternatives, FNAs provide an efficient tool in fighting antibiotic resistance and extending the application of antibacterial agents.

3.2.3 Phytochemicals

Phytochemicals are extracted from natural fruits, vegetables, legumes, and plants, especially medicinal plants. These phytochemicals are usually small molecular agents with certain pharmacological effects that can be exploited in the treatment of various diseases, as natural agents for promoting human health. For instance, resveratrol from grapes shows excellent ability to alleviate oxidative stress and inflammation [56]; quercetin, which is ubiquitous in fruits and vegetables, is a clinically applied antioxidant and antiaging agent [57]; capsaicin from chili pepper has significant anticancer activity [58]; and artemisinin extracted from the Chinese herb Qinghao has been approved as an antimalarial drug [59]. However, these compounds have poor pharmacokinetics, weak cell/tissue targeting, and penetration, and are unable to cross biological barriers, such as B-BB and alimentary mucosa, thereby motivating advanced delivery approaches for their application. It has been demonstrated that phytochemicals complex strongly with DNA by binding to minor and/or major grooves of DNA duplexes [60, 61]. Benefiting from this unique interaction, FNAs demonstrate great potential for the transport of these small-molecule phytochemicals (Figure 3.2d).

Yunfeng Lin's group at Sichuan University has focused on the delivery of phytochemical compounds. They thoroughly studied tFNAs with a side length of 21 bp as the vehicle for efficient drug loading and intelligent delivery. Through intercalation binding, tFNAs can efficiently load and transport wogonin, a type of flavonoid, into chondrocytes to suppress the inflammation process and promote cartilage regeneration during the development of osteoarthritis [62]. The tFNA/wogonin complex showed high entrapment efficiency, enhanced chondrocyte internalization, and improved biological effects. Other flavonoids, such as curcumin, quercetin, and nobiletin, have also been loaded onto tFNAs [63]. Similar to wogonin, the cargo tank effectively offset the deficiencies of these flavonoids, enhanced their therapeutic effects, and facilitated their application in clinical translation. Resveratrol, a type of polyphenol, can be loaded into tFNAs by binding to the grooves of DNA duplexes. tFNA vehicles significantly improve the solubility and stability of free resveratrol [64] and also enhance its cellular uptake and biological effects. In addition, because of the ROS scavenging ability of DNA, tFNAs and resveratrol may have a synergistic effect, which may be a reasonable explanation for the sufficient therapeutic efficacy of the complex. Compactly folded DNA nanostructures may provide higher entrapment efficiency because of the intercalation of DNA and phytochemicals; however, further investigation is needed. These studies have laid the foundation for the construction of new Chinese medicine and are of far-reaching significance for the inheritance and development of traditional Chinese medicine.

3.3 Merits of FNA Delivery Systems in Biomedical Application

3.3.1 Efficient Drug Delivery

FNAs have been extensively utilized in the transport of small molecules because of their abundant interspaces for the intercalation of small molecules, outstanding structural stability, excellent biocompatibility, and prominent cell/tissue penetration, thus providing a prospective approach for efficient drug transport. Based on the different binding affinities of different drugs for DNA, along with the differences in their influence on the stability of DNA structures, specific FNAs can be constructed to meet diverse drug delivery needs, enabling significant improvements in the therapeutic effects of the drugs. FNA vehicles also possess drug retention capabilities, which allow the loaded drugs to diffuse out of the vehicle into the targets while precluding immediate release upon transfer to an environment with low drug concentration. Thus, because of their high loading efficiency and cell internalization, FNA vehicles can significantly advance the delivery of cargo drugs.

For example, tFNA vehicles significantly enhanced the internalization of loaded agents [63], from a 5.6% uptake rate to 32.8% in normal macrophages and 12–82.9% in the inflammatory phenotype of macrophages, suggesting dramatically efficient delivery of curcumin, mediated by the FNA vehicles. FNA delivery systems can be used as carriers for a variety of therapeutic agents, including the drugs mentioned above. Vitamin B12 (VB12), a potential therapeutic option for Parkinson's disease (PD), has been loaded into tFNAs [65]. The VB12-loaded nanostructures dramatically accelerate the transport and utilization of VB12 in the brain, leading to abnormal accumulation in proteins and effective therapy for PD or other neurodegenerative diseases. Moreover, DNA-based nanostructures can deliver not only a single molecule but also two or more types of agents simultaneously. An octahedral DNA origami wireframe was constructed for the delivery of a chemotherapeutic and photothermal agent and siRNA [66]. The vehicle significantly enhanced the internalization of the payload (Figure 3.3). Once taken up by tumor cells, the nanoplatform gradually released these cargo drugs, leading to enhanced tumor elimination mediated by the synergy of chemotherapy, PTT, and gene therapy. It is beyond doubt that the nanoscale spatial heterogeneity and precise localization of FNAs present unique advantages in efficient drug delivery.

3.3.2 Targeted Drug Delivery

The special physicochemical properties of FNAs provide numerous possibilities for the modification of targeted monomers. Commonly used monomers include biotin, aptamers, polypeptides, and antibodies. Ligand-modified FNAs can selectively enter specific tissues/cells, enabling greater accumulation of therapeutic agents in the targets, resulting in enhanced treatment potency and reduced side effects. Bhatia's team attached folate, galectin-3, and Shiga toxin B-subunit to the external side of DNA icosahedra. They verified that endocytic ligand-modified DNA cages can be taken up by specific cells via a certain internalization pathway [67]. Functionalized FNAs have drawn considerable attention for targeted drug delivery. Folate, a biotin, is usually utilized in tumor targeting and can selectively conjugate with the overexpressed folate receptor on the surface of tumor cells with very high affinity [68]. Ko successfully modified DNA nanotubes with

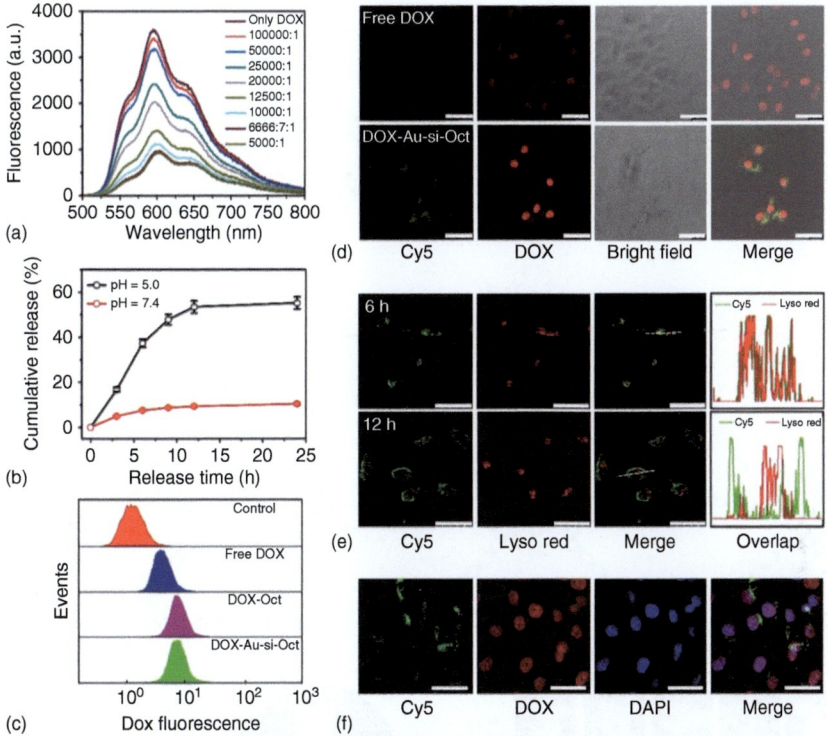

Figure 3.3 Sufficient drug delivery mediated by DNA origami frameworks. Source: Xu et al. [66], © 2021/John Wiley & Sons. (a) Fluorescence spectroscopic monitoring of the DOX loading onto Au-si-Oct at different mole ratios. (b) DOX release curves from OctDOFs in physiological (pH 7.4) and acidic (pH 5.0) environments. (c) Flow cytometric analysis of MDA-MB-231 cells treated with free DOX, DOX-Oct, and DOX-Au-si-Oct, respectively. The untreated group is set as a control. CLSM images of (d) MDA-MB-231 cells treated with free DOX and Cy5-DOX-Au-si-Oct for 12 hours, respectively; (e) MDA-MB-231 cells treated with Cy5-DOX-Au-si-Oct for 6 and 12 hours, followed by staining with LysoTracker Red; (f) MDA-MB-231 cells treated with Cy5-DOX-Au-si-Oct for 12 hours, followed by staining with DAPI. Scale bars: 50 μm. Copyright 2021, Wiley-VCH GmbH.

folate through NHS chemistry [69]. The combinatorial vehicle had much higher cellular uptake than free DNA nanotubes in a folate content-dependent manner. In another approach, a triangular DNA origami was assembled and chemically modified with folate for the targeted delivery of DOX [70]. The nanostructure displayed specific internalization and enhanced cytotoxicity (Figure 3.4). Based on conjugation between DNA and folate, folate-overhung octahedral DNA nanocages have been generated for efficient cancer therapy [72]. Furthermore, the aforementioned study also demonstrated that a distance of 2.5 nm between folate and DNA, which can be predicted by accurate design, significantly enhanced the binding affinity between the cells and nanostructures. Other nanostructures, such as folate-appended multiple-armed tFNAs, have also been developed for tumor cell targeting and drug delivery [71, 73]. Hence, folate modification provides a simple but efficient method for FNA-mediated targeted drug delivery and is feasible and highly advantageous for precise therapy.

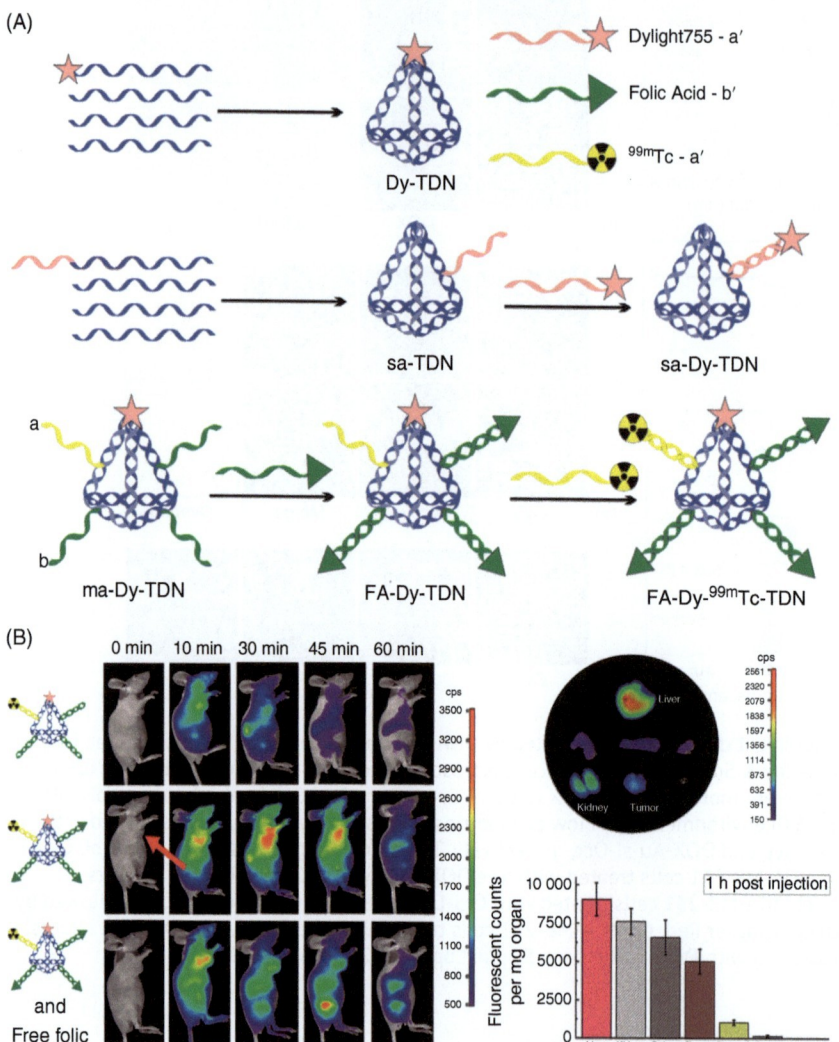

Figure 3.4 (A) Schematic design of dual-modality imaging probe based on DNA nanostructures. Source: Jiang et al. [71], © 2016/American Chemical Society. Fluorescent dye (Dylight 755, pink stars) was assembled together with multiple-armed tetrahedral DNA nanostructure (ma-TDN) to prepare ma-Dy-TDN. Then, folic acid (FA)-conjugated ssDNA (green triangles) was hybridized with ma-Dy-TDN for *in vivo* tumor targeting; last, radioactive isotope technetium-99 m (99mTc)-labeled ssDNA (trefoil symbols) was hybridized to construct the dual-labeled molecular imaging probe. (B) Dynamic biodistribution of Dy-99mTc-TDN, FA-Dy-99mTc-TDN, and the mixture of FA-Dy-99mTc-TDN and free folic acid monitored in the KB tumor-bearing nude mice for continuous one hour through fluorescent imaging. And the *ex vivo* fluorescent imaging of mice tissues, including tumors, liver, kidneys, heart, lung, stomach, and spleen. The tumor showed relatively high signals when compared with the stomach, spleen, lung, and heart. And the biodistribution profile of FA-Dy-99mTc-TDN in tumor-bearing mice one hour post-injection. Copyright 2016, American Chemical Society.

Aptamers are another widely used targeted ligand. An aptamer is a type of short ssDNA selected through the systematic evolution of ligands by the exponential enrichment (SELEX) process [74, 75]. Aptamers can selectively bind to specific targets with high affinity. The natural DNA of the aptamer allows for a simple self-assembly route and strong linking with DNA nanostructures. Aptamers have received increasing interest as they represent an alternative to antibodies and are also known as "chemical antibodies." Aptamers have many advantages, such as a wide variety of targets (from molecules to cells), facile synthesis, flexible chemical modification, excellent stability and biocompatibility, low immunogenicity, and high specificity and affinity. Because of the DNA nature of both FNAs and aptamers, there is immediate compatibility between the two, enabling aptamers to attach to the DNA sequence of FNAs via base pairing or sequence extension. AS1411, a tumor cell target aptamer, is widely used in the targeted delivery of antitumor agents [76]. AS1411 can specifically bind to nucleolin, which is overexpressed on the surface of tumor cells. AS1411 has been modified at the apex of tFNAs for tumor cell targeting [32]. The AS1411-tFNA nanostructures show specific and enhanced uptake by breast cancer cells, providing an effective platform for targeted drug delivery. An AS1411-based dynamic DNA tube was also designed as a nanorobot to specifically deliver antitumor enzymes [77]. With the guidance of AS1411, the nanorobot can specifically accumulate in tumors, control drug release, and powerfully trigger tumor necrosis. Other aptamers that specifically recognize certain cells have also been utilized in the construction of FNA-based targeted delivery systems, such as S2.2 aptamer-modified DNA icosahedra [78] and Sgc8c aptamer-functionalized dynamic DNA nanorobots [79]. Aptamer-mediated precise delivery has also been widely utilized in the treatment of other diseases and in fields such as tissue regeneration, with examples including a BACE1-specific aptamer for targeted therapy of Alzheimer's disease [80], a chondrocyte-specific aptamer for the targeted treatment of osteoarthritis [81], and a bone marrow stem cell-specific aptamer for targeted bone regeneration [82]. Although these aptamers have not been attached to DNA nanostructures, they highlight the broad potential for the development of aptamer-based FNA delivery systems for application in the biomedical field.

Polypeptides can also be used for targeted drug delivery. Fan and coworkers designed glioma-specific peptide-modified tFNAs for B-BB crossing and brain tumor targeting [83]. Another tumor-penetrating and targeting peptide, iRGD, has been chemically modified onto the surface of tFNAs for DOX delivery [40]. The prepared nanocomplexes exhibited superior tissue penetration and accumulation in the tumor, as well as selective internalization and enhanced tumor inhibition. RGD tripeptide is one of the most prominent targeting peptides and has been widely applied in targeted drug delivery for transporting antitumor agents as well as certain phytochemicals, such as curcumin [84, 85]. However, research on peptide-decorated FNAs and their applications is not sufficiently thorough. Hence, further studies are needed to extend the application of peptide-based nucleic acid nanomaterials.

3.3.3 Controlled Drug Release

DNA structures may undergo structural changes or disintegration under the influence of environmental stimuli such as pH, light, temperature, protein, and ion concentration [86]. Based on this characteristic, a series of FNAs that are sensitive to external stimuli was constructed to achieve controlled drug release. Ijäs et al. monitored the release of the intercalated cargo from DNA origami via DNA degradation under the action of deoxyribonuclease (DNase) [35]. This research demonstrated that cargo drugs are released from vehicles in the presence of DNase I, which exists under physiological conditions. The study also confirmed that the DNA degradation and drug release profile depend on the superstructure of the DNA origami. The study provides a rational design strategy

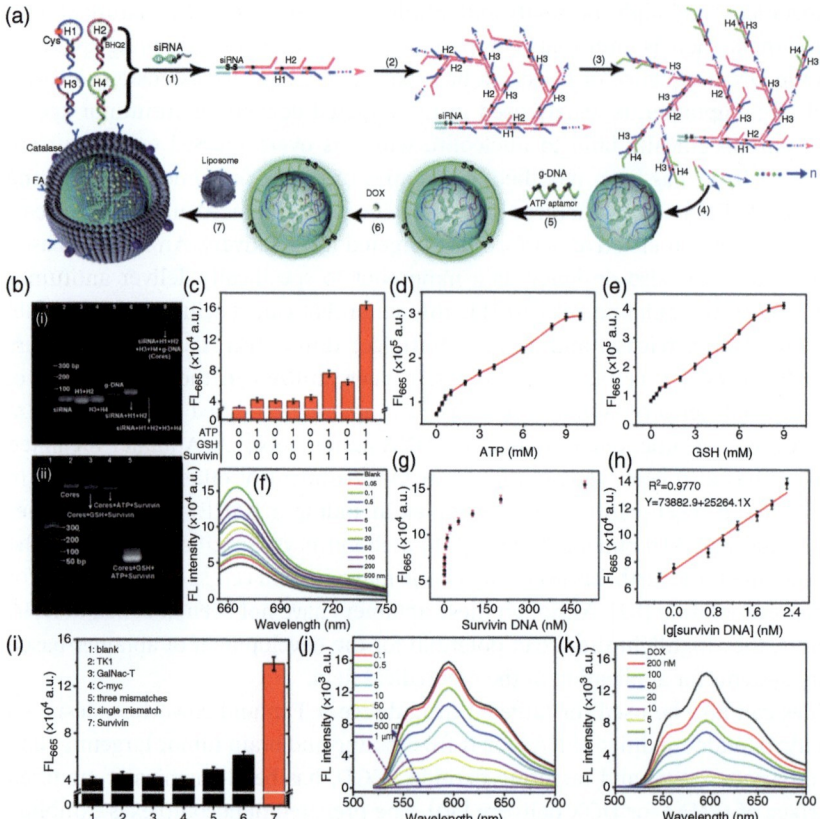

Figure 3.5 Design of siRNA-templated 3D spherical FNAs (ST-SFNAs). Source: Li et al. [87], © 2021/Royal Society of Chemistry. (a) Illustration of synthesis. (b) Agarose gel electrophoresis images: (i) ST-SFNA core construction. Lanes 1–8: DNA marker, siRNA, H1 + H2, H3 + H4, g-DNA, siRNA + H1 + H2, ST-FNA cores, and g-DNA-covered ST-SFNA cores. (ii) Cascaded-logical responses of ST-SFNA cores. Lane (1): DNA marker; (2): ST-SFNA cores; (3): ST-SFNA cores with added GSH and survivin; (4): ST-SFNA cores with added ATP and survivin; (5): ST-SFNA cores with added GSH, ATP, and survivin. (c) FL responses under all input conditions. (d) FL responses to different concentrations of ATP. GSH: 5 mM; survivin: 200 nM. (e) FL responses to different concentrations of GSH. ATP: 5 mM; survivin: 200 nM. (f) Fluorescence spectra at different concentrations of survivin. (g) FL responses versus survivin concentration. (h) Linear relationship of (f) (0.5–200 nM, $R2 = 0.9770$). (i) Selectivity. (j) FL spectra of DOX (2 μM) loaded into different concentrations of ST-SFNAs in PBS (pH = 7.4). (k) FL spectra of DOX-loaded ST-SFNAs (2 μM) during the release of DOX with different concentrations of survivin. ATP and GSH: 1 mM. Copyright 2021, The Royal Society of Chemistry.

for achieving controlled release and optimizing the therapeutic efficacy of DNA origami in antitumor agent delivery. Moreover, researchers have covered spherical FNAs with catalase- and target ligand-modified liposome membranes for the delivery of DOX and siRNA [87] to realize catalase-responsive drug release (Figure 3.5).

The pH of damaged tissues tends to be different from that of normal tissues. For instance, the pH of tumors is usually lower than that of normal tissues. Based on this aspect, pH-sensitive FNA design has been widely researched. The aforementioned RGD peptide-modified tFNAs have been reported to be pH-sensitive [40], where dramatically higher release of the payloads was achieved under low pH conditions than under neutral conditions. In another work, tFNAs were attached to drug-preloaded octahedra through a pH-sensitive DNA sequence [88]. The tFNAs act as a gate to retain the cargo drugs in the octahedra, whereas in the tumor microenvironment with low pH, the tFNAs separate from the octahedral, triggering sufficient release of the drugs, thereby inhibiting tumor growth. Other stimuli, such as temperature, have also been utilized for controlled drug release. Octahedral DNA cages with a truncated corner have been designed for thermoresponsive drug release [89, 90]. These stimuli-sensitive FNAs allow controlled release of the payloads in targets, resulting in precise therapy.

3.3.4 Overcoming Drug Resistance

Drug resistance is one of the most common problems limiting the application of small-molecule drugs, especially in antitumor and antimicrobial therapy. FNA carriers offer a prospective strategy for overcoming drug resistance, as FNAs can escape the drug efflux mechanism of tumor cells or bacteria [91, 92]. Triangular DNA origami loaded with DOX exhibit enhanced cellular internalization and prominent cytotoxicity to DOX-resistant cancer cells [91]. Circumvention of DOX resistance is presumably owing to the increased drug intake and inhibition of lysosomal acidification mediated by the special size and shape of the adopted DNA origami (Figure 3.6). Another rod-shaped DNA origami has been demonstrated to bypass efflux pump resistance mechanisms and overcome daunorubicin resistance in leukemia [52]. tFNAs have also been reported to circumvent antitumor drug resistance by inhibiting P-glycoprotein-mediated drug efflux [43]. It has also been demonstrated that FNAs can overcome antitumor agent resistance as well as antibiotic resistance. Ampicillin-loaded tFNA nanoplatforms have been employed to combat MRSA and overcome ampicillin resistance [53]. Aptamer-functionalized DNA origami delivery nanostructures have been demonstrated to target Gram-negative bacteria and fight against antibiotic resistance [55]. Furthermore, for multidrug resistance, synergistic RNAi/chemotherapy was achieved using a triangular DNA origami delivery system [93]. A targeted shRNA and a therapeutic shRNA, which can inhibit the gene of drug efflux pump protein, were appended to the DOX-preloaded DNA triangle. By combining shRNA-mediated RNAi therapy and DOX-mediated chemotherapy, the DNA nanoplatform could synergistically suppress tumor growth and overcome multidrug resistance. Similar strategies have been developed using sheet-type DNA origami [94] and tubular DNA origami [95], demonstrating the tremendous potential of FNAs in anti-resistance therapy.

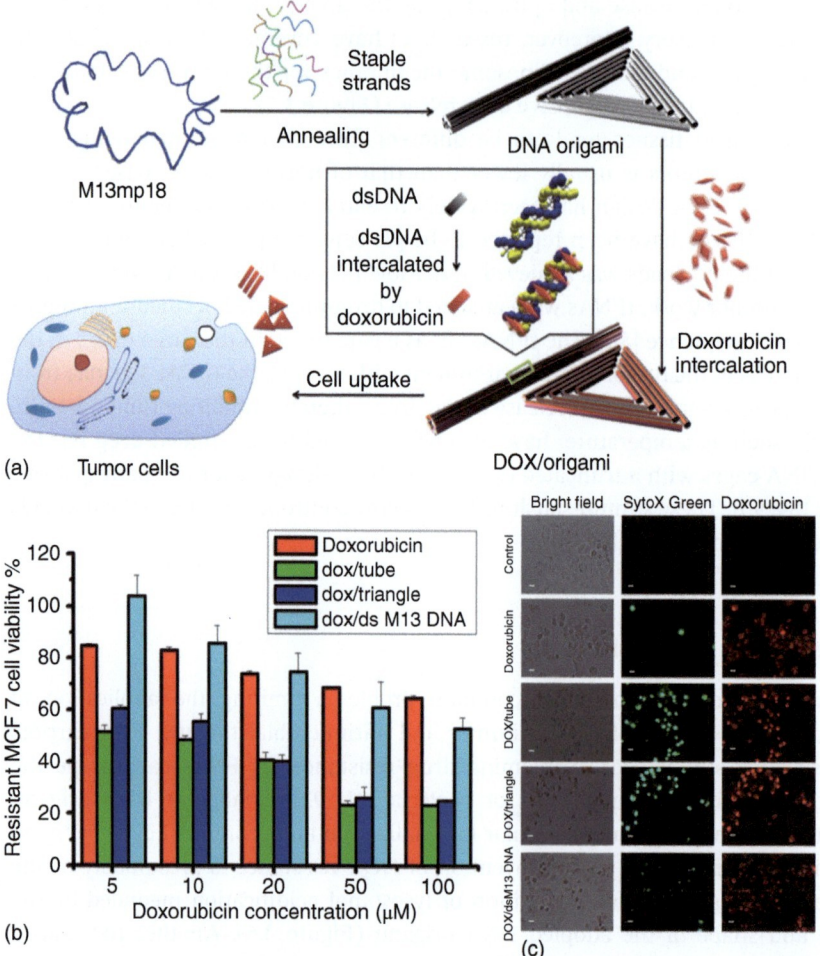

Figure 3.6 DOX-loaded DNA origami nanosystem for circumvention of drug resistance. Source: Jiang et al. [91], © 2012/American Chemical Society. (a) Schematic of the fabrication of DNA origami and doxorubicin origami delivery system. (b) Cell viability of doxorubicin res-MCF 7 cells after administration with equal concentrations of free doxorubicin and drug-loaded origami for 48 hours. Error bars represent strand deviation of three independent experiments in triplicate wells of cells. (c) Bright-field and fluorescence images of SYTOX Green dye (invitrogen) stained res-MCF 7 cells treated with 100 μM free doxorubicin, doxorubicin-triangle, and doxorubicin-tube structures (with 100 μM doxorubicin), doxorubicin–ds M13 DNA, and vehicle solvent (control) for 48 hours. Scale bars are 20 μm. Copyright 2012, American Chemical Society.

3.4 Conclusions and Prospects

Over the last 40 years, there has been tremendous development in DNA nanotechnology, and various self-assembled FNAs have been widely researched in the drug delivery field as well as in other biomedical fields. Notably, the main merits of FNAs for smart drug delivery are: (i) precise and controllable size and shape, (ii) adequate drug loading and chemical modification sites, (iii) superior biocompatibility, and (iv) excellent tissue/cell penetration. These advantages have led to breakthroughs in the intelligent transport of small molecules for myriad biological applications.

First, successful drug delivery enables drugs to cross biological barriers. Because of their distinctive spatial structure, ingeniously designed FNAs can easily cross biological barriers, ranging from cell and intracellular membranes to tissue barriers. Lin's group has clarified that specially sized tFNAs with a side length of 21 bp can cross B-BB to the full extent [28, 96], and they also demonstrated the great potential of tFNAs and their delivery system in the treatment of brain diseases. Second, precision therapy can be achieved with FNAs. FNA-based targeted delivery nanoplatforms have been demonstrated to selectively enter certain tissues, cells, and cellular organelles [97, 98], enabling accurately controlled drug release. The refined design and fabrication provide an excellent strategy for precise therapy with excellent therapeutic efficacy and reduced side effects. Third, a treatment strategy has been developed for oxidative stress-related diseases. FNAs display excellent capacity to protect or even reverse ROS damage in cells [99, 100]. In the delivery of antioxidants, FNA vehicles may synergize with the cargo drugs, leading to enhanced therapeutic efficiency and enabling the extensive application of traditional antioxidants. Fourth, these systems have been used for immunotherapy. DNA nanostructures can be recognized and taken up by immune cells in a DNA density-dependent manner [101], and are prospective tools for immune cell targeting. Although the immunogenicity of synthetic FNAs remains rare and inconsistent, DNA nanostructures have been reported to affect the immune system [102]. A Y-shaped DNA origami has been reported to stimulate innate immune cells through the activation of Toll-like receptors (TLRs) [103]. CpG-coated DNA origami has been shown to trigger an immune response through TLR-mediated immune cell activation [7], where the tFNA-resveratrol delivery system has an excellent immunomodulatory capacity for alleviating inflammation [64]. Finally, these systems have been applied to tissue regeneration. FNAs can promote various cellular processes, such as proliferation, migration, and differentiation. In addition, free tFNAs can facilitate the functions of various stem cells as well as somatic cells [6]. Moreover, drug-loaded tFNAs can promote tissue regeneration [104], providing a simple but efficient approach to tissue regeneration. Some bioactive small molecules, after delivery by the FNAs, can locally promote tissue regeneration by affecting cell behavior and the microenvironment [105, 106]. Thus, FNA vehicles may offer a powerful tool for achieving targeted transport and improved regenerative outcomes.

Although many challenging problems have been solved by FNAs and FNA-based delivery systems, it is still too early for their application in clinical trials. The aspects of high production cost, compromised yield, and difficulty in purification are still to be addressed. Pharmacokinetics and pharmacodynamic studies in living animals, especially large mammals, are essential. In addition, the long-term toxicity of these DNA vehicles is of paramount importance. Many research teams are devoted to exploring these aspects and have made certain improvements. Nevertheless, the area of FNA-based delivery is still in the initial stage; hence, breakthroughs and long-term progress are needed. FNA nanocarriers represent a promising tool for intelligent drug delivery, which may pave the way for further development of smart drug delivery and other biomedical applications, and eventual translation to clinics.

References

1 Seeman, N.C. (1982). Nucleic acid junctions and lattices. *J. Theor. Biol.* 99 (2): 237–247.
2 Winfree, E., Liu, F., Wenzler, L.A., and Seeman, N.C. (1998). Design and self-assembly of two-dimensional DNA crystals. *Nature* 394 (6693): 539–544.
3 Yan, H., Park, S.H., Finkelstein, G. et al. (2003). DNA-templated self-assembly of protein arrays and highly conductive nanowires. *Science* 301 (5641): 1882–1884.
4 Zheng, J., Birktoft, J.J., Chen, Y. et al. (2009). From molecular to macroscopic via the rational design of a self-assembled 3D DNA crystal. *Nature* 461 (7260): 74–77.
5 Goodman, R.P., Schaap, I.A., Tardin, C.F. et al. (2005). Rapid chiral assembly of rigid DNA building blocks for molecular nanofabrication. *Science* 310 (5754): 1661–1665.
6 Zhang, T., Tian, T.R., and Lin, Y.F. (2022). Functionalizing framework nucleic-acid-based nanostructures for biomedical application. *Adv. Mater.*
7 Schüller, V.J., Heidegger, S., Sandholzer, N. et al. (2011). Cellular immunostimulation by CpG-sequence-coated DNA origami structures. *ACS Nano* 5 (12): 9696–9702.
8 Shen, X., Jiang, Q., Wang, J. et al. (2012). Visualization of the intracellular location and stability of DNA origami with a label-free fluorescent probe. *Chem. Commun.* 48 (92): 11301–11303.
9 Rothemund, P.W. (2006). Folding DNA to create nanoscale shapes and patterns. *Nature* 440 (7082): 297–302.
10 Douglas, S.M., Dietz, H., Liedl, T. et al. (2009). Self-assembly of DNA into nanoscale three-dimensional shapes. *Nature* 459 (7245): 414–418.
11 Andersen, E.S., Dong, M., Nielsen, M.M. et al. (2009). Self-assembly of a nanoscale DNA box with a controllable lid. *Nature* 459 (7243): 73–76.
12 Benson, E., Mohammed, A., Gardell, J. et al. (2015). DNA rendering of polyhedral meshes at the nanoscale. *Nature* 523 (7561): 441–444.
13 Wei, B., Dai, M., and Yin, P. (2012). Complex shapes self-assembled from single-stranded DNA tiles. *Nature* 485 (7400): 623–626.
14 Tikhomirov, G., Petersen, P., and Qian, L. (2017). Fractal assembly of micrometre-scale DNA origami arrays with arbitrary patterns. *Nature* 552 (7683): 67–71.
15 Douglas, S.M., Marblestone, A.H., Teerapittayanon, S. et al. (2009). Rapid prototyping of 3D DNA-origami shapes with caDNAno. *Nucleic Acids Res.* 37 (15): 5001–5006.
16 Gustafson, R. and Källmén, H. (1989). Alcohol effects on cognitive and personality style in women with special reference to primary and secondary process. *Alcohol.: Clin. Exp. Res.* 13 (5): 644–648.
17 Liu, Z., Li, Y., Tian, C., and Mao, C. (2013). A smart DNA tetrahedron that isothermally assembles or dissociates in response to the solution pH value changes. *Biomacromolecules* 14 (6): 1711–1714.
18 Ge, Z., Gu, H., Li, Q., and Fan, C. (2018). Concept and development of framework nucleic acids. *JACS* 140 (51): 17808–17819.
19 Veneziano, R., Ratanalert, S., Zhang, K. et al. (2016). Designer nanoscale DNA assemblies programmed from the top down. *Science* 352 (6293): 1534.
20 Mokhtarzadeh, A., Vahidnezhad, H., Youssefian, L. et al. (2019). Applications of spherical nucleic acid nanoparticles as delivery systems. *Trends Mol. Med.* 25 (12): 1066–1079.

21 Kollmann, F., Ramakrishnan, S., Shen, B. et al. (2018). Superstructure-dependent loading of DNA origami nanostructures with a groove-binding drug. *ACS Omega* 3 (8): 9441–9448.
22 Ramakrishnan, S., Ijäs, H., Linko, V., and Keller, A. (2018). Structural stability of DNA origami nanostructures under application-specific conditions. *Comput. Struct. Biotechnol. J.* 16: 342–349.
23 Castro, C.E., Kilchherr, F., Kim, D.N. et al. (2011). A primer to scaffolded DNA origami. *Nat. Methods* 8 (3): 221–229.
24 Perrault, S.D. and Shih, W.M. (2014). Virus-inspired membrane encapsulation of DNA nanostructures to achieve in vivo stability. *ACS Nano* 8 (5): 5132–5140.
25 Mikkilä, J., Eskelinen, A.P., Niemelä, E.H. et al. (2014). Virus-encapsulated DNA origami nanostructures for cellular delivery. *Nano Lett.* 14 (4): 2196–2200.
26 Ahmadi, Y., De Llano, E., and Barišić, I. (2018). (Poly)cation-induced protection of conventional and wireframe DNA origami nanostructures. *Nanoscale* 10 (16): 7494–7504.
27 Zeng, Y., Liu, J., Yang, S. et al. (2018). Time-lapse live cell imaging to monitor doxorubicin release from DNA origami nanostructures. *J. Mater. Chem. B* 6 (11): 1605–1612.
28 Shi, S., Li, Y., Zhang, T. et al. (2021). Biological effect of differently sized tetrahedral framework nucleic acids: endocytosis, proliferation, migration, and biodistribution. *ACS Appl. Mater. Interfaces* 13 (48): 57067–57074.
29 Tian, T., Zhang, C., Li, J. et al. (2021). Proteomic exploration of endocytosis of framework nucleic acids. *Small* 17 (23): e2100837.
30 Liang, L., Li, J., Li, Q. et al. (2014). Single-particle tracking and modulation of cell entry pathways of a tetrahedral DNA nanostructure in live cells. *Angew. Chem.* 53 (30): 7745–7750.
31 Tian, T., Zhang, T., Zhou, T. et al. (2017). Synthesis of an ethyleneimine/tetrahedral DNA nanostructure complex and its potential application as a multi-functional delivery vehicle. *Nanoscale* 9 (46): 18402–18412.
32 Li, Q., Zhao, D., Shao, X. et al. (2017). Aptamer-modified tetrahedral DNA nanostructure for tumor-targeted drug delivery. *ACS Appl. Mater. Interfaces* 9 (42): 36695–36701.
33 Wiraja, C., Zhu, Y., Lio, D.C.S. et al. (2019). Framework nucleic acids as programmable carrier for transdermal drug delivery. *Nat. Commun.* 10 (1): 1147.
34 Hasanzadeh, M. and Shadjou, N. (2016). Pharmacogenomic study using bio- and nanobioelectrochemistry: drug-DNA interaction. *Mater. Sci. Eng. C* 61: 1002–1017.
35 Ijäs, H., Shen, B., Heuer-Jungemann, A. et al. (2021). Unraveling the interaction between doxorubicin and DNA origami nanostructures for customizable chemotherapeutic drug release. *Nucleic Acids Res.* 49 (6): 3048–3062.
36 Zhang, Q., Jiang, Q., Li, N. et al. (2014). DNA origami as an in vivo drug delivery vehicle for cancer therapy. *ACS Nano* 8 (7): 6633–6643.
37 Zhao, Y.X., Shaw, A., Zeng, X. et al. (2012). DNA origami delivery system for cancer therapy with tunable release properties. *ACS Nano* 6 (10): 8684–8691.
38 Ge, Z., Guo, L., Wu, G. et al. (2020). DNA origami-enabled engineering of ligand-drug conjugates for targeted drug delivery. *Small* 16 (16): e1904857.

39 Sun, P., Zhang, N., Tang, Y. et al. (2017). SL2B aptamer and folic acid dual-targeting DNA nanostructures for synergic biological effect with chemotherapy to combat colorectal cancer. *Int. J. Nanomed.* 12: 2657–2672.

40 Liu, M., Ma, W., Zhao, D. et al. (2021). Enhanced penetrability of a tetrahedral framework nucleic acid by modification with iRGD for DOX-targeted delivery to triple-negative breast cancer. *ACS Appl. Mater. Interfaces* 13 (22): 25825–25835.

41 Xia, Z., Wang, P., Liu, X. et al. (2016). Tumor-penetrating peptide-modified DNA tetrahedron for targeting drug delivery. *Biochemistry* 55 (9): 1326–1331.

42 Shi, S., Fu, W., Lin, S. et al. (2019). Targeted and effective glioblastoma therapy via aptamer-modified tetrahedral framework nucleic acid-paclitaxel nanoconjugates that can pass the blood brain barrier. *Nanomed. Nanotechnol. Biol. Med.* 21: 102061.

43 Xie, X., Shao, X., Ma, W. et al. (2018). Overcoming drug-resistant lung cancer by paclitaxel loaded tetrahedral DNA nanostructures. *Nanoscale* 10 (12): 5457–5465.

44 Zhong, Y.F., Cheng, J., Liu, Y. et al. (2020). DNA nanostructures as Pt(IV) prodrug delivery systems to combat chemoresistance. *Small* 16 (38): e2003646.

45 Jorge, A.F., Aviñó, A., Pais, A. et al. (2018). DNA-based nanoscaffolds as vehicles for 5-fluoro-2′-deoxyuridine oligomers in colorectal cancer therapy. *Nanoscale* 10 (15): 7238–7249.

46 Ma, W., Yang, Y., Zhu, J. et al. (2022). Biomimetic nanoerythrosome-coated aptamer-DNA tetrahedron/maytansine conjugates: pH-responsive and targeted cytotoxicity for HER2-positive breast cancer. *Adv. Mater.* e2109609.

47 Zhuang, X., Ma, X., Xue, X. et al. (2016). A photosensitizer-loaded DNA origami nanosystem for photodynamic therapy. *ACS Nano* 10 (3): 3486–3495.

48 Shaukat, A., Anaya-Plaza, E., Julin, S. et al. (2020). Phthalocyanine-DNA origami complexes with enhanced stability and optical properties. *Chem. Commun.* 56 (53): 7341–7344.

49 Kim, K.R., Bang, D., and Ahn, D.R. (2016). Nano-formulation of a photosensitizer using a DNA tetrahedron and its potential for in vivo photodynamic therapy. *Biomater. Sci.* 4 (4): 605–609.

50 Wright, G.D. (2016). Antibiotic adjuvants: rescuing antibiotics from resistance. *Trends Microbiol.* 24 (11): 862–871.

51 Bhattacharya, P., Mukherjee, S., and Mandal, S.M. (2020). Fluoroquinolone antibiotics show genotoxic effect through DNA-binding and oxidative damage. *Spectrochim. Acta, Part A* 227: 117634.

52 Halley, P.D., Lucas, C.R., McWilliams, E.M. et al. (2016). Daunorubicin-loaded DNA origami nanostructures circumvent drug-resistance mechanisms in a leukemia model. *Small* 12 (3): 308–320.

53 Sun, Y., Li, S., Zhang, Y. et al. (2020). Tetrahedral framework nucleic acids loading ampicillin improve the drug susceptibility against methicillin-resistant *Staphylococcus aureus*. *ACS Appl. Mater. Interfaces* 12 (33): 36957–36966.

54 Sun, Y., Liu, Y., Zhang, B. et al. (2021). Erythromycin loaded by tetrahedral framework nucleic acids are more antimicrobial sensitive against *Escherichia coli* (*E. coli*). *Bioact. Mater.* 6 (8): 2281–2290.

55 Mela, I., Vallejo-Ramirez, P.P., Makarchuk, S. et al. (2020). DNA nanostructures for targeted antimicrobial delivery. *Angew. Chem.* 59 (31): 12698–12702.

- 56 Pastor, R.F., Restani, P., Di Lorenzo, C. et al. (2019). Resveratrol, human health and winemaking perspectives. *Crit. Rev. Food Sci. Nutr.* 59 (8): 1237–1255.
- 57 Xu, M., Pirtskhalava, T., Farr, J.N. et al. (2018). Senolytics improve physical function and increase lifespan in old age. *Nat. Med.* 24 (8): 1246–1256.
- 58 Srinivasan, K. (2016). Biological activities of red pepper (*Capsicum annuum*) and its pungent principle capsaicin: a review. *Crit. Rev. Food Sci. Nutr.* 56 (9): 1488–1500.
- 59 Ma, N., Zhang, Z., Liao, F. et al. (2020). The birth of artemisinin. *Pharmacol. Ther.* 216: 107658.
- 60 Hussain, I., Fatima, S., Siddiqui, S. et al. (2021). Exploring the binding mechanism of β-resorcylic acid with calf thymus DNA: insights from multi-spectroscopic, thermodynamic and bioinformatics approaches. *Spectrochim. Acta, Part A* 260: 119952.
- 61 Platella, C., Mazzini, S., Napolitano, E. et al. (2021). Plant-derived stilbenoids as DNA-binding agents: from monomers to dimers. *Chemistry* 27 (34): 8832–8845.
- 62 Sirong, S., Yang, C., Taoran, T. et al. (2020). Effects of tetrahedral framework nucleic acid/wogonin complexes on osteoarthritis. *Bone Res.* 8: 6.
- 63 Zhang, M., Zhang, X., Tian, T. et al. (2022). Anti-inflammatory activity of curcumin-loaded tetrahedral framework nucleic acids on acute gouty arthritis. *Bioact. Mater.* 8: 368–380.
- 64 Li, Y., Gao, S., Shi, S. et al. (2021). Tetrahedral framework nucleic acid-based delivery of resveratrol alleviates insulin resistance: from innate to adaptive immunity. *Nano Micro Lett.* 13 (1): 86.
- 65 Cui, W.T., Yang, X., Chen, X.Y. et al. (2021). Treating LRRK2-related Parkinson's disease by inhibiting the mTOR signaling pathway to restore autophagy. *Adv. Funct. Mater.* 31 (38): 13.
- 66 Xu, T., Yu, S., Sun, Y. et al. (2021). DNA origami frameworks enabled self-protective siRNA delivery for dual enhancement of chemo-photothermal combination therapy. *Small* 17 (46): e2101780.
- 67 Bhatia, D., Arumugam, S., Nasilowski, M. et al. (2016). Quantum dot-loaded monofunctionalized DNA icosahedra for single-particle tracking of endocytic pathways. *Nat. Nanotechnol.* 11 (12): 1112–1119.
- 68 Leamon, C.P. and Reddy, J.A. (2004). Folate-targeted chemotherapy. *Adv. Drug Delivery Rev.* 56 (8): 1127–1141.
- 69 Ko, S., Liu, H., Chen, Y., and Mao, C. (2008). DNA nanotubes as combinatorial vehicles for cellular delivery. *Biomacromolecules* 9 (11): 3039–3043.
- 70 Pal, S. and Rakshit, T. (2021). Folate-functionalized DNA origami for targeted delivery of doxorubicin to triple-negative breast cancer. *Front. Chem.* 9: 721105.
- 71 Jiang, D., Sun, Y., Li, J. et al. (2016). Multiple-armed tetrahedral DNA nanostructures for tumor-targeting, dual-modality in vivo imaging. *ACS Appl. Mater. Interfaces* 8 (7): 4378–4384.
- 72 Raniolo, S., Vindigni, G., Ottaviani, A. et al. (2018). Selective targeting and degradation of doxorubicin-loaded folate-functionalized DNA nanocages. *Nanomed. Nanotechnol. Biol. Med.* 14 (4): 1181–1190.
- 73 Bu, Y.Z., Xu, J.R., Luo, Q. et al. (2020). A precise nanostructure of folate-overhung mitoxantrone DNA tetrahedron for targeted capture leukemia. *Nanomaterials* 10 (5).

74 Zhu, G. and Chen, X. (2018). Aptamer-based targeted therapy. *Adv. Drug Delivery Rev.* 134: 65–78.

75 Wu, L., Wang, Y., Xu, X. et al. (2021). Aptamer-based detection of circulating targets for precision medicine. *Chem. Rev.* 121 (19): 12035–12105.

76 Tong, X., Ga, L., Ai, J., and Wang, Y. (2022). Progress in cancer drug delivery based on AS1411 oriented nanomaterials. *J. Nanobiotechnol.* 20 (1): 57.

77 Li, H., Liu, J., and Gu, H. (2019). Targeting nucleolin to obstruct vasculature feeding with an intelligent DNA nanorobot. *J. Cell. Mol. Med.* 23 (3): 2248–2250.

78 Chang, M., Yang, C.S., and Huang, D.M. (2011). Aptamer-conjugated DNA icosahedral nanoparticles as a carrier of doxorubicin for cancer therapy. *ACS Nano* 5 (8): 6156–6163.

79 Douglas, S.M., Bachelet, I., and Church, G.M. (2012). A logic-gated nanorobot for targeted transport of molecular payloads. *Science* 335 (6070): 831–834.

80 Liang, H., Shi, Y., Kou, Z. et al. (2015). Inhibition of BACE1 activity by a DNA aptamer in an Alzheimer's disease cell model. *PLoS One* 10 (10): e0140733.

81 Ji, M.L., Jiang, H., Wu, F. et al. (2020). Precise targeting of miR-141/200c cluster in chondrocytes attenuates osteoarthritis development. *Ann. Rheum. Dis.*

82 Luo, Z.W., Li, F.X., Liu, Y.W. et al. (2019). Aptamer-functionalized exosomes from bone marrow stromal cells target bone to promote bone regeneration. *Nanoscale* 11 (43): 20884–20892.

83 Tian, T., Li, J., Xie, C. et al. (2018). Targeted imaging of brain tumors with a framework nucleic acid probe. *ACS Appl. Mater. Interfaces* 10 (4): 3414–3420.

84 Tian, T., Zhang, H.X., He, C.P. et al. (2018). Surface functionalized exosomes as targeted drug delivery vehicles for cerebral ischemia therapy. *Biomaterials* 150: 137–149.

85 Bari, E., Serra, M., Paolillo, M. et al. (2021). Silk fibroin nanoparticle functionalization with Arg-Gly-Asp cyclopentapeptide promotes active targeting for tumor site-specific delivery. *Cancers* 13 (5).

86 Zhang, Y., Pan, V., Li, X. et al. (2019). Dynamic DNA structures. *Small* 15 (26): e1900228.

87 Li, J., Zhang, Y., Sun, J. et al. (2021). SiRNA-templated 3D framework nucleic acids for chemotactic recognition, and programmable and visualized precise delivery for synergistic cancer therapy. *Chem. Sci.* 12 (46): 15353–15361.

88 Zhang, P., Ouyang, Y., Sohn, Y.S. et al. (2021). pH- and miRNA-responsive DNA-tetrahedra/metal-organic framework conjugates: functional sense-and-treat carriers. *ACS Nano* 15 (4): 6645–6657.

89 Juul, S., Iacovelli, F., Falconi, M. et al. (2013). Temperature-controlled encapsulation and release of an active enzyme in the cavity of a self-assembled DNA nanocage. *ACS Nano* 7 (11): 9724–9734.

90 Franch, O., Iacovelli, F., Falconi, M. et al. (2016). DNA hairpins promote temperature controlled cargo encapsulation in a truncated octahedral nanocage structure family. *Nanoscale* 8 (27): 13333–13341.

91 Jiang, Q., Song, C., Nangreave, J. et al. (2012). DNA origami as a carrier for circumvention of drug resistance. *JACS* 134 (32): 13396–13403.

92 Kim, K.R., Kim, D.R., Lee, T. et al. (2013). Drug delivery by a self-assembled DNA tetrahedron for overcoming drug resistance in breast cancer cells. *Chem. Commun.* 49 (20): 2010–2012.

93 Liu, J., Song, L., Liu, S. et al. (2018). A tailored DNA nanoplatform for synergistic RNAi-/chemotherapy of multidrug-resistant tumors. *Angew. Chem.* 57 (47): 15486–15490.

94 Pan, Q., Nie, C., Hu, Y. et al. (2020). Aptamer-functionalized DNA origami for targeted codelivery of antisense oligonucleotides and doxorubicin to enhance therapy in drug-resistant cancer cells. *ACS Appl. Mater. Interfaces* 12 (1): 400–409.

95 Wang, Z., Song, L., Liu, Q. et al. (2021). A tubular DNA nanodevice as a siRNA/chemo-drug co-delivery vehicle for combined cancer therapy. *Angew. Chem.* 60 (5): 2594–2598.

96 Zhu, J., Yang, Y., Ma, W. et al. (2022). Antiepilepticus effects of tetrahedral framework nucleic acid via inhibition of gliosis-induced downregulation of glutamine synthetase and increased AMPAR internalization in the postsynaptic membrane. *Nano Lett.* 22 (6): 2381–2390.

97 Gao, Y., Chen, X., Tian, T. et al. (2022). A lysosome-activated tetrahedral nanobox for encapsulated siRNA delivery. *Adv. Mater.* 34: e2201731.

98 Zhang, B.W., Tian, T.R., Xiao, D.X. et al. (2022). Facilitating in situ tumor imaging with a tetrahedral DNA framework-enhanced hybridization chain reaction probe. *Adv. Funct. Mater.* 32 (16): 11.

99 Wang, Y., Li, Y., Gao, S. et al. (2022). Tetrahedral framework nucleic acids can alleviate taurocholate-induced severe acute pancreatitis and its subsequent multiorgan injury in mice. *Nano Lett.* 22 (4): 1759–1768.

100 Zhou, M., Zhang, T., Zhang, B. et al. (2021). A DNA nanostructure-based neuroprotectant against neuronal apoptosis via inhibiting toll-like receptor 2 signaling pathway in acute ischemic stroke. *ACS Nano* 16: 1456–1470.

101 Maezawa, T., Ohtsuki, S., Hidaka, K. et al. (2020). DNA density-dependent uptake of DNA origami-based two-or three-dimensional nanostructures by immune cells. *Nanoscale* 12 (27): 14818–14824.

102 Tseng, C.Y., Wang, W.X., Douglas, T.R., and Chou, L.Y.T. (2022). Engineering DNA nanostructures to manipulate immune receptor signaling and immune cell fates. *Adv. Healthcare Mater.* 11 (4): e2101844.

103 Yang, G., Koo, J.E., Lee, H.E. et al. (2019). Immunostimulatory activity of Y-shaped DNA nanostructures mediated through the activation of TLR9. *Biomed. Pharmacother.* 112: 108657.

104 Li, J., Yao, Y., Wang, Y. et al. (2022). Modulation of the crosstalk between Schwann cells and macrophages for nerve regeneration: a therapeutic strategy based on multifunctional tetrahedral framework nucleic acids system. *Adv. Mater.* 34 (46): 2202513.

105 Qin, H., Zhao, A., and Fu, X. (2017). Small molecules for reprogramming and transdifferentiation. *Cell. Mol. Life Sci.* 74 (19): 3553–3575.

106 Li, J., Lai, Y.X., Li, M.X. et al. (2022). Repair of infected bone defect with clindamycin-tetrahedral DNA nanostructure complex-loaded 3D bioprinted hybrid scaffold. *Chem. Eng. J.* 435: 13.

4

The Application of Framework Nucleic Acid-Based Nanomaterials in the Treatment of Mitochondrial Dysfunction

Lan Yao and Tao Zhang

Sichuan University, State Key Laboratory of Oral Diseases, West China Hospital of Stomatology, 3rd Sec, Ren Min Nan Road, Chengdu 610041, PR China

4.1 Introduction

Mitochondria are eukaryotic cell organelles that can reproduce independently, and their chromosomes are circular mitochondrial DNA (mtDNA). Mitochondria with great autonomy are essential for oxidative phosphorylation, intracellular calcium control, apoptosis, and cell metabolism [1, 2]. At the same time, the mitochondria are constantly undergoing dynamic changes such as fusion division [3], movement [4], and clearance [5, 6], so as to maintain low heterogeneity and adapt to the different physiological needs of cells. When the above functional disorders are manifested as hyperoxidation, calcium overload, mtDNA mutations, mitochondrial autophagy, and apoptosis, it comes to metabolic diseases, tumors, neurodegenerative lesions, and other severe diseases [7–12].

Currently, there is no perfect treatment for mitochondrial dysfunction, and severe mitochondrial dysfunction even requires mitochondrial transplantation [12–14]. It is of great significance to seek a less invasive, less costly, and more effective therapy [15]. For the treatment of targeted mitochondria, the drug or delivery molecule is required to have a more exquisite structure and adaptability. It needs hierarchical targeting and high enrichment. Compared with the broad spectrum of chemical synthetic drugs, nucleic acid nanomaterial drugs have the potential of "precision targeting." Furthermore, numerous studies have successfully demonstrated the important reversal role of nucleic acids or delivery structures in advance [16, 17] and have found that exogenous nucleic acids can enter intracellular mitochondria to exert biological activity [18, 19]. In this review, we summarize the therapeutic role of nucleic acid-based delivery systems in mitochondrial dysfunction and shed light on opportunities for future development.

4.2 Treatment Mechanisms in Mitochondrial Dysfunction

Current drugs for mitochondrial treatment include chemical synthetic drugs targeting proteins and nucleic acid drugs targeting genes, in which the nucleic acid drugs can change the

Nucleic Acid-Based Nanomaterials: Stabilities and Applications, First Edition.
Edited by Yunfeng Lin and Shaojingya Gao.
© 2024 WILEY-VCH GmbH. Published 2024 by WILEY-VCH GmbH.

replication and translation functions of target genes after entering the cell, thereby bringing longer-term stability and specific efficacy.

Due to the different sizes and structures of nucleic acid drugs, the efficiency of cell targeting, cell entry, and stability *in vivo* are uneven [20]. Therefore, the modification of nucleic acid drugs is the basis for ensuring their efficacy, mainly based on chemical modifications [21–23] and others like bioconjugation [24–26]. In recent years, the United States Food and Drug Administration (FDA) and the European Medicines Agency (EMA) have approved more than a dozen *in vivo* and *in vitro* therapies based on nucleic acid drugs, including spinal muscular atrophy [27, 28] and COVID-19 [29], which currently lack effective treatment options. Mitochondrial dysfunction plays a primary or promoting role in the pathogenesis of many diseases, but nucleic acid drugs for mitochondrial diseases are still in the early stages of research, even the nucleic acid sequence has been modified, the efficiency and accuracy of cell entrance are still lacking, and the use of transfection reagents or some delivery systems such as liposomes to smoothly transport to the cell is essential. The harsh design requirements have led to the fact that current relevant *in vivo* and clinical trials are very few. Nevertheless, based on the superiority of nucleic acid drugs themselves, their therapeutic potential in targeted mitochondrial diseases cannot be denied (Figure 4.1).

4.2.1 Treating in mtDNA

MtDNA is more susceptible to mutation than nuclear DNA (nDNA) due to its long exposure to ROS [30]. Mutated mtDNA is constantly mismatched and copied. Until the ratio of mutated mtDNA to wild-type mtDNA (WT-mtDNA) is out of balance, dysfunction happens, which is also called heterogeneity [31, 32]. Since the introduction of exogenous WT-mtDNA into mitochondria also faces the problem of genetic mutations, only two directions are discussed below to reduce heterogeneity: clearing and inhibiting mutant mtDNA replication.

4.2.1.1 Clearing Mutations

Clearing mutant mtDNA primarily relies on mitochondrial-targeting nucleic acid endonucleases, including restriction enzymes, zinc-finger nucleases (ZFNs), transcription activator-like effector nucleases (TALEN), and customized endonucleases. Although restriction endonucleases work only against m8993T>G mutations [33], Reddy et al. [34] found that they were effective in reducing mutant mtDNA in germ cells and embryonic cells to reduce mitochondrial gene mutation disease at the genetic level. By design, ZFNs can shear almost all DNA sequences. Minczuk's team [35] designed and modified ZFNs that target mitochondria and observed co-localization with mitochondria in human osteosarcoma cells, while the nucleus did not detect signals for these ZFNs. The mandatory heterodimer ZFNs carry the mitochondrial targeting sequence (MTS) and nuclear localization sequence (NLS), which can successfully identify and eliminate the mutated site in mtDNA and reduce heterogeneity. TALEN designed by Bacman et al. [36] can also successfully target the mitochondria of human osteosarcoma cells and reduce heterogeneity. Yang et al. [37] reported that TALEN reduced the heterogeneity of the mitochondria of human pluripotent stem cells and porcine oocytes and restored the normal phosphorylation function of mitochondria, proposing more possibilities for

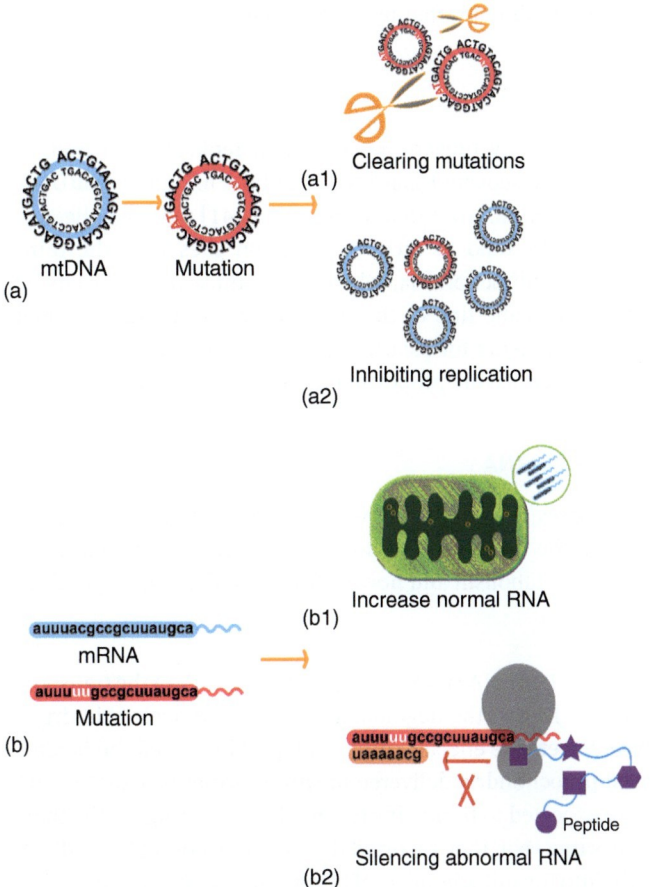

Figure 4.1 Treatment mechanisms in mitochondrial dysfunction. (a) Therapeutic strategies for mitochondrial mutated DNA. (a1) Mutated mtDNA is cleared using endonuclease targeting mitochondria like restriction enzymes, zinc-finger nucleases (ZFNs), transcription activator-like effector nucleases (TALEN), and customized endonucleases. (a2) Hinders the replication of mutant mtDNA and reduces the proportion of mutated and wild mtDNA. (b) Therapeutic strategies for mitochondrial mutated mRNA. (b1) Normal sequences of mutant mRNAs are introduced into the mitochondria to correct their proportions in the mitochondria. (b2) Oligonucleotide fragments bind to mutation sites that prevent mutant mRNAs from carrying out downstream physiological activities.

TALEN. Subsequently, McCann et al. [38] reported on the introduction of ZFNs into mouse embryos and reduced heterogeneity. However, as a nucleic acid endonuclease, it inevitably has the disadvantages of large volume and poor flexibility to identify single-base variations. Zekonyte et al. [39] designed a large range of nucleases with monomeric properties that can recognize single bases against mt-tRNA^{Ala} mutations based on natural homodimerases, which are encapsulated in adenoviruses and transported in mice, targeting mouse skeletal muscle, heart, and liver organs, effectively eliminating the heterogeneity of mtDNA *in vivo* and restoring mt-tRNA^{Ala} to normal levels. The use of nucleases is the preferred way to reduce mitochondrial gene heterogeneity, which can identify multiple mutation sites according to design but is currently mainly limited by the efficiency of transport *in vivo* and the off-target effects of nDNA.

4.2.1.2 Inhibiting Replication

The heterogeneity can also be adjusted by preventing continued replication of mutant mtDNA. 20 years ago, Taylor et al. [40] discovered that PNAs can bind to mutated mtDNA templates *in vitro* to inhibit their replication. After that, Comte et al. [41] replaced oligonucleotide fragments that can specifically bind to mtDNA mutation sites with the third motif of 5s rRNA. Then it imports to mitochondria, which can reduce the intracellular mutant mtDNA by 15–35%. With the same line of thought, Loutre et al. [42] designed 5s rRNA-carrying anti-replication fragments for Kearns–Sayre Syndrome can effectively enter intracellular mitochondria and reduce the proportion of mutant mtDNA.

4.2.2 Treating in mRNA, tRNA, and rRNA

MtDNA has a total of 37 pairs of bases, encoding 13 mRNA, 22 tRNA, and 2 rRNA [43]. These coding RNAs are involved in the synthesis of mitochondrial proteins and have significant implications for the energy metabolism and normal function of the mitochondria.

4.2.2.1 Increase Normal RNA

The transport of wild-type mtRNA (WT-mRNA) is more precise and safer than the transport of WT-mtDNA to the mitochondria. Due to the degradability of RNA, nucleic acid drugs that enter the mitochondria have fewer side effects. Yuma et al. [44] improved the heterogeneity of 12s rRNA in fibroblast mitochondria delivered *in vitro* to patients with A1555G mitochondrial mutations and are expected to be an effective nucleic acid drug for the therapy of hereditary deafness. In the same year, they reported that the transport of WT-mRNA encoding the ND3 protein *in vitro* into the mitochondria of fibroblasts contributes to the formation of normal respiratory chains and improves mitochondrial respiratory productivity [45]. In addition, the high heterogeneity caused by mutational abnormalities in tRNA can also be treated through this idea. Kawamura et al. [46] reversed the function of mitochondria by delivering pre-WT-tRNAs *in vitro* to mitochondria, reducing the ratio of mutant tRNAs.

4.2.2.2 Silencing Abnormal RNA

As the core of the energy supply, the mRNA of the mitochondrial gene encoding the protein of the oxidative phosphorylation system is very important. Mutations and mismatches of their corresponding genes during replication or translation may cause disorders in protein synthesis and lead to mitochondrial dysfunction. The application of antisense oligonucleotides (ASOs) can accurately identify erroneous mRNAs in the mitochondria and degrade the target one. For example, Furukawa et al. [47] and Kawamura et al. [48] both successfully introduced ASO targeting COX II into mitochondria *in vitro*, reducing the level of mitochondrial oxidative phosphorylation and inhibiting cell growth. Subsequently, Luis et al. [49] created more complex delivery systems for RNA transportation, which successfully introduced jac-1-oligonucleotides into purified mitochondria, reducing COX1 expression. In addition to the proteins of the respiratory chain, mitochondrial dynamic stability is also extremely important. Balanced mitochondrial fusion and division help maintain a normal mutation rate and can make the mitochondria change shape into balls, rods, etc., adapting to the different physiological needs of cells [50]. ASOs targeting the outer membrane fission factor Drp1 of the mitochondria can effectively reduce the fragmentation of mitochondria in cellular models and save the respiratory function of the mitochondria [51]. Gao et al. [18] found that RNA greater than 77nt in length has difficulty entering mitochondria through holes in the mitochondrial membrane that do not consume energy, whereas small RNAs, including synthesized siRNAs, have the potential to enter mitochondria. They then verified that siRNAs entered Hela cells and C2C12 cells, down-regulating transcription of ND1 and COX I.

4.2.2.3 Treating in Noncoding RNA

LncRNAs are a kind of regulatory noncoding RNA found in genes that contribute in information transfer between mitochondria and nuclei and are required to produce mitochondrial proteins and the assembly of ribosomes in mitochondria [52]. In recent years, as the research on lncRNA has become clearer, there has been increasing evidence that its tumorigenesis associated with mitochondrial dysfunction is closely related to metastasis [53]. In this regard, *in vivo* and *in vivo* studies knocking out antisense noncoding mitochondrial RNA (ASlncmtRNA) have shown inhibition of both human and murine cancer cells [54–56]. Lorena et al. [57] treated mouse melanoma with ASO-1560S, which effectively decreased the growth rate of subcutaneous tumors and the number of metastatic nodules in the lungs; furthermore, they used ASO-1537S to knock out breast cancer cells ASlncmtRNA causing tumor cell death and reducing tumor cell secretion of extracellular vesicle-mediated signaling agents for tumor metastasis, further exploring the feasibility of oligonucleotide-targeted ASlncmtRNA for tumors [58].

Mitochondrial circRNA (mito-circRNA) is involved in the splicing regulation of mRNA in mito-circRNA in mito-mitochondria, the relationship between microRNAs and proteins, and the synthesis of peptides [59, 60]. Zhao et al. [61] found that three mito-circRNAs were downregulated in fibroblasts of alcoholic hepatitis (NASH) patients and subsequently established mito-NPs that specifically target mitochondria with PH-responsive function. It is equipped with the steatohepatitis-related regulator SCAR into the mitochondria, which can reduce the activation of NASH's fibroblasts *in vivo* and *in vitro*, thereby reducing liver inflammation.

4.3 Nucleic Acid Nanomaterial-Based Delivery System in Mitochondrial Treatment

Due to the structure of double mitochondrial membranes and the selective permeability of cell membranes, chemical synthetic and nucleic acid drugs that target mitochondria have difficulties in entering the cell and mitochondria, so the drugs rarely reach the target organelles of the designated tissue. Obviously, these drugs need a carrier. For intelligent delivery systems [62], they are expected to have good tissue specificity, cell targeting, internal environmental stability, precise release, and low cytotoxicity. In recent years, studies of diseases related to mitochondrial dysfunction have found that traditional delivery systems, including liposomes and inorganic nanoparticles [63, 64], polymers [65, 66], and other delivery structures, face difficulty with application *in vivo*. For example, MITO-porter, the most popular delivery system based on liposomes, can be loaded with a large number of drugs while the tissue targeting and cell specificity of this system are poor, and corresponding modifications may be required for different cells to increase cell targeting and cell entry efficiency [67–69]. Complex aptamer connections decrease the uniformity of synthesis and increase the synthesis steps associated with cytotoxic unpredictability.

Therefore, it is necessary to explore nano-transfection systems with simpler structures, fewer synthetic by-products, better cell penetration and targeting, and less cytotoxicity [62]. Among them, the strong editability and controllability of nucleic acid materials make them more likely to carry drugs and precisely target release [70], after a well-defined structure changes with the environmental denaturation [71]. Moreover, nucleic acid nanomaterials have excellent cell internalization ability and good biocompatibility [72, 73]. There have been many studies reporting that nucleic acid nanomaterial-based delivery systems are used to transport drugs to treat cancer, autoimmune diseases, neurodegenerative diseases, etc. [74]. Similarly, nucleic acid nanomaterial-based delivery systems have the potential to be seen in mitochondrial-related diseases.

4.3.1 Cell and Mitochondria Targeting

4.3.1.1 Cell Targeting
Because of their stable structure and controllable size and diameter, nucleic acid nanomaterials can reach almost all organs in the body through bloodstream, including the brain, lungs, liver, lymph nodes, and bones [75]. In the case of organ or tissue damage, nucleic acid nanomaterials also exhibit excellent passive enrichment [76], which can accumulate in large quantities at the site of injury. Based on the editability of the nucleic acid structure, functional groups such as aptamers [77], small molecules [78], and proteins [79] can be attached to the endpoints of nucleic acid nanomaterials, thereby giving nucleic acid nanomaterials excellent cell targeting.

4.3.1.2 Mitochondria Targeting
When nucleic acid nanomaterials enter cells intact, they need to exert "secondary targeting" capabilities to locate the mitochondria. Due to the characteristics of charge and protein on the mitochondrial membrane, there are several types of structures that have a stronger affinity for mitochondria, including peptides [71], small molecules [70], and transition metal complexes [80]. The nucleated short peptides need to be fed into the mitochondria, and these short peptides have a specific MTS that is sheared after entering the mitochondria and is an irreversible process. Small molecules are represented by out-of-situ lipophilic cations, particularly triphenyl phosphonium (TPP) and rhodamine 123, which can be enriched in large quantities and enter the mitochondria without the need for transporters but may have dose-dependent mitochondrial toxicity. The transition metal complexes, represented by ruthenium (Ru) and iridium (Ir) complexes, have been shown to specifically target mitochondria and exert anticancer effects. Nucleic acid nanomaterials have been reported to form stable connections with peptide chains, small molecules, metals, etc., which lays the foundation for the connection of mitochondrial-targeted structures (Figure 4.2).

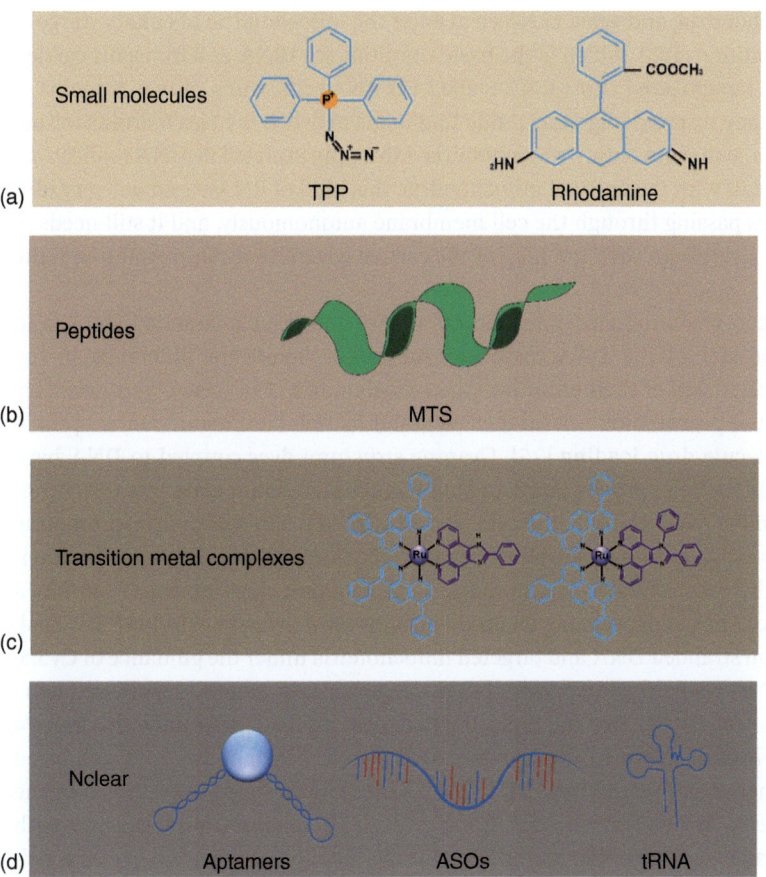

Figure 4.2 Available ways to target mitochondria. (a) Chemical small molecule drugs target mitochondrial membranes through electrostatic interactions like TPP and rhodamine. (b) Mitochondrial proteins have mitochondrial targeting sequences at the N-terminus for recognition by the outer mitochondrial membrane for transport. (c) Metal complexes target mitochondria. (d) The idea of nucleic acid materials as targeted delivery to mitochondria, including by some aptamers, various oligonucleotides, and 5sRNA and tRNA fractional structures.

In addition to modifying the targeting sequence of drugs that need to enter the mitochondria, there are also delivery pathways for targeted, separated mitochondria designed according to the autonomous transport mechanism of the mitochondria. 5s rRNA is the most naturally occurring RNA in mammalian cells that enters mitochondria [81]. It enters the mitochondria from the cytoplasm with the help of two proteins, matrix-localized rhodanese and mitochondrial ribosomal protein L18, and is involved in the assembly of ribosomes [82]. There are three motifs on the 5s rRNA structure, and the deletion and replacement of the third motif do not affect its entry into the mitochondria. It is also a site for replacing therapeutic oligonucleotides or small-molecule drugs [82]. For example, the third motif of the 5s rRNA can be replaced with a nucleotide sequence that inhibits DNA replication, and after replacing 13–15 nucleotides in varying numbers, the material can enter the separated mitochondria in large quantities [42]. In mammalian cells, tRNAs$^{(Gln)}$ encoded by nDNA can be fed into the mitochondria, and yeast tRNAs can enter the mitochondria of eukaryotic cells under natural conditions [83], which is the basic condition for tRNA as a transport carrier. Caroline et al. [41] engineered short RNA (called FD-RNA) consisting of two domains of yeast tRNA, and they inserted oligonucleotide fragments between the two hairpins to successfully target the sequences within mitochondria, inhibiting mutated mtDNA replication. Although it has dealt with the entry of mitochondria, this kind of RNA-based delivery idea still has difficulties passing through the cell membrane autonomously, and it still needs to be released in the cytoplasm with the help of the carrier system that can pass through the cell membrane.

Mitochondria can be introduced into natural single- and double-stranded DNA, and depending on the size of the DNA molecule, there is a significant difference in the efficiency and mechanism of their entry into the mitochondria [19]. Base-complementary paired DNA has the potential of a natural carrier, and its rich base region is the optimal site for small-molecule drug loading [75]. Cyanine structural dyes coupled to DNA have been demonstrated to preferentially target the mitochondria of tumor cells, and the carrier delivering such colors may penetrate the cells and localize to the mitochondria, prompting researchers to investigate mitochondrial delivery methods employing DNA nanocarbonates. As one of the most efficient chemotherapy medications, doxorubicin (Dox) has a limited effect on cell targeting, making it simple to harm normal organ cells [84]. Tao et al. [85] carried Dox on stranded DNA and targeted mitochondria under the guidance of Cy5.5, inhibiting p-AKT to mediate mitochondrial-derived cancer cell apoptosis while reducing damage to normal cells. After that, by replacing the cargo on the DNAtrain with TMPyP4 and Cy3, the DNAtrain drove to the hypoxic MCF-7 cell mitochondria and enhanced the efficacy of photodynamic therapy (PDT) [86]. Curcumin has also been linked to the regulation of energy metabolism in cells. Yang et al. [87] combine stranded DNA with FeS_2 to form a stable nanotube-like material under the junction of dopamine, with a rich GC base pair becoming an ideal location for curcumin loading. The complex then enters the cells to target the mitochondria, and curcumin is successfully released to reduce the energy metabolism of MCF-7 cells.

Even though the modified double-stranded DNA can partially enter the cell and be localized to the mitochondria through targeted sequences, the stability of the double-stranded DNA nanostructures under physiological conditions is far less than that of the framework DNA nanomaterials [88]. At the same time, there are fewer sites for modification of stranded DNA, which may not meet the requirements for transport routes with high targeting accuracy and complex targeting processes.

4.3.2 Framework Nucleic Acid-Based Delivery System in Mitochondria Treatment

DNA with good editability makes it more imaginative in structure, and long enough strands of DNA can fold into any two- or even three-dimensional structure [89–91]. DNA from the chain to the frame structure increases the rigidity of the material itself and also increases the possibility of carrying drugs. Numerous studies have shown that tetrahedral framework DNA or spherical DNA nanostructures are mostly through foveal-mediated encapsulation [92, 93], and there are also reports of the use of functionalized vertical silicon nanoneedle

(SiNN) arrays to directly transfer frame DNA into cytoplasm without encapsulation [94], which makes it common for DNA framework material to enter cells. They can enter almost all cells. By the way, spatial stereostructure allows DNA nanomaterials to have more editable and replaceable sites, including alternative aptamers, RNA, PNA, and other structures [95–99], giving the delivery framework richer functionality. Frame nucleic acids are more resistant to degradation by nucleases, serum, acidity, and other conditions than other nucleic acid monomers, which explains why they can exhibit good delivery effects *in vivo* [72]. After DNA enters the cytoplasm, in order to increase its targeting and efficiency of entering mitochondria, it is usually used to connect some targeted peptides, cations, etc. Then, modified DNA framework structures can be used to carry therapeutic drugs into the mitochondria [86, 100–102] (Figure 4.3).

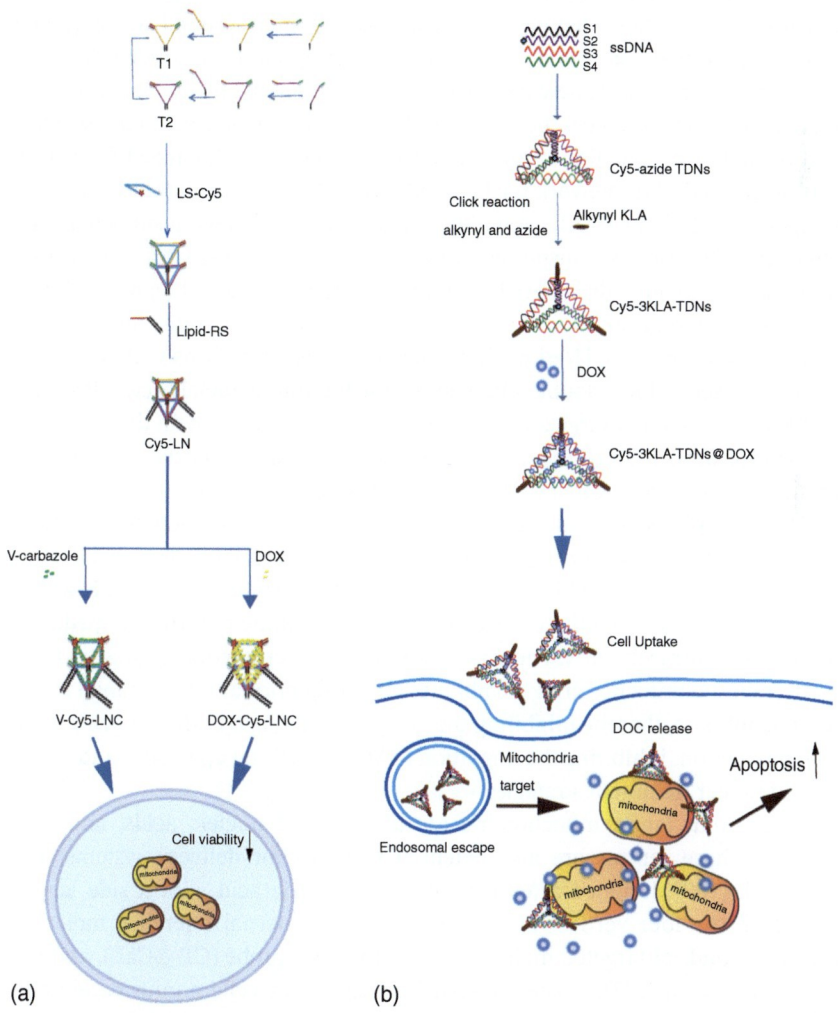

Figure 4.3 (a) The structure of 3D lipid-functionalized DNA nanocages and its application. Source: Adapted from Chan et al. [103]. (b) The design of tetrahedral framework nucleic acids with DOX loading.

Connecting MTS and NLS at the apex of DNA cage structure can successfully localize this framework nucleic acid to the mitochondria and nucleus [94], which lays the foundation for its use as a nanocarrier for locating organelles. They then modified the 3D framework using double-tailed C18 lipid chain-based DNA amphiphiles [103], allowing it to efficiently target into the cytoplasm of MCF-7 cells and transfer to mitochondria. At the same time as the degradation of the DNA framework, drugs such as V-carbazole or Dox carried within the framework are released (Figure 4.3b), causing significant cytotoxicity. While loaded DOX successfully targets mitochondria and performs biological functions, the design of such framework nucleic acids is complex, and its manufacturing process and technical sensitivity may limit its wider application.

As the simplest and most synthetically stable stereonucleic acid nanocarrier [104], tetrahedral framework nucleic acids were first demonstrated in 2014 to be transported into lysosomes by encapsulating vesicles. They can eventually escape from lysosomal and be guided to the nucleus through their connected nuclear localization signals [93], which provides a theoretical basis for tetrahedral framework nucleic acids to target organelles. Subsequently, by replacing the targeting sequence, the tetrahedral framework nucleic acid targets mitochondria and regulates the function of mitochondria. Yan et al. [105] carried four DNA single strands of tetrahedral framework nucleic acids formed by base complementary pairings on the apex with D-(KLAKLAK)$_2$ peptide modifications, enhancing mitochondrial targeting after cell entry. Dox is transported to the mitochondria along with the GC-rich regions of DNA, achieving high drug enrichment in the mitochondria and high lethality of tumor cells. At the same time, another way of synthesizing tetrahedral frame nucleic acids has also been reported. Wu et al. [106] prefabricated four 3-way junctions without stems, which were then crossed to form double-bundled tetrahedral frame nucleic acids. It simultaneously binds to an ASO that targets COXII and a KillerRed (KR) protein drug for PDT and attaches AS1411 at its four vertices. This tool targets the mitochondria of Hela cells, causing significant downregulation of COXII and apoptosis of cells (Figure 4.4).

Tetrahedral framework nucleic acids can also regulate mitochondrial function by forming dynamic polymers intracellularly through sequence design. Li et al. [107] designed a tetrahedral framework nucleic acid in which one vertex carries triphenylphosphine TPP targets mitochondria, and the remaining three vertices are rich in guanine to form K+-mediated tetrahedral aggregates intracellularly. These negatively charged tetrahedral aggregates form a shielding effect on the mitochondrial membrane, hindering normal ion exchange in the mitochondrial membrane, thereby inhibiting the energy metabolism of the mitochondria, reducing ATP production, inhibiting the migration of MCF-7 cells by up to 50%, and is one of the promising cancer treatment strategies.

Imaging systems that target mitochondria are also framed nucleic acids, and it is not excluded that these reports are also potential methods for delivery systems. For example, Liu et al. [108] designed a tetrahedral frame nucleic acid with a side length of 12 nm, at which the four vertices are connected to the internal reference molecule (AF660), the mitochondrial-targeted molecule (TPP), the Ca^{2+} probe (CD@CaL), and the PH-responsive fluorescein (FITC). The nanoparticles can enter cells noninvasively and target mitochondria, enabling simultaneous imaging and sensing of two biological signals within mitochondria.

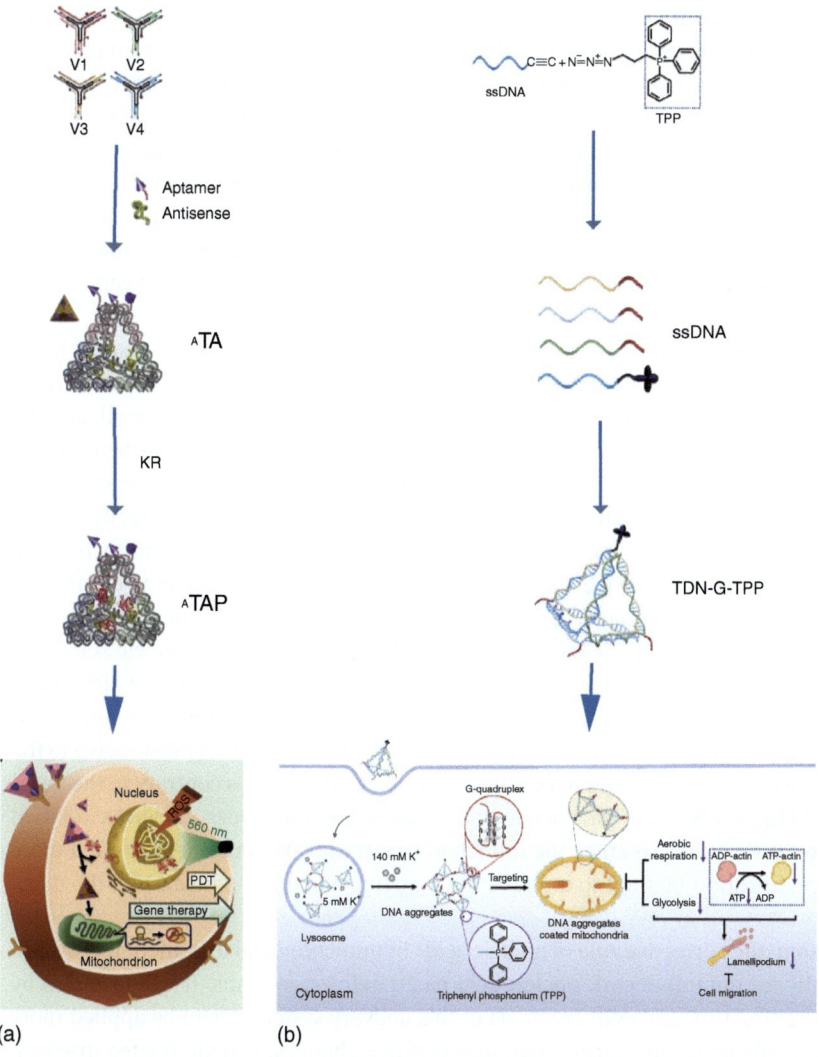

Figure 4.4 (a) The design of tetrahedral framework nucleic acids loading with functional nucleic acid and protein. Source: Reproduced with permission Wu et al. [106]. Copyright 2020, American Chemical Society. (b) The form of dynamic aggregates with tetrahedral framework nucleic acids on mitochondrial membrane. Source: Reproduced with permission Li et al. [107]. Copyright 2022, American Chemical Society.

4.4 Challenges and Prospectives

In summer, whether it is a simple nucleic acid therapeutic drug or the development of a framework nucleic acid-based delivery system, it provides a guiding inquiry into the treatment of mitochondrial dysfunction, although most studies are still stayed *in vitro*. Due to the limited knowledge of the mechanism of action of nucleic acids entering the mitochondria, there is a clear obstacle to the optimization of nucleic acid structure.

Traditional drug delivery systems exhibit poor permeability when passing through biological intrinsic barriers such as the blood–brain barrier and the blood–placental barrier due to their size and structure, resulting in low efficiency in drug treatment of serious diseases, particularly central nervous system diseases. Mitochondrial dysfunction is linked to a number of neurodegenerative disorders. Framework nucleic acid nanoparticles are effective in transmitting physiological barriers. Taking the tetrahedral framework nucleic acid as an example, its ability to cross the blood–brain barrier and transport drugs into nerve cells [109, 110], combined with the neuroprotective effect and editable targeting of the tetrahedral framework nucleic acid itself, will be expected to make it an efficient delivery system for neuropathy caused by mitochondrial dysfunction [111, 112].

On the other hand, framework nucleic acid-based delivery systems have made a lot of evidence in targeting cells, including *in vivo* and *in vitro* work, but they seem to be less powerful in the study of targeted organelles; the number of studies in this area is small, and the mechanism is not clearly elaborated. At present, the treatment of mitochondrial dysfunction diseases is mostly focused on cancer-related exploration, and there are fewer studies on other dysfunctions such as calcium ion overload and abnormal mitochondrial dynamics. Framework nucleic acids are a very promising delivery system that targets mitochondria and may also have certain mitochondrial function regulation effects, such as tetrahedral framework nucleic acids that regulate oxidative stress of cells. However, the specific mechanism by which framework nucleic acids carry small-molecule drugs or nucleic acids into the mitochondria through two layers of mitochondrial membranes is unclear, which is also a major limitation of the use of framework nucleic acids in the treatment of mitochondrial dysfunction. Secondly, the special structure of mitochondria is more conducive to the targeting of positively charged and lipophilic materials [113], which is contrary to the negative electrical characteristics of the nucleic acids, and for the framework nucleic acid-based delivery systems that wish to enter the mitochondria, it may be necessary to find a more systematic modification method to improve the targeting and entry efficiency of the mitochondria.

In conclusion, the corrective role of nucleic acids in mitochondrial dysfunction is unlimited, especially for the framework nucleic acid-based delivery systems. It is hoped that by optimizing the structure and delivery mode of this delivery system, it can be applied more accurately and efficiently to cancer, neurodegenerative changes, genetic-related diseases, and other diseases related to mitochondria.

Funding

This study was supported by the National Natural Science Foundation of China (81800947), Sichuan Science and Technology Program (Grant 2020YFS0176), and the Postdoctoral Science Foundation of China (Grants 2018M640930 and 2020T130443).

References

1 Friedman, J.R. and Nunnari, J. (2014). Mitochondrial form and function. *Nature* 505 (7483): 335–343.

- 2 Bock, F.J. and Tait, S.W.G. (2020). Mitochondria as multifaceted regulators of cell death. *Nat. Rev. Mol. Cell Biol.* 21 (2): 85–100.
- 3 Wong, Y.C., Ysselstein, D., and Krainc, D. (2018). Mitochondria-lysosome contacts regulate mitochondrial fission via RAB7 GTP hydrolysis. *Nature* 554 (7692): 382–386.
- 4 Chen, Y. and Sheng, Z.H. (2013). Kinesin-1-syntaphilin coupling mediates activity-dependent regulation of axonal mitochondrial transport. *J. Cell Biol.* 202 (2): 351–364.
- 5 Ding, W.X. and Yin, X.M. (2012). Mitophagy: mechanisms, pathophysiological roles, and analysis. *Biol. Chem.* 393 (7): 547–564.
- 6 Youle, R.J. and Narendra, D.P. (2011). Mechanisms of mitophagy. *Nat. Rev. Mol. Cell Biol.* 12 (1): 9–14.
- 7 Wang, X.L., Feng, S.T., Wang, Z.Z. et al. (2021). Role of mitophagy in mitochondrial quality control: mechanisms and potential implications for neurodegenerative diseases. *Pharmacol. Res.* 165: 105433.
- 8 Anzell, A.R., Fogo, G.M., Gurm, Z. et al. (2021). Mitochondrial fission and mitophagy are independent mechanisms regulating ischemia/reperfusion injury in primary neurons. *Cell Death Dis.* 12 (5): 475.
- 9 Ogrodnik, M., Miwa, S., Tchkonia, T. et al. (2017). Cellular senescence drives age-dependent hepatic steatosis. *Nat. Commun.* 8: 15691.
- 10 Mehta, A.R., Gregory, J.M., Dando, O. et al. (2021). Mitochondrial bioenergetic deficits in C9orf72 amyotrophic lateral sclerosis motor neurons cause dysfunctional axonal homeostasis. *Acta Neuropathol.* 141 (2): 257–279.
- 11 Cassim, S., Vucetic, M., Zdralevic, M., and Pouyssegur, J. (2020). Warburg and beyond: the power of mitochondrial metabolism to collaborate or replace fermentative glycolysis in cancer. *Cancers (Basel).* 12 (5).
- 12 Grunewald, A., Kumar, K.R., and Sue, C.M. (2019). New insights into the complex role of mitochondria in Parkinson's disease. *Prog. Neurobiol.* 177: 73–93.
- 13 Lleonart, M.E., Grodzicki, R., Graifer, D.M., and Lyakhovich, A. (2017). Mitochondrial dysfunction and potential anticancer therapy. *Med. Res. Rev.* 37 (6): 1275–1298.
- 14 Prakash, Y.S., Pabelick, C.M., and Sieck, G.C. (2017). Mitochondrial dysfunction in airway disease. *Chest* 152 (3): 618–626.
- 15 Wen, R., Banik, B., Pathak, R.K. et al. (2016). Nanotechnology inspired tools for mitochondrial dysfunction related diseases. *Adv. Drug Delivery Rev.* 99 (Pt A): 52–69.
- 16 Kulkarni, J.A., Witzigmann, D., Thomson, S.B. et al. (2021). The current landscape of nucleic acid therapeutics. *Nat. Nanotechnol.* 16 (6): 630–643.
- 17 Vaughan, H.J., Green, J.J., and Tzeng, S.Y. (2020). Cancer-targeting nanoparticles for combinatorial nucleic acid delivery. *Adv. Mater.* 32 (13): e1901081.
- 18 Gao, K., Cheng, M., Zuo, X. et al. (2021). Active RNA interference in mitochondria. *Cell Res.* 31 (2): 219–228.
- 19 Verechshagina, N.A., Konstantinov, Y.M., Kamenski, P.A., and Mazunin, I.O. (2018). Import of proteins and nucleic acids into mitochondria. *Biochemistry (Mosc).* 83 (6): 643–661.
- 20 Goodchild, J., Kim, B., and Zamecnik, P.C. (1991). The clearance and degradation of oligodeoxynucleotides following intravenous injection into rabbits. *Antisense Res. Dev.* 1 (2): 153–160.

21 Deleavey, G.F. and Damha, M.J. (2012). Designing chemically modified oligonucleotides for targeted gene silencing. *Chem. Biol.* 19 (8): 937–954.

22 Shen, W., De Hoyos, C.L., Migawa, M.T. et al. (2019). Chemical modification of PS-ASO therapeutics reduces cellular protein-binding and improves the therapeutic index. *Nat. Biotechnol.* 37 (6): 640–650.

23 Manoharan, M. (1999). 2′-carbohydrate modifications in antisense oligonucleotide therapy: importance of conformation, configuration and conjugation. *Biochim. Biophys. Acta* 1489 (1): 117–130.

24 Springer, A.D. and Dowdy, S.F. (2018). GalNAc-siRNA conjugates: leading the way for delivery of RNAi therapeutics. *Nucleic Acid Ther.* 28 (3): 109–118.

25 McNally, E.M. and Leverson, B.D. (2019). Better living through peptide-conjugated chemistry: next-generation antisense oligonucleotides. *J. Clin. Invest.* 129 (11): 4570–4571.

26 McClorey, G. and Banerjee, S. (2018). Cell-penetrating peptides to enhance delivery of oligonucleotide-based therapeutics. *Biomedicines* 6 (2).

27 Finkel, R.S., Mercuri, E., Darras, B.T. et al. (2017). Nusinersen versus sham control in infantile-onset spinal muscular atrophy. *N. Engl. J. Med.* 377 (18): 1723–1732.

28 Mercuri, E., Darras, B.T., Chiriboga, C.A. et al. (2018). Nusinersen versus sham control in later-onset spinal muscular atrophy. *N. Engl. J. Med.* 378 (7): 625–635.

29 Sun, L., Li, P., Ju, X. et al. (2021). In vivo structural characterization of the SARS-CoV-2 RNA genome identifies host proteins vulnerable to repurposed drugs. *Cell* 184 (7): 1865–83 e20.

30 Wallace, D.C. (2005). A mitochondrial paradigm of metabolic and degenerative diseases, aging, and cancer: a dawn for evolutionary medicine. *Annu. Rev. Genet.* 39: 359–407.

31 Russell, O. and Turnbull, D. (2014). Mitochondrial DNA disease-molecular insights and potential routes to a cure. *Exp. Cell. Res.* 325 (1): 38–43.

32 Gorman, G.S., Chinnery, P.F., DiMauro, S. et al. (2016). Mitochondrial diseases. *Nat. Rev. Dis. Primers* 2: 16080.

33 Alexeyev, M.F., Venediktova, N., Pastukh, V. et al. (2008). Selective elimination of mutant mitochondrial genomes as therapeutic strategy for the treatment of NARP and MILS syndromes. *Gene Ther.* 15 (7): 516–523.

34 Reddy, P., Ocampo, A., Suzuki, K. et al. (2015). Selective elimination of mitochondrial mutations in the germline by genome editing. *Cell* 161 (3): 459–469.

35 Gammage, P.A., Rorbach, J., Vincent, A.I. et al. (2014). Mitochondrially targeted ZFNs for selective degradation of pathogenic mitochondrial genomes bearing large-scale deletions or point mutations. *EMBO Mol. Med.* 6 (4): 458–466.

36 Bacman, S.R., Williams, S.L., Pinto, M. et al. (2013). Specific elimination of mutant mitochondrial genomes in patient-derived cells by mitoTALENs. *Nat. Med.* 19 (9): 1111–1113.

37 Yang, Y., Wu, H., Kang, X. et al. (2018). Targeted elimination of mutant mitochondrial DNA in MELAS-iPSCs by mitoTALENs. *Protein Cell* 9 (3): 283–297.

38 McCann, B.J., Cox, A., Gammage, P.A. et al. (2018). Delivery of mtZFNs into early mouse embryos. *Methods Mol. Biol.* 1867: 215–228.

39 Zekonyte, U., Bacman, S.R., Smith, J. et al. (2021). Mitochondrial targeted meganuclease as a platform to eliminate mutant mtDNA in vivo. *Nat. Commun.* 12 (1): 3210.

40 Taylor, R.W., Chinnery, P.F., Turnbull, D.M., and Lightowlers, R.N. (1997). Selective inhibition of mutant human mitochondrial DNA replication in vitro by peptide nucleic acids. *Nat. Genet.* 15: 212–215.

41 Comte, C., Tonin, Y., Heckel-Mager, A.M. et al. (2013). Mitochondrial targeting of recombinant RNAs modulates the level of a heteroplasmic mutation in human mitochondrial DNA associated with Kearns Sayre Syndrome. *Nucleic Acids Res.* 41 (1): 418–433.

42 Loutre, R., Heckel, A.M., Jeandard, D. et al. (2018). Anti-replicative recombinant 5S rRNA molecules can modulate the mtDNA heteroplasmy in a glucose-dependent manner. *PLoS One* 13 (6): e0199258.

43 Chan, D.C. (2006). Mitochondria: dynamic organelles in disease, aging, and development. *Cell* 125 (7): 1241–1252.

44 Yamada, Y., Maruyama, M., Kita, T. et al. (2020). The use of a MITO-Porter to deliver exogenous therapeutic RNA to a mitochondrial disease's cell with a A1555G mutation in the mitochondrial 12S rRNA gene results in an increase in mitochondrial respiratory activity. *Mitochondrion* 55: 134–144.

45 Yamada, Y., Somiya, K., Miyauchi, A. et al. (2020). Validation of a mitochondrial RNA therapeutic strategy using fibroblasts from a Leigh syndrome patient with a mutation in the mitochondrial ND3 gene. *Sci. Rep.* 10 (1): 7511.

46 Kawamura, E., Maruyama, M., Abe, J. et al. (2020). Validation of gene therapy for mutant mitochondria by delivering mitochondrial RNA using a MITO-porter. *Mol. Ther. Nucleic Acids* 20: 687–698.

47 Furukawa, R., Yamada, Y., Kawamura, E., and Harashima, H. (2015). Mitochondrial delivery of antisense RNA by MITO-Porter results in mitochondrial RNA knockdown, and has a functional impact on mitochondria. *Biomaterials* 57: 107–115.

48 Kawamura, E., Hibino, M., Harashima, H., and Yamada, Y. (2019). Targeted mitochondrial delivery of antisense RNA-containing nanoparticles by a MITO-Porter for safe and efficient mitochondrial gene silencing. *Mitochondrion* 49: 178–188.

49 Cruz-Zaragoza, L.D., Dennerlein, S., Linden, A. et al. (2021). An in vitro system to silence mitochondrial gene expression. *Cell* 184 (23): 5824–5837. e15.

50 Chan, D.C. (2020). Mitochondrial dynamics and its involvement in disease. *Annu. Rev. Pathol.* 15: 235–259.

51 Garcia, D.A., Powers, A.F., Bell, T.A. 3rd, et al. (2022). Antisense oligonucleotide-mediated silencing of mitochondrial fusion and fission factors modulates mitochondrial dynamics and rescues mitochondrial dysfunction. *Nucleic Acid Ther.* 32 (1): 51–65.

52 Siira, S.J., Rossetti, G., Richman, T.R. et al. (2018). Concerted regulation of mitochondrial and nuclear non-coding RNAs by a dual-targeted RNase Z. *EMBO Rep.* 19 (10).

53 Zhao, Y., Sun, L., Wang, R.R. et al. (2018). The effects of mitochondria-associated long noncoding RNAs in cancer mitochondria: New players in an old arena. *Crit. Rev. Oncol. Hematol.* 131: 76–82.

54 Varas-Godoy, M., Lladser, A., Farfan, N. et al. (2018). In vivo knockdown of antisense non-coding mitochondrial RNAs by a lentiviral-encoded shRNA inhibits melanoma tumor growth and lung colonization. *Pigm. Cell Melanoma Res.* 31 (1): 64–72.

55 Borgna, V., Villegas, J., Burzio, V.A. et al. (2017). Mitochondrial ASncmtRNA-1 and ASncmtRNA-2 as potent targets to inhibit tumor growth and metastasis in the RenCa murine renal adenocarcinoma model. *Oncotarget.* 8 (27): 43692–43708.

56 Vidaurre, S., Fitzpatrick, C., Burzio, V.A. et al. (2014). Down-regulation of the antisense mitochondrial non-coding RNAs (ncRNAs) is a unique vulnerability of cancer cells and a potential target for cancer therapy. *J. Biol. Chem.* 289 (39): 27182–27198.

57 Lobos-Gonzalez, L., Silva, V., Araya, M. et al. (2016). Targeting antisense mitochondrial ncRNAs inhibits murine melanoma tumor growth and metastasis through reduction in survival and invasion factors. *Oncotarget* 7 (36): 58331–58350.

58 Lobos-Gonzalez, L., Bustos, R., Campos, A. et al. (2020). Exosomes released upon mitochondrial ASncmtRNA knockdown reduce tumorigenic properties of malignant breast cancer cells. *Sci. Rep.* 10 (1): 343.

59 Wilusz, J.E. (2018). A 360 degrees view of circular RNAs: From biogenesis to functions. *Wiley Interdiscip. Rev.: RNA* 9 (4): e1478.

60 Kristensen, L.S., Andersen, M.S., Stagsted, L.V.W. et al. (2019). The biogenesis, biology and characterization of circular RNAs. *Nat. Rev. Genet.* 20 (11): 675–691.

61 Zhao, Q., Liu, J., Deng, H. et al. (2020). Targeting mitochondria-located circRNA SCAR alleviates NASH via reducing mROS output. *Cell* 183 (1): 76–93. e22.

62 Hossen, S., Hossain, M.K., Basher, M.K. et al. (2019). Smart nanocarrier-based drug delivery systems for cancer therapy and toxicity studies: a review. *J. Adv. Res.* 15: 1–18.

63 Yuan, P., Mao, X., Wu, X. et al. (2019). Mitochondria-targeting, intracellular delivery of native proteins using biodegradable silica nanoparticles. *Angew. Chem. Int. Ed. Engl.* 58 (23): 7657–7661.

64 Zhang, J., Li, C., Xue, Q. et al. (2021). An efficient carbon-based drug delivery system for cancer therapy through the nucleus targeting and mitochondria mediated apoptotic pathway. *Small Methods* 5 (12): e2100539.

65 Velichkovska, M., Surnar, B., Nair, M. et al. (2019). Targeted mitochondrial COQ10 delivery attenuates antiretroviral-drug-induced senescence of neural progenitor cells. *Mol. Pharmaceutics* 16 (2): 724–736.

66 Liu, H.N., Guo, N.N., Wang, T.T. et al. (2018). Mitochondrial targeted doxorubicin-triphenylphosphonium delivered by hyaluronic acid modified and pH responsive nanocarriers to breast tumor: in vitro and in vivo studies. *Mol. Pharmaceutics* 15 (3): 882–891.

67 Abe, J., Yamada, Y., Takeda, A., and Harashima, H. (2018). Cardiac progenitor cells activated by mitochondrial delivery of resveratrol enhance the survival of a doxorubicin-induced cardiomyopathy mouse model via the mitochondrial activation of a damaged myocardium. *J. Controlled Release* 269: 177–188.

68 Katayama, T., Kinugawa, S., Takada, S. et al. (2019). A mitochondrial delivery system using liposome-based nanocarriers that target myoblast cells. *Mitochondrion* 49: 66–72.

69 Yamada, Y., Tabata, M., Yasuzaki, Y. et al. (2014). A nanocarrier system for the delivery of nucleic acids targeted to a pancreatic beta cell line. *Biomaterials* 35 (24): 6430–6438.

70 Tan, X., Jia, F., Wang, P., and Zhang, K. (2020). Nucleic acid-based drug delivery strategies. *J. Controlled Release* 323: 240–252.

71 Banerjee, A., Bhatia, D., Saminathan, A. et al. (2013). Controlled release of encapsulated cargo from a DNA icosahedron using a chemical trigger. *Angew. Chem. Int. Ed. Engl.* 52 (27): 6854–6857.

72 Li, J., Pei, H., Zhu, B. et al. (2011). Self-assembled multivalent DNA nanostructures for noninvasive intracellular delivery of immunostimulatory CpG oligonucleotides. *ACS Nano* 5 (11): 8783–8789.

73 Walsh, A.S., Yin, H., Erben, C.M. et al. (2011). DNA cage delivery to mammalian cells. *ACS Nano* 5 (7): 5427–5432.

74 Wang, Z., Song, L., Liu, Q. et al. (2021). A tubular DNA nanodevice as a siRNA/chemo-drug co-delivery vehicle for combined cancer therapy. *Angew. Chem. Int. Ed. Engl.* 60 (5): 2594–2598.

75 Hu, Q., Li, H., Wang, L. et al. (2019). DNA nanotechnology-enabled drug delivery systems. *Chem. Rev.* 119 (10): 6459–6506.

76 Zhang, Q., Jiang, Q., Li, N. et al. (2014). DNA origami as an in vivo drug delivery vehicle for cancer therapy. *ACS Nano* 8 (7): 6633–6643.

77 Zhu, G., Zheng, J., Song, E. et al. (2013). Self-assembled, aptamer-tethered DNA nanotrains for targeted transport of molecular drugs in cancer theranostics. *PNAS* 110 (20): 7998–8003.

78 Lee, H., Lytton-Jean, A.K., Chen, Y. et al. (2012). Molecularly self-assembled nucleic acid nanoparticles for targeted in vivo siRNA delivery. *Nat. Nanotechnol.* 7 (6): 389–393.

79 Modi, S., Nizak, C., Surana, S. et al. (2013). Two DNA nanomachines map pH changes along intersecting endocytic pathways inside the same cell. *Nat. Nanotechnol.* 8 (6): 459–467.

80 Liu, J., Chen, Y., Li, G. et al. (2015). Ruthenium(II) polypyridyl complexes as mitochondria-targeted two-photon photodynamic anticancer agents. *Biomaterials* 56: 140–153.

81 Smirnov, A.V., Entelis, N.S., Krasheninnikov, I.A. et al. (2008). Specific features of 5S rRNA structure – its interactions with macromolecules and possible functions. *Biochemistry (Mosc).* 73 (13): 1418–1437.

82 Smirnov, A., Entelis, N., Martin, R.P., and Tarassov, I. (2011). Biological significance of 5S rRNA import into human mitochondria: role of ribosomal protein MRP-L18. *Genes Dev.* 25 (12): 1289–1305.

83 Schneider, A. (2011). Mitochondrial tRNA import and its consequences for mitochondrial translation. *Annu. Rev. Biochem.* 80: 1033–1053.

84 Tacar, O., Sriamornsak, P., and Dass, C.R. (2013). Doxorubicin: an update on anticancer molecular action, toxicity and novel drug delivery systems. *J. Pharm. Pharmacol.* 65 (2): 157–170.

85 Jiang, T., Zhou, L., Liu, H. et al. (2019). Monitorable mitochondria-targeting DNAtrain for image-guided synergistic cancer therapy. *Anal. Chem.* 91 (11): 6996–7000.

86 Liu, J., Ding, G., Chen, S. et al. (2021). Multifunctional programmable DNA nanotrain for activatable hypoxia imaging and mitochondrion-targeted enhanced photodynamic therapy. *ACS Appl. Mater. Interfaces* 13 (8): 9681–9690.

87 Yang, R., Fang, X.L., Zhen, Q. et al. (2019). Mitochondrial targeting nano-curcumin for attenuation on PKM2 and FASN. *Colloids Surf., B* 182: 110405.

88 Keum, J.W. and Bermudez, H. (2009). Enhanced resistance of DNA nanostructures to enzymatic digestion. *Chem. Commun. (Camb).* 45: 7036–7038.

89 Shih, W.M., Quispe, J.D., and Joyce, G.F. (2004). A 1.7-kilobase single-stranded DNA that folds into a nanoscale octahedron. *Nature* 427 (6975): 618–621.

90 Rothemund, P.W. (2006). Folding DNA to create nanoscale shapes and patterns. *Nature* 440 (7082): 297–302.

91 Goodman, R.P., Schaap, I.A., Tardin, C.F. et al. (2005). Rapid chiral assembly of rigid DNA building blocks for molecular nanofabrication. *Science* 310 (5754): 1661–1665.

92 Choi, C.H., Hao, L., Narayan, S.P. et al. (2013). Mechanism for the endocytosis of spherical nucleic acid nanoparticle conjugates. *PNAS* 110 (19): 7625–7630.

93 Liang, L., Li, J., Li, Q. et al. (2014). Single-particle tracking and modulation of cell entry pathways of a tetrahedral DNA nanostructure in live cells. *Angew. Chem. Int. Ed. Engl.* 53 (30): 7745–7750.

94 Chan, M.S. and Lo, P.K. (2014). Nanoneedle-assisted delivery of site-selective peptide-functionalized DNA nanocages for targeting mitochondria and nuclei. *Small* 10 (7): 1255–1260.

95 Qin, X., Xiao, L., Li, N. et al. (2022). Tetrahedral framework nucleic acids-based delivery of microRNA-155 inhibits choroidal neovascularization by regulating the polarization of macrophages. *Bioact. Mater.* 14: 134–144.

96 Ma, W., Yang, Y., Zhu, J. et al. (2022). Biomimetic nanoerythrosome-coated aptamer-DNA tetrahedron/maytansine conjugates: pH-responsive and targeted cytotoxicity for HER2-positive breast cancer. *Adv. Mater.* e2109609.

97 Li, S., Liu, Y., Tian, T. et al. (2021). Bioswitchable delivery of microRNA by framework nucleic acids: application to bone regeneration. *Small* 17 (47): e2104359.

98 Green, C.M., Hastman, D.A., Mathur, D. et al. (2021). Direct and efficient conjugation of quantum dots to DNA nanostructures with peptide-PNA. *ACS Nano* 15 (5): 9101–9110.

99 Zhang, T., Tian, T., and Lin, Y. (2021). Functionalizing framework nucleic-acid-based nanostructures for biomedical application. *Adv. Mater.* e2107820.

100 Chai, X., Fan, Z., Yu, M.M. et al. (2021). A redox-activatable DNA nanodevice for spatially-selective, AND-gated imaging of ATP and glutathione in mitochondria. *Nano Lett.* 21 (23): 10047–10053.

101 Luo, L., Wang, M., Zhou, Y. et al. (2021). Ratiometric fluorescent DNA nanostructure for mitochondrial ATP imaging in living cells based on hybridization chain reaction. *Anal. Chem.* 93 (17): 6715–6722.

102 Liu, J., Yang, L., Xue, C. et al. (2021). Reductase and light programmatical gated DNA nanodevice for spatiotemporally controlled imaging of biomolecules in subcellular organelles under hypoxic conditions. *ACS Appl. Mater. Interfaces* 13 (29): 33894–33904.

103 Chan, M.S., Tam, D.Y., Dai, Z. et al. (2016). Mitochondrial delivery of therapeutic agents by amphiphilic DNA nanocarriers. *Small* 12 (6): 770–781.

104 He, Y., Ye, T., Su, M. et al. (2008). Hierarchical self-assembly of DNA into symmetric supramolecular polyhedra. *Nature* 452 (7184): 198–201.

105 Yan, J., Chen, J., Zhang, N. et al. (2020). Mitochondria-targeted tetrahedral DNA nanostructures for doxorubicin delivery and enhancement of apoptosis. *J. Mater. Chem. B* 8 (3): 492–503.

106 Wu, T., Liu, Q., Cao, Y. et al. (2020). Multifunctional double-bundle DNA tetrahedron for efficient regulation of gene expression. *ACS Appl. Mater. Interfaces* 12 (29): 32461–32467.

107 Li, F., Liu, Y., Dong, Y. et al. (2022). Dynamic assembly of DNA nanostructures in living cells for mitochondrial interference. *JACS* 144 (10): 4667–4677.

108 Liu, Z., Pei, H., Zhang, L., and Tian, Y. (2018). Mitochondria-targeted DNA nanoprobe for real-time imaging and simultaneous quantification of Ca(2+) and pH in neurons. *ACS Nano* 12 (12): 12357–12368.

109 Zhou, M., Zhang, T., Zhang, B. et al. (2021). A DNA nanostructure-based neuroprotectant against neuronal apoptosis via inhibiting toll-like receptor 2 signaling pathway in acute ischemic stroke. *ACS Nano* 16 (1): 1456–1470.

110 Tian, T., Li, J., Xie, C. et al. (2018). Targeted imaging of brain tumors with a framework nucleic acid probe. *ACS Appl. Mater. Interfaces* 10 (4): 3414–3420.

111 Ma, W., Shao, X., Zhao, D. et al. (2018). Self-assembled tetrahedral DNA nanostructures promote neural stem cell proliferation and neuronal differentiation. *ACS Appl. Mater. Interfaces* 10 (9): 7892–7900.

112 Shao, X., Ma, W., Xie, X. et al. (2018). Neuroprotective effect of tetrahedral DNA nanostructures in a cell model of Alzheimer's disease. *ACS Appl. Mater. Interfaces* 10 (28): 23682–23692.

113 Saminathan, A., Zajac, M., Anees, P., and Krishnan, Y. (2021). Organelle-level precision with next-generation targeting technologies. *Nat. Rev. Mater.* 7 (5): 355–371.

5

Regeneration of Bone-Related Diseases by Nucleic Acid-Based Nanomaterials: Perspectives from Tissue Regeneration and Molecular Medicine

Xiaoru Shao

Jining Medical University, Affiliated Hospital of Jining Medical University, Department of Stomatology, Guhuai Road, Jining, Shandong 272029, PR China

5.1 Introduction

There are many types of bone-related diseases that seriously affect the quality of life of patients. The dysfunction or loss of bone tissue function caused by inflammation, injury, trauma, disease, aging, or genetic disease not only brings great pain to the patient but also brings a series of socioeconomic problems. Common bone-related diseases include osteoporosis, osteoarthritis, bone metastases, osteomyelitis, myeloma, and bone defects [1–5]. Bone-related diseases can seriously affect the quality of life of patients; therefore, it is very important to seek effective treatment methods. Traditional bone tissue regeneration treatment methods mainly include transplantation of autologous bone, allogeneic bone, and artificial bone. However, all of the above treatment methods have certain drawbacks and limitations, such as increased risk of infection, rejection of recipients, and insufficient donors, which cannot fully meet clinical needs [6–8]. In addition, the patient's economic status and their acceptance of materials for repairing bone defects are also important issues for surgeons to consider when repairing bone tissue defects [9]. Screening out new materials with a better ability to repair bone tissue defects to meet clinical needs has become an urgent problem for researchers and clinicians to solve. Bone tissue engineering avoids the shortcomings of traditional biomaterials and explores new biomaterials through continuous research on emerging technologies. The focus of bone tissue engineering research includes stem cells, biomaterials, and growth factors. Bioscaffold materials are designed to mimic the extracellular matrix, providing a suitable microenvironment to support and promote bone tissue growth. The bioscaffold materials used should have good biocompatibility, biodegradability, and low immunogenicity. Therefore, methods for discovering simple synthetic biomaterials and improving the biosafety of materials need to be explored. It is urgent to carry out in-depth research on biomaterials that can effectively repair bone tissue regeneration.

Nanotechnology has attracted widespread attention since its emergence. Nanotechnology refers to the study of the properties and technologies of materials in the range of 0.1–100 nm. The unique small-scale effect, surface effect, tunnel effect, etc., make nanomaterials have

Nucleic Acid-Based Nanomaterials: Stabilities and Applications, First Edition.
Edited by Yunfeng Lin and Shaojingya Gao.
© 2024 WILEY-VCH GmbH. Published 2024 by WILEY-VCH GmbH.

a wide range of application prospects and have achieved rapid development in the past 20 years [10, 11]. In particular, the application of nanotechnology in the fields of drug carriers, biosensing, bioimaging, gene delivery, disease diagnosis, and treatment has been deepened day by day. Nanomaterials, as new carriers for drug delivery, have many advantages compared with traditional free drugs, including increased drug stability, controllable drug release rate, reduced dosage, and improved bioavailability. It will become a hot research field in future medical development [12–14].

Nucleic acid nanotechnology is a new type of nanotechnology developed in recent years. The principle of nucleic acid nanotechnology is to use nucleic acid and its derivatives as basic materials. By designing highly specific and programmable base sequences, nucleic acid molecules are folded and paired to form nanomaterials with specific shapes, geometries, and scales [15, 16]. Due to the biomolecular properties of nucleic acid itself, it can be used as a new type of transport carrier and may also carry specific genetic information. Therefore, nucleic acid-based functional nanomaterials have special application prospects and research value. Nucleic acid-based functional nanomaterials use the inherent nucleic acid of the human body as the raw material, so they have many advantages, such as good biocompatibility and degradation ability, being nontoxic to organisms, and having rich functional modification sites. Nucleic acid nanotechnology has been widely used in molecular diagnosis, biological imaging, molecular delivery, and targeted drug delivery [17, 18]. Therefore, how to effectively regenerate bone tissue with nucleic acid-based nanomaterials has become an interesting research direction. This chapter will review the role of nucleic acid-based nanomaterials in bone-related disease regeneration from the perspective of tissue regeneration and molecular medicine.

5.2 The Development Process of Functional Nucleic Acid

The development of nucleic acid nanotechnology has made nucleic acid widely used as a material. By applying base pairing with highly specific and programmable properties, a large number of framework nucleic acid nanomaterials with different dimensions, shapes, and geometric structures can be perfectly designed and synthesized [19, 20]. Among the many potential application directions of nucleic acid nanomaterials, biological and biomedical applications are very attractive due to the biomolecular properties of nucleic acids. Nucleic acid nanomaterials generally have low cytotoxicity and strong resistance to enzymatic degradation in biological environments, which are favorable for the *in vivo* application of nucleic acid-based functional nanomaterials. Since Watson and Crick proposed the double helix structure model, the biochemical properties, structure, and morphology of DNA molecules have been well studied. Nucleic acid nanotechnology refers to nanotechnology based on the physical and chemical properties of DNA, which is mainly used for molecular assembly to produce functional assembly aggregates. Therefore, the research of nucleic acid nanotechnology occupies an extremely important position not only in the field of life sciences but also in the field of nanomaterials [21, 22]. The essence of nucleic acid nanotechnology is DNA self-assembly, and the core of self-assembly is the principle of complementary base pairing. As a natural biological macromolecule, DNA has unique advantages in constructing functional structures at the nanoscale, including high specificity, predictability, programmability, high mechanical rigidity, and high physicochemical stability [23, 24]. Throughout the development of nucleic acid nanotechnology, it has been found that functional nucleic acids mainly include DNA tile, DNA origami, three-dimensional DNA self-assembly, DNA nanorobots, and DNA microchips.

5.2.1 DNA Tile

In 1982, Seeman of New York University's Department of Chemistry first proposed that two-dimensional ordered arrays could be constructed using branched DNA molecules with complementary sticky ends [25]. The basic building unit of two-dimensional DNA structure (also known as "DNA tile") is a four-arm-junction complex constructed by four single-stranded DNA, and the base sequences of the four sticky ends are specially designed, and through the complementary pairing between the sticky ends in different motifs, periodic two-dimensional DNA planar structures can be constructed [26]. Seeman's genius idea was later proven experimentally to be the first stepping stone to structural DNA nanotechnology. Early nucleic acid nanotechnology mainly studied various DNA self-assembly modules and various periodic structures generated by their assembly. Based on Seeman's original idea, in 1999, Mao Chengde's research group at Purdue University connected four Holliday junctions together to construct parallelogram DNA structural units, which could self-assemble to form a 2D DNA array with diamond-shaped holes [27]. Later, Yan Hao's group reported a cross-shaped primitive, namely a 4×4 primitive, which can be used as a template to form conductive nanowires or two-dimensional protein arrays [28, 29].

5.2.2 DNA Origami

In 2006, Rothemund's research brought an exciting advance in structural DNA nanotechnology. He first proposed "DNA origami" in Nature: using more than 200 short DNA auxiliary strands, by folding M13mp18 genomic DNA, two-dimensional structures of about 100 nm in size and arbitrary shapes, such as rectangles, squares, triangles, stars, and smiley faces [30]. In 2006, under the leadership of He Lin, an academician of the Chinese Academy of Sciences, the Bio-X Research Institute of Shanghai Jiaotong University, pioneered the creation of a "DNA China Map" with a length of 150 nm, a width of 120 nm, a thickness of 2 nm, and a resolution of 6 nm. The world's first asymmetric two-dimensional DNA pattern overcomes the stress problem caused by asymmetric patterns [31]. The experiment proves that asymmetric, complex, two-dimensional shapes can be constructed using the method of DNA origami. They further deduced that DNA origami has the ability to construct nearly any complex two-dimensional nanoscale shape. The advantages of DNA origami self-assembly are that the design method is simple, the operation is easy, the raw DNA does not need to be purified, complex patterns can be assembled in one step, and the material yield is high. Combined with other methods and techniques, more complex three-dimensional structures can be assembled through origami.

5.2.3 Three-dimensional DNA Self-assembly

When DNA self-assembly could easily build more complex planar graphics, scientists began to pay more attention to building the three-dimensional structure of DNA. In the past five years, research on 3D DNA nanostructures has achieved vigorous development, and a large number of 3D DNA motifs have been successfully constructed. Turberfield's group successfully constructed DNA tetrahedral structures of a series of sizes using four single-stranded DNAs and tried to use these DNA tetrahedra as drug delivery systems for drug-controlled release studies [32, 33]. Other 3D DNA nanostructures, including cube, tetrahedron, octahedron, dodecahedron, buckyball, and nanotube, have also been extensively studied [34–36].

5.2.4 DNA Nanobots and DNA Microchips

In a 2012 Science article, researchers used DNA origami to build a nanorobot that can deliver tiny "cargo" loads to individual cells and affect their behavior [37]. The technique folds DNA strands together into a complex shape with a cross section that resembles a hexagon. They then loaded it with nanoparticles such as gold and fluorescently labeled antibody fragments and watched their nanorobots deliver their cargo to cells in a tissue culture. These nanoparticles can be designed to respond to certain cell surface proteins or specific combinations of them and deliver a variety of different molecules that affect cell behavior in different ways. In the new study, published in Nature on January 23, 2013, the scientists "recorded" a 26-second clip of audio, a copy of a paper, a photo, and other information in artificially synthesized DNA [38]. The findings mark a milestone toward the practicality of using nucleic acids for storing information, which is more compact and durable than current hard drives or magnetic tapes.

5.3 Nucleic Acid-Based Functional Nanomaterials

In addition to the abovementioned DNA-based nanostructures, functional nucleic acids are combined with nanomaterials to form many functional nucleic acid nanomaterials. Nucleic acid-based functional nanomaterials have the following characteristics: good biocompatibility, good biodegradability, good biological stability, and editability. Given these properties of nucleic acid-based functional nanomaterials, they have promising application prospects in biomedical applications such as drug delivery, tissue engineering, bioimaging, and biosensing.

5.3.1 Nanomaterials That Can Bind to Functional Nucleic Acids

A wide variety of nanomaterials can be combined with functional nucleic acids, including metal-based nanomaterials, carbon-based nanomaterials, biological nanomaterials, quantum dots, and composite nanomaterials. Different types of nanomaterials combine with functional nucleic acids in different ways and play different roles in bioimaging, biosensing, drug delivery, tissue engineering, and other fields.

5.3.1.1 Metal-Based Nanomaterials

Metal-based nanomaterials are often combined with functional nucleic acids. This combination enables functional nucleic acids with greater stability and new catalytic activities [39]. In addition, functional nucleic acids also enabled metal-based nanomaterials to have better biocompatibility and altered their charge properties. The most basic combinatorial approach is to use metal-based nanomaterials as templates for functional nucleic acid immobilization and localization. Functional nucleic acid-metal nanomaterials have broad applications in biosensing, bioimaging, and biomedicine. At the same time, functional nucleic acids also endow metal nanomaterials with more precise targeting capabilities [40, 41]. The principle of nucleic acid template-forming metal nanoclusters (NCs) is the process of nucleic acid metallization. Because the size of NCs is very small, the surface energy is very high, and NCs are easy to gather, so a stabilizer or matrix must be used to isolate NCs. Prior to the excellent work of Dickson et al. using dendrimers as stabilizers, metal NCs were only prepared in low-temperature inert gas or solid zeolite. After their initial discovery, Dickson's team first proved that cytosine (C)-rich DNA can also stabilize fluorescent AgNCs, which establishes a connection between metal NCs and DNA. There is a strong specific coordination between the N3 position of cytosine and Ag+, while the C-rich DNA strand shows a high affinity for Ag nanomaterials [42–44]. Subsequent work showed that all nucleotides could participate in the binding of metal NCs. Each nucleotide has at least one metal binding site. After further study, it was concluded that AgNC contains approximately the same number of neutral silver atoms and silver cations, and these Ag species are arranged in a rod-like structure rather than a plane or spherical structure. The existence of different emission colors is attributed to the change in rod length caused by a specific alkali/Ag interaction [41].

5.3.1.2 Carbon-Based Nanomaterials

The carbon nanostructures most commonly associated with functional nucleic acids are graphene oxide (GO), carbon nanotubes (CNTs), graphene, and nanodiamonds [45]. Due to their large surface area, charge characteristics, and excellent electron transfer ability, functional nucleic acids can be firmly immobilized on the surface of various carbon-based nanomaterials. Functional nucleic acid-carbon nanomaterials are ideal signal recognition elements for building fluorescent biosensors [46, 47]. Functional nucleic acids (FNA) can be firmly fixed on the surface of various carbon-based nanomaterials. The interaction between FNA and carbon surface may lead to significant changes in some biosensor signals. In addition, carbon-based nanomaterials have efficient fluorescence quenching ability. Therefore, FNA nanocarbon material is an ideal signal recognition element for constructing fluorescent biosensors [45]. Mesoporous silica nanoparticles (MSNs) have ideal surface chemical properties, which make them easy to modify with polycation. For example, MSNs modified with carbosilane dendrimers are used to deliver oligonucleotides [48, 49]. In addition, the abundant hydroxyl groups on the surface of MSN make it easy to functionalize into MSN-NH2, MSN-COOH, MSN-N3, and MSN-Cl. The introduction of these groups makes the conjugation of FNA possible. The excellent surface activity of MSNs is one of the primary qualities as nanomaterials. The new FNA-MSN nanomaterial is an effective drug delivery material. The targeting ability of FNA makes the probe or delivery system based on MSN have more accurate direction *in vitro* and *in vivo* [50, 51].

5.3.1.3 Bionanomaterials

Commonly used bionanomaterials are exosomes, which originate from the fusion of multivesicular bodies with the cytoplasmic membrane and are then released into the extracellular space. Exosomes play an important role in the intercellular transport of proteins and nucleic acids, which provides a theoretical basis for the combined application of exosomes and functional nucleic acid molecules [52]. Most exosomes can carry characteristic molecules that certain cells can recognize, facilitating the selective targeting and adsorption of recipient cells. The use of exosomes to deliver siRNA has emerged as a leading approach for therapeutic gene regulation with good biocompatibility, low side effects, low immunogenicity, and reduced nucleic acid degradation [53, 54]. Most exosomes can carry some characteristic molecules that can be recognized by cells, thus promoting the selective targeting and adsorption of receptor cells. Some exosomes also showed a strong ability to abscond and identify, as well as to prevent the clearance or degradation of immune cells. Early studies have shown that small exosomes allow them to penetrate into tissues, so exosomes are suitable as nanocarriers [54–57]. In addition, the exosomes showed slightly negative zeta potential in long circulation, had a deformable cytoskeleton, and their membrane properties were almost the same as those of the cells themselves. Using exosomes to deliver siRNA has become the main method for gene regulation. Because exosomes have good biocompatibility, no obvious side effects have been found in exosome-mediated gene delivery systems used *in vivo*. This method avoids nonspecific delivery, prevents immunogenicity of the delivery vector, and reduces nucleic acid degradation [58].

5.3.1.4 Quantum Dots

Quantum dots are semiconductor nanoparticles that are very small in size. So there are quantum constraints in all three-dimensional spaces. Quantum dots can be divided into several types, including metal-ion-doped inorganic semiconductor quantum dots, metal nanoparticles and NCs, silicon quantum dots, carbon quantum dots, and graphene quantum dots [59, 60]. Quantum dots have received extensive attention due to their unique, highly sensitive optical and electronic properties [60, 61]. Functional nucleic acids can be easily introduced into the surface of quantum dots based on their abundant surface amino and carboxyl groups, and fluorescence emission and quenching occur rapidly between quantum dots and fluorophores attached to functional nucleic acids. The above properties make fluorescent functional nucleic acid-quantum dot nanomaterials suitable for bioimaging and biosensing [62, 63].

5.3.1.5 Magnetic Nanomaterials

Magnetic nanomaterials can be divided into magnetic beads (MBs), magnetic nanoenzymes, magnetosomes, and other magnetic nanomaterials according to their different structural characteristics and applications. MB is the most widely used magnetic nanomaterial in biosensors and biomedicine. Polymer surface coatings play an important role in the application of MBs with FNA. Although MB nanoparticles have no magnetic attraction due to their superparamagnetic properties, they still have a significant aggregation trend due to their high surface energy. In addition, polymer coatings provide abundant binding sites for FNA. For example, streptavidin can attach to the surface of MBs, and biotin-modified FNA can connect to MBs through streptavidin–biotin interaction [64]. Bacterial magnetosomes are synthesized under strict genetic control, and bacterial magnetosomes have uniform shape, size, dispersion, and chemical composition [65]. They allow cells to align passively along magnetic field lines, a behavior known as magnetotaxis [5]. In 2015, Borg et al. prepared a ZnO/SiO_2-encapsulated magnetosome core–shell heterostructure [66]. Compared with the component materials, the new nanostructures have the following advantages: the nanometer size of the magnetic body and the modification ability of silicon dioxide. This makes it possible to produce FNA magnetosome composite nanomaterials by a new construction method.

5.3.1.6 Composite Nanomaterials

The combination of composite nanomaterials is rich, and the core principle is to play on the unique advantages of all nanomaterials [67]. For example, Jurado Sanchez and his colleagues created a Janus micromotor, a composite nanomaterial composed of the semiconducting polymer polycaprolactone, graphene quantum dots, and Pt/Fe_3O_4. This Janus micromotor can be driven only in the presence of H_2O_2 or magnetic force without adding other fuels. By coupling Janus micromotors with functional nucleic acids, a micromotor probe with autonomous motion and capture capability can be synthesized, which can be widely used in environmental and life sciences [68, 69].

5.3.2 Combination of Functional Nucleic Acids and Nanomaterials

Nucleic acid-based functional nanomaterials are composed of functional nucleic acids (DNA tile, DNA origami, three-dimensional DNA self-assembly, DNA nanorobots, and DNA microchips) and nanomaterials (metal-based nanomaterials, carbon-based nanomaterials, biological nanomaterials, quantum dots, and composite nanomaterials) formed by self-assembly (Figure 5.1). There are various ways of combining functional nucleic acid with

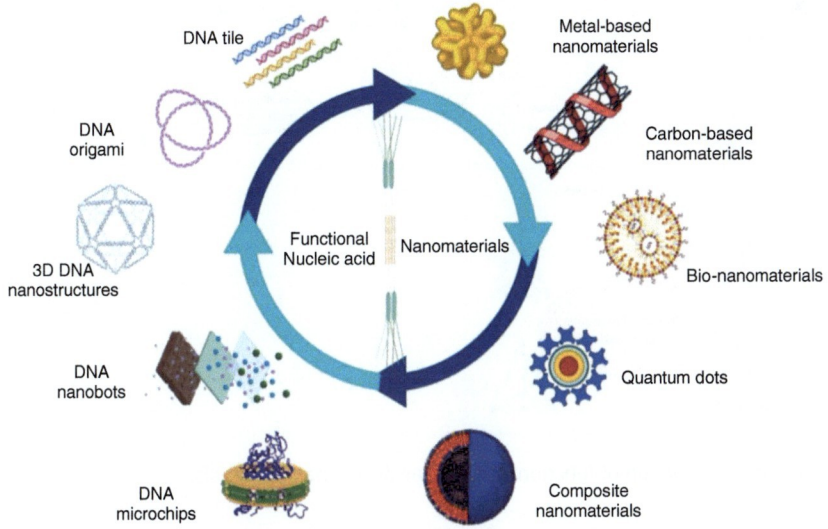

Figure 5.1 Schematic diagram of the composition of nucleic acid-based functional nanomaterials.

nanomaterials, including biological conjugation [70, 71], adsorption interaction [72, 73], functional nucleic acid-template assembly [74, 75], functional nucleic acid-nanomaterial heterojunction [76], and functional nucleic acid self-assembly [77–79], of which functional nucleic acid self-assembly is the most noteworthy. Self-assembled DNA nanotechnology constructs nucleic acid nanomaterials of different shapes and sizes, including tetrahedron, octahedron, icosahedron, and other more complex DNA origami structures, based on the "bottom-up" synthesis principle. Self-assembled nucleic acid nanostructures inherit many advantages of functional nucleic acids, including small-scale effects, high programmability, and excellent biocompatibility [80]. Self-assembled nucleic acid nanostructures take the inherent DNA of human body as the raw material. Self-assembled nucleic acid nanostructures have been widely used in molecular diagnosis, bioimaging, molecular delivery, and targeted drug delivery due to their good biocompatibility and degradation ability, nontoxic effect on organisms, rich functional modification sites, and many other advantages. Self-assembled framework nucleic acid nanostructures are modified in a variety of ways. For example, tetrahedral framework nucleic acid nanostructures (tFNAs) are modified in vertex type, mosaic type, capsule type, and suspension type. There are many kinds of groups and drugs that can be used to modify tetrahedral framework nucleic acid nanostructures, including chemical polymers, aptamers, micro-RNA, PNA, LNA, and traditional Chinese medicine monomers [81, 82]. These groups or drugs are connected with tetrahedral framework nucleic acid nanostructures in different ways to construct new nanomaterials with tFNAs as carriers, which can play a significant role in the repair of bone tissue defects [83] (Figure 5.2).

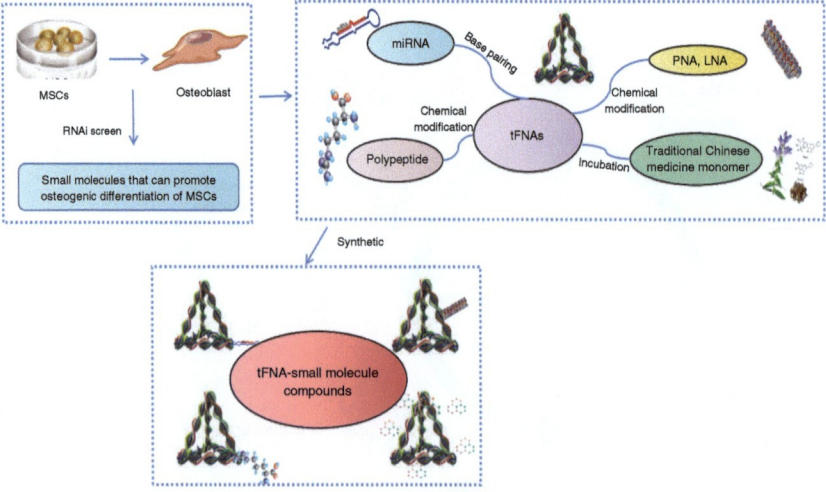

Figure 5.2 Different combination of functional nucleic acids and nanomaterials.

5.4 Multiple Roles of Nucleic Acid-Based Functional Nanomaterials in Bone Tissue Repair and Regeneration

In view of the good molecular recognition and drug delivery capabilities of functional nucleic acid nanomaterials, they have broad application prospects in the field of biomedicine, including the diagnosis and treatment of diseases. The molecular recognition and drug delivery capabilities of nucleic acid nanomaterials can be widely used in the field of biomedicine, mainly in the diagnosis and treatment of diseases. Nanomaterials, such as metallic nanomaterials and magnetic nanomaterials, have well-known advantages and optical properties for synthesis, bioconjugation, injury imaging, and photothermal therapy. Self-assembled 3D functional nucleic acid nanostructures, such as DNA origami and DNA hydrogels, have good biocompatibility, stability, flexibility, and precise programmability, and are easy to synthesize and modify. In view of the above advantages of nucleic acid nanomaterials, they have potential application prospects in the field of bone tissue disease repair and regeneration (Figure 5.3).

5.4.1 Sustained Release

The sustained-release effect of nucleic acid nanomaterials can avoid the sudden release of drugs, slow down the rate of drug release, and achieve long-term treatment for better efficacy. The nucleic acid nanomaterial sustained-release system has the following

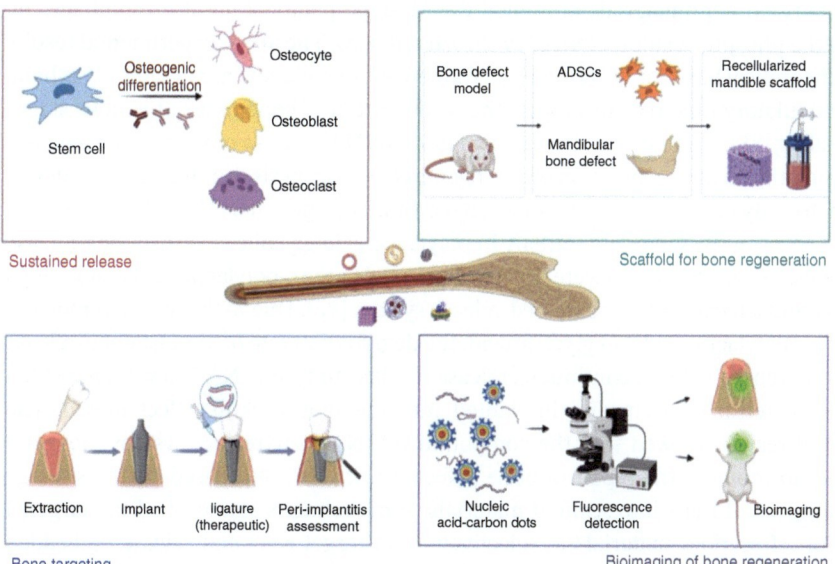

Figure 5.3 Schematic illustration of the role of nucleic acid-based functional nanomaterials in bone regeneration.

advantages: First, the use of the nucleic acid nanomaterial sustained-release system can reduce the frequency of administration and provide a more comfortable treatment process. Second, the sustained-release system can achieve continuous release of drugs in the body, thereby ensuring that the serum drug concentration has been maintained in a relatively stable state. Third, some nucleic acid nano sustained-release systems have localized effects, which can improve the effect of local treatment and reduce other systemic side effects. To sum up, the nucleic acid nano sustained-release system can greatly improve the therapeutic effect of drugs. Cui et al. applied lipid nanoparticles to carry TNF-α siRNA to treat rheumatoid arthritis in mice. The results of the study found that the nucleic acid nanosystem was an ideal sustained-release drug carrier because of its relatively high siRNA encapsulation efficiency (>90%) and low siRNA burst release rate (<5%). After intravenous administration, the inflammatory response in arthritic mice was significantly reduced, confirming that the sustained-release system has great application potential in the treatment of rheumatoid arthritis [84]. The results of Zhao et al. showed that the tetrahedral framework nucleic acid could enter mouse monocyte/macrophage cells without a carrier, reduce the M1-type marker inducible nitric oxide synthase after lipopolysaccharide combined with interferon-α treatment, and promote the induction of nitric oxide synthase. The production of inflammatory factors, nitric oxide, and reactive oxygen species increased, and at the same time, the expression of M2-type markers arginase 1 and interleukin-10, transforming growth factor-β increased, suggesting that tetrahedral framework nucleic acid nanomaterials promoted post-injury healing and bone tissue remodeling while organizing the inflammatory response process. Further studies showed that the tetrahedral framework nucleic acid nanomaterials promoted the M2-type polarization of macrophages by regulating the phosphorylation of STAT *in vitro* and *in vivo*. The above experimental results provide a theoretical basis for the tetrahedral framework nucleic acid nanomaterials as bone immunomodulatory materials to support the regeneration of bone tissue at various stages of bone healing [85]. Some studies have shown that miRNA-26a can promote bone regeneration through the positive regulation of angiogenesis–osteogenesis coupling. Li et al. developed a low-toxicity polymer (LP series) and a hyperbranched polymer (HP series) as miRNA transfection system. By wrapping miRNA in polylactic acid glycolic acid sustained-release microspheres, the microspheres were further connected to the L-polylactic acid nanoporous scaffold to build a local miRNA sustained-release system [86]. Due to the different molecular weights of polylactic acid and glycolic acid, the degradation rate of the sustained-release system is different to achieve continuous release. In this study, miRNA-26a was embedded into this slow-release system and implanted into the mouse skull defect model. The experimental results showed that the compound miRNA slow-release system showed a good ability to promote the repair of bone defects *in situ* [87, 88]. Based on the strategy of embedding small RNA into scaffold materials, some scholars also studied the targeted modification of scaffold materials to selectively bind target cells and release embedded miRNA/siRNA, which is of great significance for the treatment of bone-related diseases. It has been reported that polyurethane nanomicelles modified with osteoblast-targeting peptides can be used to encapsulate miRNA/siRNA and then used for osteoblast release [89]. The experimental results show that the SDSSD-PU delivery system not only selectively targets the bone formation surface but also selectively targets osteoblasts without significant toxicity and inducing an immune response. Anti-miR-214 was delivered to osteoblasts using

SDSSD-PU delivery system, and the results showed that bone formation, bone microstructure, and bone mass increased in ovariectomized osteoporosis mouse models [90].

5.4.2 Bone Targeting

Targeted drug delivery systems play an important role in the medical field. Targeted drug delivery systems refer to nucleic acid nanomaterials that deliver appropriate concentrations of drugs to target tissues, organs, and cells through local or systemic blood circulation. Traditional drug delivery systems easily affect the structure and function of normal tissues. In comparison, targeted drug delivery systems can increase the specificity and targeting of treatment and improve drug efficacy through the selection of targeted sites. In view of the above advantages of bone-targeted nucleic acid nano-drug delivery systems, they are of great significance in the treatment of diseases such as bone metastases, myeloma, and rheumatoid arthritis. The research results of Li et al. confirmed that the microRNA bioswitchable delivery system based on tetrahedral framework nucleic acid nanomaterials has significant advantages in biosafety compared with traditional carriers and confirmed that the delivery system greatly improves the stability of nucleic acid drugs. The miRNA-2861 carried by the tetrahedral framework nucleic acid was successfully unloaded and screened in bone marrow mesenchymal stem cells, and the ideal bone regeneration effect was observed *in vitro* and *in vivo*. The huge number of miRNAs, as a potential member of nucleic acid drugs, also enables this nucleic acid nanomaterial-based microRNA bioswitchable delivery strategy to play a more powerful role in biomedical fields such as tumor-targeted therapy and antibacterial therapy [91]. Xavier et al. fixed multiple oligonucleotides on gold nanoparticles through mercaptan bonds to form spherical nucleic acid (SNA). This structure can not only detect bone stem cells based on mRNA expression but also quickly isolate bone stem cells from human bone marrow without affecting the long-term survival of bone cells. This is the key to using bone stem cells to help bone regeneration in tissue engineering strategies. Although the number of bone stem cells in bone marrow is limited, this study shows that for every 200 bone stem cells in bone marrow, SNA can specifically detect and enrich one bone stem cell, which is equivalent to a 50~500-fold enrichment. Combined with the speed and simplicity of this strategy, this method has broad application potential [92]. After Shao et al. exposed adipose mesenchymal stem cells to tetrahedral framework nucleic acid nanomaterials, alkaline phosphatase and calcium deposition experiments showed that tetrahedral framework nucleic acid nanomaterials played an important role in inducing bone formation by promoting alkaline phosphatase activity and matrix mineralization. Tetrahedral framework nucleic acid nanomaterials induce osteoblastic differentiation of adipose-derived mesenchymal stem cells by activating classical Wnt/β-Catenin signaling pathway, which is closely related to bone formation and development. Tetrahedral framework nucleic acid nanomaterials can provide a unique cell environment, promote the osteogenic differentiation and proliferation of adipose-derived mesenchymal stem cells, and provide future opportunities for their potential applications in bone tissue regeneration therapy [93] (Figure 5.4). Xing et al. utilized siRNA-modified particles to design layered nanostructured coatings on clinically used titanium implants for synergistic regeneration of bone and vascular tissue. They combined siRNA, which can modulate cathepsin K expression, into nanoparticles and assembled the functionalized

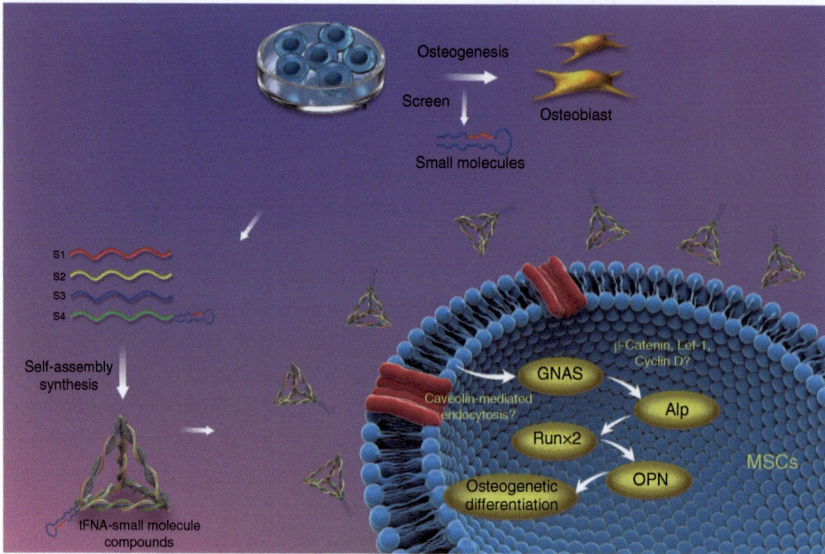

Figure 5.4 Schematic diagram of the role of tetrahedral frame nucleic acid nanomaterials in promoting osteogenic differentiation of MSCs.

nanoparticles onto bone implants to form a layered nanostructured coating. By regulating mRNA transcription, the coating significantly promoted cell viability and the release of growth factors associated with angiogenesis. Furthermore, microchip-based experiments showed that the nanostructured coating promotes macrophage-induced synergy to upregulate at least seven bone and vascular growth factors. Ovariectomized rat and comprehensive beagle dog models confirmed that this siRNA-integrated nanostructured coating significantly promotes bone regeneration in *in vivo* animal models and is expected to be an ideal nucleic acid nanomaterial for bone tissue regeneration [94].

5.4.3 Scaffold Material for Bone Regeneration

Trauma, tumor, congenital deformity, etc., can all cause defects in bone tissue. A bone defect refers to damage to the structural integrity of the bone. The repair and regeneration of bone tissue is a complex process in which many growth factors play an important role. Therefore, designing bioscaffolds with load-related growth factors or specific sequences that can regulate the expression of growth factors is an important research direction in the field of bone tissue regeneration. Nucleic acid nanomaterials used for bone tissue scaffolds need to have the following characteristics: First, nucleic acid nanoscaffold materials should have good biocompatibility and surface properties, avoid cytotoxicity and infection, and provide good microstructure for cell growth, migration, and differentiation. environment. Second, nucleic acid nanomaterials as scaffold materials should have good osteoconductivity and osteoinductivity, promote the differentiation of bone marrow mesenchymal stem cells into osteoblasts and osteocytes, and promote bone regeneration. Third, an ideal nucleic acid nanoscaffold material should have a rich porous structure that can provide three-dimensional space for cell survival, migration, and differentiation and facilitate the transport of nutrients and regulatory factors. Finally, an ideal nucleic acid nanoscaffold material should possess suitable mechanical strength and controllable degradability. To promote the regeneration of alveolar bone, Yang and his colleagues developed a chitosan-based injectable thermosensitive hydrogel scaffold structure. The scaffold structure combines chitosan nanoparticles with plasmid DNA loaded with BMP-2. In animal experiments, it has been found that it can significantly promote bone formation in rat calvaria defects and bone defect healing in beagle dogs [95]. Hydrogels combined with DNA can be loaded with bioactive substances that are difficult to deliver into mammalian cells or easily degraded in animals, thereby playing a corresponding role in bone tissue engineering to achieve bone regeneration and bone repair [96, 97]. In their study of osteoporosis, Ignatius and colleagues reported a biocompatible and biodegradable hybrid protein–DNA hydrogel carrying Rho-inhibiting C3 toxin. The hydrogel can target inhibit the formation and activity of osteoclasts. The protein is hybridized into DNA when the DNA hydrogel is mixed, without the use of other organic reagents or catalysts [98]. Nucleic acid nanomaterials as scaffolds for bone tissue regeneration can combine growth factors, functional proteins, nucleic acids, or nucleic acid analogs with bioscaffold structures to provide ideal repair materials for bone tissue regeneration.

5.4.4 Bioimaging of Bone Tissue Regeneration

With the development of materials science, fluorescent pigments, carbon dots (CDs), and some metal nanoparticles have been applied in bioimaging, and carbon quantum dots have become one of the most attractive materials in bioimaging. Significant advantages of these imageable bioimaging materials are their chemical inertness, water solubility, high fluorescence, biocompatibility, and nontoxicity. Nucleic acid nanomaterials can play a good role in bioimaging in the process of bone tissue regeneration. Bu et al. created ascorbic acid-PEI CDs capable of carrying miR-2861 by a microwave-assisted pyrolysis method. The results show that CD has excellent fluorescence stability and enables good fluorescence imaging *in vitro* and *in vivo*. CDs are efficiently internalized into bone marrow stromal cells (BMSCs) via a clathrin-mediated endocytic pathway. CD and miR-2861 can synergistically promote osteogenic differentiation *in vitro* and new bone regeneration *in vivo* [99]. Chen et al. found that citrate-based CDs could not only track but also promote osteogenic differentiation of mesenchymal stem cells. Cell uptake experiments showed that CDs formed obvious intracellular fluorescence after three hours of interaction with rat bone marrow mesenchymal stem cells (rBMSCs). In addition, co-incubation of CDs and rBMSCs can significantly promote osteogenic differentiation. The results of genetic testing showed that under the influence of CDs, ALP, runt-related transcription factor 2 (RUNX2), osteocalcin (cyanate), and bone sialoprotein (BSP) were significantly increased [100]. In addition, Leblanc et al. found that their CDs could serve as sensitive detectors of calcified zebrafish bones. There is no specific functional ligand on CD, but CD can target calcified zebrafish bone and serve as a highly specific biologic for bone imaging and diagnosis [101]. Therefore, nucleic acid nanomaterials can play a good role in bioimaging and drug delivery in the process of bone tissue regeneration.

5.5 Conclusion and Perspectives

To sum up, in view of the good physical and chemical properties of nucleic acid-based functional nanomaterials, they have a good application prospect in bone tissue regeneration. In terms of bone tissue regeneration, nucleic acid-based functional nanomaterials can be used as drug carriers, bone tissue scaffolds, and coating materials for implants. As a drug carrier, nucleic acid nanomaterials can reduce the burst release effect of drugs and increase the targeting and effectiveness of drug therapy. In terms of bone tissue engineering applications, nucleic acid nanomaterials have suitable mechanical strength and porous structure to provide support for cell adhesion, proliferation, migration, and differentiation. At the same time, nucleic acid nanomaterials can be linked to growth factors that promote the expression of bone regeneration or be linked to specific sequences that can regulate the expression of osteogenesis-related growth factors to promote bone tissue regeneration. In addition, nucleic acid-based functional nanomaterials can be used as coating materials for implants, coated on metal alloys and bioceramics, to promote the healing of bone defects after implantation. At the same time, benefiting from the continuous exploration of nanomaterials such as carbon quantum dots, nucleic acid-based functional nanomaterials can play a better role in bioimaging in the process of bone tissue regeneration. The above-mentioned role of nucleic acid-based functional nanomaterials in bone tissue regeneration expands the scope of its treatment of bone-related diseases, improves the efficiency of bone tissue regeneration, and reduces the side effects during bone tissue repair and regeneration. Nucleic acid-based functional nanomaterials have great potential for application in bone repair and regeneration.

Although nucleic acid-based functional nanomaterials have been reported to have great potential in the field of bone regeneration, the application of nucleic acid-based functional nanomaterials in bone tissue repair and regeneration remains challenging. Many studies are still in their infancy, and further exploration is urgently needed for practical application. In the future, it is necessary to further fine-tune the structure of nucleic acid-based functional nanomaterials. In terms of the composition and biological functions of nucleic acid nanomaterials, it is necessary to combine bionic properties to create a real microenvironment for bone regeneration and further endow nucleic acid-based functional nanomaterials with more efficient and effective bone regeneration capabilities.

References

1 Khosla, S. and Hofbauer, L.C. (2017). Osteoporosis treatment: recent developments and ongoing challenges. *Lancet Diabetes Endocrinol.* 5 (11): 898–907.
2 Berenbaum, F., Griffin, T.M., and Liu-Bryan, R. (2017). Review: metabolic regulation of inflammation in osteoarthritis. *Arthr. Rheumatol.* 69 (1): 9–21.
3 Jin, K., Li, T., van Dam, H. et al. (2017). Molecular insights into tumour metastasis: tracing the dominant events. *J. Pathol.* 241 (5): 567–577.
4 Nandi, S.K., Bandyopadhyay, S., Das, P. et al. (2016). Understanding osteomyelitis and its treatment through local drug delivery system. *Biotechnol. Adv.* 34 (8): 1305–1317.

5 Frankel, R.B. and Bazylinski, D.A. (2006). How magnetotactic bacteria make magnetosomes queue up. *Trends Microbiol.* 14 (8): 329–331.
 6 Tang, D., Tare, R.S., Yang, L.Y. et al. (2016). Biofabrication of bone tissue: approaches, challenges and translation for bone regeneration. *Biomaterials* 83: 363–382.
 7 Asafo-Adjei, T.A., Chen, A.J., Najarzadeh, A., and Puleo, D.A. (2016). Advances in controlled drug delivery for treatment of osteoporosis. *Curr. Osteoporos. Rep.* 14 (5): 226–238.
 8 Stevenson, S. (1998). Enhancement of fracture healing with autogenous and allogeneic bone grafts. *Clin. Orthop. Relat. Res.* 355 (Suppl): S239–S246.
 9 Murphy, P.S. and Evans, G.R. (2012). Advances in wound healing: a review of current wound healing products. *Plast. Surg. Int.* 2012: 190436.
 10 Bawa, R. (2011). Regulating nanomedicine – can the FDA handle it? *Curr. Drug Delivery* 8 (3): 227–234.
 11 Davis, M.E., Chen, Z.G., and Shin, D.M. (2008). Nanoparticle therapeutics: an emerging treatment modality for cancer. *Nat. Rev. Drug Discovery* 7 (9): 771–782.
 12 Han, D., Pal, S., Nangreave, J. et al. (2011). DNA origami with complex curvatures in three-dimensional space. *Science* 332 (6027): 342–346.
 13 Seeman, N.C. (2003). DNA in a material world. *Nature* 421 (6921): 427–431.
 14 Fu, Y., Zeng, D., Chao, J. et al. (2013). Single-step rapid assembly of DNA origami nanostructures for addressable nanoscale bioreactors. *JACS* 135 (2): 696–702.
 15 Li, J., Pei, H., Zhu, B. et al. (2011). Self-assembled multivalent DNA nanostructures for noninvasive intracellular delivery of immunostimulatory CpG oligonucleotides. *ACS Nano* 5 (11): 8783–8789.
 16 Benenson, Y., Gil, B., Ben-Dor, U. et al. (2004). An autonomous molecular computer for logical control of gene expression. *Nature* 429 (6990): 423–429.
 17 Wang, F., Willner, B., and Willner, I. (2014). DNA-based machines. *Top. Curr. Chem.* 354: 279–338.
 18 Heuer-Jungemann, A., Kirkwood, R., El-Sagheer, A.H. et al. (2013). Copper-free click chemistry as an emerging tool for the programmed ligation of DNA-functionalised gold nanoparticles. *Nanoscale* 5 (16): 7209–7212.
 19 Zhang, Y., Tu, J., Wang, D. et al. (2018). Programmable and multifunctional DNA-based materials for biomedical applications. *Adv. Mater.* 30 (24): e1703658.
 20 Zhang, M., Zhang, X., Tian, T. et al. (2022). Anti-inflammatory activity of curcumin-loaded tetrahedral framework nucleic acids on acute gouty arthritis. *Bioact. Mater.* 8: 368–380.
 21 Garzon, R., Marcucci, G., and Croce, C.M. (2010). Targeting microRNAs in cancer: rationale, strategies and challenges. *Nat. Rev. Drug Discovery* 9 (10): 775–789.
 22 Qin, X., Xiao, L., Li, N. et al. (2022). Tetrahedral framework nucleic acids-based delivery of microRNA-155 inhibits choroidal neovascularization by regulating the polarization of macrophages. *Bioact. Mater.* 14: 134–144.
 23 Zhang, Y., Ma, W., Zhu, Y. et al. (2018). Inhibiting methicillin-resistant *Staphylococcus aureus* by tetrahedral DNA nanostructure-enabled antisense peptide nucleic acid delivery. *Nano Lett.* 18 (9): 5652–5659.

24 Zhou, M., Gao, S., Zhang, X. et al. (2021). The protective effect of tetrahedral framework nucleic acids on periodontium under inflammatory conditions. *Bioact. Mater.* 6 (6): 1676–1688.
25 Seeman, N.C. (1982). Nucleic acid junctions and lattices. *J. Theor. Biol.* 99 (2): 237–247.
26 Yan, H., Park, S.H., Finkelstein, G. et al. (2003). DNA-templated self-assembly of protein arrays and highly conductive nanowires. *Science* 301 (5641): 1882–1884.
27 Winfree, E., Liu, F., Wenzler, L.A., and Seeman, N.C. (1998). Design and self-assembly of two-dimensional DNA crystals. *Nature* 394 (6693): 539–544.
28 Yao, G., Zhang, F., Wang, F. et al. (2020). Meta-DNA structures. *Nat. Chem.* 12 (11): 1067–1075.
29 Liu, X., Zhang, F., Jing, X. et al. (2018). Complex silica composite nanomaterials templated with DNA origami. *Nature* 559 (7715): 593–598.
30 Rothemund, P.W. (2006). Folding DNA to create nanoscale shapes and patterns. *Nature* 440 (7082): 297–302.
31 Qian, L., Zhao, J., Shi, Y. et al. (2007). Brain-derived neurotrophic factor and risk of schizophrenia: an association study and meta-analysis. *Biochem. Biophys. Res. Commun.* 353 (3): 738–743.
32 Goodman, R.P., Schaap, I.A., Tardin, C.F. et al. (2005). Rapid chiral assembly of rigid DNA building blocks for molecular nanofabrication. *Science* 310 (5754): 1661–1665.
33 Dunn, K.E., Dannenberg, F., Ouldridge, T.E. et al. (2015). Guiding the folding pathway of DNA origami. *Nature* 525 (7567): 82–86.
34 Haley, N.E.C., Ouldridge, T.E., Mullor Ruiz, I. et al. (2020). Design of hidden thermodynamic driving for non-equilibrium systems via mismatch elimination during DNA strand displacement. *Nat. Commun.* 11 (1): 2562.
35 Fu, J., Yang, Y.R., Dhakal, S. et al. (2016). Assembly of multienzyme complexes on DNA nanostructures. *Nat. Protoc.* 11 (11): 2243–2273.
36 Derr, N.D., Goodman, B.S., Jungmann, R. et al. (2012). Tug-of-war in motor protein ensembles revealed with a programmable DNA origami scaffold. *Science* 338 (6107): 662–665.
37 Douglas, S.M., Bachelet, I., and Church, G.M. (2012). A logic-gated nanorobot for targeted transport of molecular payloads. *Science* 335 (6070): 831–834.
38 Church, G. (2013). Improving genome understanding. *Nature* 502 (7470): 143.
39 Braun, E., Eichen, Y., Sivan, U., and Ben-Yoseph, G. (1998). DNA-templated assembly and electrode attachment of a conducting silver wire. *Nature* 391 (6669): 775–778.
40 Zhu, D., Song, P., Shen, J. et al. (2016). PolyA-mediated DNA assembly on gold nanoparticles for thermodynamically favorable and rapid hybridization analysis. *Anal. Chem.* 88 (9): 4949–4954.
41 Park, J.E., Jung, Y., Kim, M., and Nam, J.M. (2018). Quantitative nanoplasmonics. *ACS Cent. Sci.* 4 (10): 1303–1314.
42 Chen, Z., Liu, C., Cao, F. et al. (2018). DNA metallization: principles, methods, structures, and applications. *Chem. Soc. Rev.* 47 (11): 4017–4072.
43 Zheng, J. and Dickson, R.M. (2002). Individual water-soluble dendrimer-encapsulated silver nanodot fluorescence. *JACS* 124 (47): 13982–13983.
44 Petty, J.T., Zheng, J., Hud, N.V., and Dickson, R.M. (2004). DNA-templated Ag nanocluster formation. *JACS* 126 (16): 5207–5212.

45 Rasheed, P.A. and Sandhyarani, N. (2017). Carbon nanostructures as immobilization platform for DNA: a review on current progress in electrochemical DNA sensors. *Biosens. Bioelectron.* 97: 226–237.

46 Bosi, S., Da Ros, T., Spalluto, G., and Prato, M. (2003). Fullerene derivatives: an attractive tool for biological applications. *Eur. J. Med. Chem.* 38 (11-12): 913–923.

47 Balasubramanian, K. and Burghard, M. (2005). Chemically functionalized carbon nanotubes. *Small* 1 (2): 180–192.

48 Bhattarai, S.R., Muthuswamy, E., Wani, A. et al. (2010). Enhanced gene and siRNA delivery by polycation-modified mesoporous silica nanoparticles loaded with chloroquine. *Pharm. Res.* 27 (12): 2556–2568.

49 Martinez, A., Fuentes-Paniagua, E., Baeza, A. et al. (2015). Mesoporous silica nanoparticles decorated with carbosilane dendrons as new non-viral oligonucleotide delivery carriers. *Chemistry* 21 (44): 15651–15666.

50 Chen, C., Pu, F., Huang, Z. et al. (2011). Stimuli-responsive controlled-release system using quadruplex DNA-capped silica nanocontainers. *Nucleic Acids Res.* 39 (4): 1638–1644.

51 Chen, Z., Sun, M., Luo, F. et al. (2018). Stimulus-response click chemistry based aptamer-functionalized mesoporous silica nanoparticles for fluorescence detection of thrombin. *Talanta* 178: 563–568.

52 Kowal, J., Tkach, M., and Thery, C. (2014). Biogenesis and secretion of exosomes. *Curr. Opin. Cell Biol.* 29: 116–125.

53 Skotland, T., Sandvig, K., and Llorente, A. (2017). Lipids in exosomes: current knowledge and the way forward. *Prog. Lipid Res.* 66: 30–41.

54 Peng, Q. and Mu, H. (2016). The potential of protein-nanomaterial interaction for advanced drug delivery. *J. Controlled Release* 225: 121–132.

55 Hood, J.L. and Wickline, S.A. (2012). A systematic approach to exosome-based translational nanomedicine. *Wiley Interdiscip. Rev. Nanomed. Nanobiotechnol.* 4 (4): 458–467.

56 Malhotra, H., Sheokand, N., Kumar, S. et al. (2016). Exosomes: tunable nano vehicles for macromolecular delivery of transferrin and lactoferrin to specific intracellular compartment. *J. Biomed. Nanotechnol.* 12 (5): 1101–1114.

57 Vader, P., Mol, E.A., Pasterkamp, G., and Schiffelers, R.M. (2016). Extracellular vesicles for drug delivery. *Adv. Drug Delivery Rev.* 106 (Pt A): 148–156.

58 El Andaloussi, S., Lakhal, S., Mager, I., and Wood, M.J. (2013). Exosomes for targeted siRNA delivery across biological barriers. *Adv. Drug Delivery Rev.* 65 (3): 391–397.

59 Filali, S., Pirot, F., and Miossec, P. (2020). Biological applications and toxicity minimization of semiconductor quantum dots. *Trends Biotechnol.* 38 (2): 163–177.

60 Cai, X., Luo, Y., Zhang, W. et al. (2016). pH-Sensitive ZnO quantum dots-doxorubicin nanoparticles for lung cancer targeted drug delivery. *ACS Appl. Mater. Interfaces* 8 (34): 22442–22450.

61 Wang, L., Hui, J., Tang, J. et al. (2018). Precursor self-assembly identified as a general pathway for colloidal semiconductor magic-size clusters. *Adv. Sci.* 5 (12): 1800632.

62 Larson, D.R., Zipfel, W.R., Williams, R.M. et al. (2003). Water-soluble quantum dots for multiphoton fluorescence imaging in vivo. *Science* 300 (5624): 1434–1436.

63 Su, S., Fan, J., Xue, B. et al. (2014). DNA-conjugated quantum dot nanoprobe for high-sensitivity fluorescent detection of DNA and micro-RNA. *ACS Appl. Mater. Interfaces* 6 (2): 1152–1157.

64 Hess, K.L., Medintz, I.L., and Jewell, C.M. (2019). Designing inorganic nanomaterials for vaccines and immunotherapies. *Nano Today* 27: 73–98.

65 Jacob, J.J. and Suthindhiran, K. (2016). Magnetotactic bacteria and magnetosomes – scope and challenges. *Mater. Sci. Eng. C Mater. Biol. Appl.* 68: 919–928.

66 Borg, S., Rothenstein, D., Bill, J., and Schuler, D. (2015). Generation of multishell magnetic hybrid nanoparticles by encapsulation of genetically engineered and fluorescent bacterial magnetosomes with ZnO and SiO_2. *Small* 11 (33): 4209–4217.

67 Han, Q., Zhang, L., He, C. et al. (2012). Metal-organic frameworks with phosphotungstate incorporated for hydrolytic cleavage of a DNA-model phosphodiester. *Inorg. Chem.* 51 (9): 5118–5127.

68 Jurado-Sanchez, B., Pacheco, M., Rojo, J., and Escarpa, A. (2017). Magnetocatalytic graphene quantum dots janus micromotors for bacterial endotoxin detection. *Angew. Chem. Int. Ed. Engl.* 56 (24): 6957–6961.

69 Liu, Q., Wang, H., Shi, X. et al. (2017). Self-assembled DNA/peptide-based nanoparticle exhibiting synergistic enzymatic activity. *ACS Nano* 11 (7): 7251–7258.

70 Hill, H.D. and Mirkin, C.A. (2006). The bio-barcode assay for the detection of protein and nucleic acid targets using DTT-induced ligand exchange. *Nat. Protoc.* 1 (1): 324–336.

71 Chen, J., Zuehlke, A., Deng, B. et al. (2017). A target-triggered DNAzyme motor enabling homogeneous, amplified detection of proteins. *Anal. Chem.* 89 (23): 12888–12895.

72 Chen, L., Chao, J., Qu, X. et al. (2017). Probing cellular molecules with polyA-based engineered aptamer nanobeacon. *ACS Appl. Mater. Interfaces* 9 (9): 8014–8020.

73 Xu, Y., Wu, Q., Sun, Y. et al. (2010). Three-dimensional self-assembly of graphene oxide and DNA into multifunctional hydrogels. *ACS Nano* 4 (12): 7358–7362.

74 Chao, J., Wang, J., Wang, F. et al. (2019). Solving mazes with single-molecule DNA navigators. *Nat. Mater.* 18 (3): 273–279.

75 Zhang, S., Wang, K., Li, K.B. et al. (2017). A DNA-stabilized silver nanoclusters/graphene oxide-based platform for the sensitive detection of DNA through hybridization chain reaction. *Biosens. Bioelectron.* 91: 374–379.

76 Huang, F., Chen, M., Zhou, Z. et al. (2021). Spatiotemporal patterning of photoresponsive DNA-based hydrogels to tune local cell responses. *Nat. Commun.* 12 (1): 2364.

77 Ma, W., Yang, Y., Zhu, J. et al. (2022). Biomimetic nanoerythrosome-coated aptamer-DNA tetrahedron/maytansine conjugates: pH-responsive and targeted cytotoxicity for HER2-positive breast cancer. *Adv. Mater.* e2109609.

78 Zhang, T., Tian, T., and Lin, Y. (2021). Functionalizing framework nucleic-acid-based nanostructures for biomedical application. *Adv. Mater.* e2107820.

79 Zhou, M., Zhang, T., Zhang, B. et al. (2021). A DNA nanostructure-based neuroprotectant against neuronal apoptosis via inhibiting toll-like receptor 2 signaling pathway in acute ischemic stroke. *ACS Nano* .

80 Wang, Y., Li, Y., Gao, S. et al. (2022). Tetrahedral framework nucleic acids can alleviate taurocholate-induced severe acute pancreatitis and its subsequent multiorgan injury in mice. *Nano Lett.* 22 (4): 1759–1768.

81 Haque Bhuyan, M.Z., Tamura, Y., Sone, E. et al. (2017). The intra-articular injection of RANKL-binding peptides inhibits cartilage degeneration in a murine model of osteoarthritis. *J. Pharmacol. Sci.* 134 (2): 124–130.

82 Ye, Y., Jing, X., Li, N. et al. (2017). Icariin promotes proliferation and osteogenic differentiation of rat adipose-derived stem cells by activating the RhoA-TAZ signaling pathway. *Biomed. Pharmacother.* 88: 384–394.

83 Abuna, R.P.F., Oliveira, F.S., Adolpho, L.F. et al. (2020). Frizzled 6 disruption suppresses osteoblast differentiation induced by nanotopography through the canonical Wnt signaling pathway. *J. Cell. Physiol.* 235 (11): 8293–8303.

84 Aldayel, A.M., O'Mary, H.L., Valdes, S.A. et al. (2018). Lipid nanoparticles with minimum burst release of TNF-alpha siRNA show strong activity against rheumatoid arthritis unresponsive to methotrexate. *J. Controlled Release* 283: 280–289.

85 Zhao, D., Cui, W., Liu, M. et al. (2020). Tetrahedral framework nucleic acid promotes the treatment of bisphosphonate-related osteonecrosis of the jaws by promoting angiogenesis and M2 polarization. *ACS Appl. Mater. Interfaces* 12 (40): 44508–44522.

86 Li, Y., Fan, L., Liu, S. et al. (2013). The promotion of bone regeneration through positive regulation of angiogenic-osteogenic coupling using microRNA-26a. *Biomaterials* 34 (21): 5048–5058.

87 Li, Y., Fan, L., Hu, J. et al. (2015). MiR-26a rescues bone regeneration deficiency of mesenchymal stem cells derived from osteoporotic mice. *Mol. Ther.* 23 (8): 1349–1357.

88 Li, Y., Chen, J., Liu, J. et al. (2013). Estimation of the reliability of all-ceramic crowns using finite element models and the stress-strength interference theory. *Comput. Biol. Med.* 43 (9): 1214–1220.

89 Nguyen, M.K., Jeon, O., Dang, P.N. et al. (2018). RNA interfering molecule delivery from in situ forming biodegradable hydrogels for enhancement of bone formation in rat calvarial bone defects. *Acta Biomater.* 75: 105–114.

90 Yamada, Y., Wakao, S., Kushida, Y. et al. (2018). S1P-S1PR2 axis mediates homing of muse cells into damaged heart for long-lasting tissue repair and functional recovery after acute myocardial infarction. *Circ. Res.* 122 (8): 1069–1083.

91 Li, S., Liu, Y., Tian, T. et al. (2021). Bioswitchable delivery of microRNA by framework nucleic acids: application to bone regeneration. *Small* 17 (47): e2104359.

92 Xavier, M., Kyriazi, M.E., Lanham, S. et al. (2021). Enrichment of skeletal stem cells from human bone marrow using spherical nucleic acids. *ACS Nano* 15 (4): 6909–6916.

93 Shao, X.R., Lin, S.Y., Peng, Q. et al. (2017). Effect of tetrahedral DNA nanostructures on osteogenic differentiation of mesenchymal stem cells via activation of the Wnt/beta-catenin signaling pathway. *Nanomedicine* 13 (5): 1809–1819.

94 Xing, H., Wang, X., Xiao, G. et al. (2020). Hierarchical assembly of nanostructured coating for siRNA-based dual therapy of bone regeneration and revascularization. *Biomaterials* 235: 119784.

95 Li, H., Ji, Q., Chen, X. et al. (2017). Accelerated bony defect healing based on chitosan thermosensitive hydrogel scaffolds embedded with chitosan nanoparticles for the delivery of BMP2 plasmid DNA. *J. Biomed. Mater. Res. A* 105 (1): 265–273.

96 Zinchenko, A., Miwa, Y., Lopatina, L.I. et al. (2014). DNA hydrogel as a template for synthesis of ultrasmall gold nanoparticles for catalytic applications. *ACS Appl. Mater. Interfaces* 6 (5): 3226–3232.

97 Li, J., Wang, Y., Zhou, T. et al. (2015). Nanoparticle superlattices as efficient bifunctional electrocatalysts for water splitting. *JACS* 137 (45): 14305–14312.

98 Gacanin, J., Kovtun, A., Fischer, S. et al. (2017). Spatiotemporally controlled release of rho-inhibiting C3 toxin from a protein-DNA hybrid hydrogel for targeted inhibition of osteoclast formation and activity. *Adv. Healthcare Mater.* 6 (21).

99 Bu, W., Xu, X., Wang, Z. et al. (2020). Ascorbic acid-PEI carbon dots with osteogenic effects as miR-2861 carriers to effectively enhance bone regeneration. *ACS Appl. Mater. Interfaces* 12 (45): 50287–50302.

100 Shao, D., Lu, M., Xu, D. et al. (2017). Carbon dots for tracking and promoting the osteogenic differentiation of mesenchymal stem cells. *Biomater. Sci.* 5 (9): 1820–1827.

101 Li, S., Skromne, I., Peng, Z. et al. (2016). "Dark" carbon dots specifically "light-up" calcified zebrafish bones. *J. Mater. Chem. B* 4 (46): 7398–7405.

6

In Situ Fluorescence Imaging and Biotherapy of Tumor Based on Hybridization Chain Reaction

Ye Chen, Songhang Li, and Taoran Tian

West China Hospital of Stomatology, Sichuan University, State Key Laboratory of Oral Diseases & National Clinical Research Center for Oral Diseases & Department of Oral and Maxillofacial Surgery, 14#, 3rd, Section of Renmin South Road, Chengdu, Sichuan 610041, PR China

With fluorescence probe development, the detection and imaging of specific nuclear acids, proteins, molecules in tumor cells, tissue slices, and intact organic matter can be achieved, which provides detailed information on tumor tissue for diagnosis and prognosis. Fluorescence-based detection facilitates multiple target identifications and real-time molecular monitoring [1]. However, the fluorescence of the traditional molecular beacons is rather faint, limiting the detection of slight targets. Besides, detecting a mutated gene or similar targets by molecular beacon can often cause false-positive results due to off-target binding [2]. Therefore, it is necessary to develop a fluorescence magnificent strategy to strengthen detection sensitivity and specificity.

Due to the self-assembly and disassembly properties, nucleic acid-based amplification reaction becomes an indispensable method for molecular detection and imaging, which can amplify the fluorescence signaling of the target rather than that of the scrambled substance [3]. The nucleic acid-based amplification reaction can be summarized into two categories, including enzyme-based and nonenzymic methods. The former includes helicase-dependent amplification [4] and rolling circle amplification [5], while the latter includes strand displacement amplification and entropy-driven DNA machine [6]. Therefore, *in situ* tumor fluorescence imaging based on nucleic acid amplification reaction can visualize tumor cellular genome and protein structure and function.

Besides, nucleic acid-based antitumor treatment has aroused wide attention due to its low cost, biocompatibility, and programmability. The nucleic acid can become a nanocarrier for antitumor drugs, photosensitizers, and small noncoding RNA. Moreover, nucleic acid can also recognize tumor microenvironments and cells with high specificity and sensitivity. However, single DNA strands have limited loading capacity, whereas nucleic acid amplification can extend nucleic acid length and strengthen loading capacity. Therefore, biotherapy based on nucleic acid amplification can become a kind of novel therapy method for antitumor treatment.

Hybridization chain reaction (HCR) is a new nonenzymic DNA amplification signal amplification technology proposed in 2004 by Dirks and Pierce [7]. As an efficient amplification platform, HCR has been used to sensitively detect a variety of analytes, including nucleic acids [8], proteins [9], and small molecules [10]. Therefore, HCR can be applied to diagnose cancer genome [11] and antigen [12], while monitoring various enzymes such as DNA methyltransferase [13] and telomerase level [14]. In addition, HCR has also attracted significant attention in biological therapy, including targeted drug delivery [15] and photodynamic therapy (PTD) [16]. Based on these merits, we summarized the application of HCR in *in situ* fluorescence imaging and biotherapy of tumors.

6.1 Hybridization Chain Reaction

The essential components of HCR include an initiator and a pair of hairpins [17]. The hairpin consists of a toehold, a complementary stem, and loop. The stem region b in hairpin1 is complementary to b* region in hairpin2. The toehold of one hairpin is complementary to the loop sequence of another hairpin. The c and b regions are the complementary sequence of target nucleic acid or the aptamer of target proteins [7]. When there is no initiator, hairpin probes remain metastable in the reaction solution [18]. When there exist initiators, the pair of hairpins are opened in turns for hybridization, eventually forming a DNA-long and nicked double-stranded structure [19]. Some scholars propose four hairpin design principles for robust HCR. (i) The toehold length keeps within 12 nt; (ii) The GC content of the toehold maintains less than 40%; (iii) The stem length is longer than the toehold length; (iv) The stem GC content keeps more than 60% [20]. The hairpin probe can be divided into always-on and switch-on/off types [21]. The former always releases fluorescence and can gather around target positions, resulting in an increased signal-to-background ratio [22]. The switch-on/off hairpin probes can release fluorescence only when encountering targets due to diminishing quenching effect [23] or fluorescence resonance energy transfer (FRET) [24]. The trace initiator can cause multiple fluorophore amplifications, which thus improves the detection sensitivity significantly [25].

Compared with polymerase chain reaction (PCR), PCR depends on enzyme assistance, which can be affected by temperature, ionic strength, pH, etc. [16]. As an enzyme-free reaction, HCR can proceed in a simple environment, even at room temperature [26]. HCR has comparable sensitivity and specificity in nucleic acid detection to PCR [27]. Furthermore, HCR saves a considerable reaction time and displays a significant difference within 30 minutes [28]. Especially for RNA detection, RNA can initiate HCR via complementary base pairing with RNA, thus avoiding reverse transcription [29]. Furthermore, PCR probes cannot discriminate DNA and mRNA due to DNA denaturation, whereas HCR can distinguish them [30]. Still, HCR depends on probe amplification rather than target amplification, which avoids nonspecific amplification results. HCR can produce result signals in varying forms, such as electrochemical signaling, fluorescence signaling, and photoacoustic signaling [31]. Therefore, due to portability, HCR can act as a promising method of point-of-care (POC) detection.

6.2 Nucleic Acid Detection

HCR has been widely applied in tumor-related nucleic acid detection involving tumor-derived miRNA [32], tumor-derived circRNA [29], tumor-derived DNA [33], and single-nucleotide variants (SNV) [34]. They play a vital role in cancer diagnosis and prognosis, thus becoming the valued cancer-related biomarkers. Many detection methods have been investigated, such as next-generation sequencing [35], deep sequencing [36], and PCR [37]. Nevertheless, current technology is limited by time-consuming, high-costing, and laborious procedures. Therefore, HCR has become a promising analytical technology due to its sensitivity and agility.

6.2.1 miRNA Detection

MicroRNAs (miRNAs), a short noncoding RNA, regulate cancerous messenger RNA (mRNA) translation, which plays a vital role in tumor diagnosis and treatment [38]. However, due to extremely low expression in cells, the detection and *in situ* imaging of miRNA are quite challenging [39]. Recently, a series of fluorescence *in situ* hybridization (FISH) methods were developed to detect miRNA, but they are confined due to undesired intracellular interference and low sensitivity [40]. HCR, a nonenzymic method that can achieve signaling amplification, and precise detection, is thus proposed for a new FISH strategy [41]. Many scholars have reported HCR-based nanotechnology to detect intracellular miRNAs, such as miR-21 [42] and let-7a [43]. Zhang et al. [28] integrated the complementary sequence of miRNA into the toehold and stem region of hairpin1. miR-21 could trigger hairpin structure to be unlocked and expose sticking ends to hybridize with hairpin2. The Cy5 fluorescence could recover due to the far distance from quencher BHQ-2. Due to weaker affinity between RNA and DNA, Pu et al. [43] designed two-step miRNA detection to achieve higher sensitivity, including toehold-mediated strand displacement (TMSD) and HCR. Let-7a could produce initiators, which could subsequently induce HCR. The two-step miRNA detection could lower detection limits down to 25pM. Besides, some methods, such as autocatalytic HCR biocircuit and nonlinear HCR, have been studied to increase the sensitivity of miRNA detection based on HCR. The description and comparison of these various HCR detection systems are summarized below.

6.2.1.1 Autocatalytic HCR Biocircuit

The detection of trace account of miRNA is often limited by faint fluorescence by HCR. Therefore, some self-sustained HCR biocircuits were designed [44]. Metal-ion-dependent DNAzymes can selectively cleave the RNA position of the binding substrate with the aid of metal ions such as Mn^{2+} and Pb^{2+} [45]. Single DNAzyme-based detection was limited by its low catalytic efficiency and the interference from the intracellular environment [46]. Therefore, some scholars developed an autocatalytic, self-sufficient amplifier detection system via the integrity of HCR and DNAzyme. As Figure 6.1 indicated, miR-21 triggered the successive cross-opening of hairpin1 and hairpin2 base complementary pairing regions to form a long and nicked DNA duplex. The hybridization pulled the separated DNAzyme subunit closer and formed many DNAzyme catalytic units. The DNAzyme could competitively bind with substrate strands, which were locked in double strands. The cleavage of the substrate strand could produce multiple amplicons, which could initiate HCR continuously. Meanwhile, the amplified DNAzyme units could cleave quencher/fluorophore-labeled substrate strand, resulting in fluorescence recovery [47]. Compared with HCR and DNAzyme alone, the autocatalytic HCR biocircuit could significantly improve the fluorescence intensity of

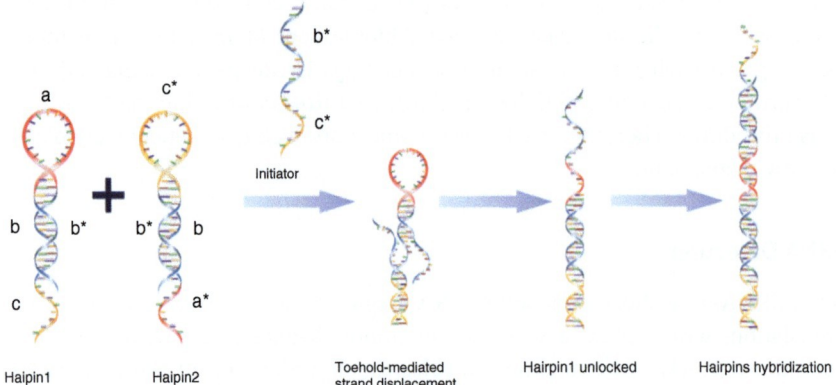

Figure 6.1 The principles of hybridization chain reaction.

the target and lower the limit of detection. The sensitive detection system also performed good specificity in recognizing mutated miRNA and cancer cells [48]. However, the activation of the autocatalytic system depends on the metal ion supplementary, because intracellular Mg^{2+} is not sufficient to drive DNAzyme catalyzing reaction [49]. Wei et al. [50] incorporated autocatalytic HCR biocircuit into GSH-responsive MnO_2 nanosponge, which could selectively deliver the HCR system into a GSH-rich tumor microenvironment.

6.2.1.2 Nonlinear HCR System

Usually, the initial strand triggers linear HCR, forming a long and nicked DNA duplex. A linear relationship can be observed between fluorescence intensity and target concentration [51]. Recently, nonlinear HCR systems have been developed, which can form dendritic or hyperbranched DNA products. Moreover, the fluorescence intensity of the nonlinear HCR system can achieve exponential amplification, thus increasing detection sensitivity [52]. Some scholars developed a dendritic HCR system, including two block strands and two assistant strands. Hairpin recognized and hybridized with target miRNA, exposing the complementary base regions of block-A. Block-B is bound separately to two repetitive sequence regions of block-A. The two assistant strands induced TMSD and produced by-products. Therefore, purification steps to remove by-products were needed to maintain reaction cycles. The hairpins were assembled on the surface of magnetic nanoparticles to facilitate the purification step. And zinc(II)-protoporphyrin IX (Zn_{II}PPIX) could be selectively intercalated into G-quadruplex structures in dendritic DNA structures to produce significant fluorescence. The exponential amplification reaction achieved the femtomolar level detection limit and demonstrated excellent selectivity toward the target [53] (Figure 6.2). Besides, Wu et al. [47] developed a two-layer cascaded HCR. The initiator of the second layer was separated and attached to 5′ and 3′ terminal ends of hairpins. The pair of hairpins hydride with target miRNA and form multiple intact and nicked trigger sequences, which induce down-streaming HCR. The two-layer cascade HCR could also produce dendritic DNA structures. The hyperbranched HCR product also caused DNAzyme amplification, which cleaved the substrate and recovered the fluorescence. The nonlinear HCR system also performed excellent specificity in detecting cancerous EV-derived miR-21

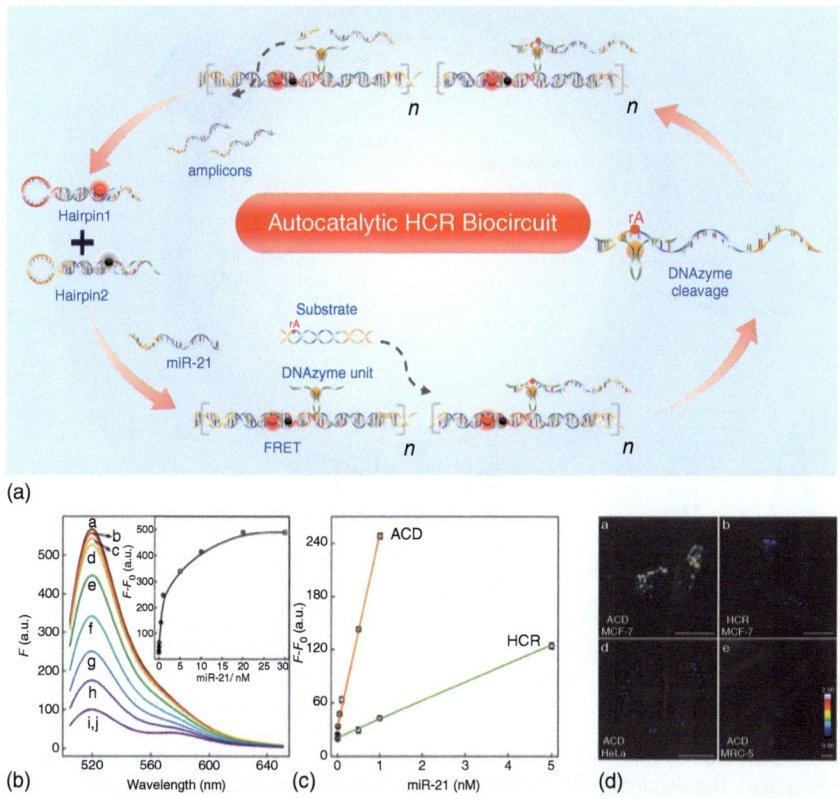

Figure 6.2 Autocatalytic HCR Biocircuit and its application in detecting miR-21. (a) miR-21 in tumor cells triggers hybridization between hairpin1 and hairpin2. Separated DNAzyme units are conjugated to 5′ and 3′ terminal ends of hairpin2. When HCR occurred, two DNAzyme units will be pulled closer and form enzyme units with the aid of metal ions. They could cleave RNA position in DNA substrates. The cleaved single strand could trigger HCR subsequently. The autocatalytic HCR biocircuit could be formed. (b) The fluorescence spectrum of autocatalytic HCR biocircuit in analyzing miR-21 with different concentrations. Source: Adapted with permission. Copyright 2020, Wiley-VCH. (c) The autocatalytic HCR biocircuit could strengthen detection signaling compared to HCR. Source: Adapted with permission. Copyright 2020, Wiley-VCH. (d) The confocal laser microscope results of different cells with autocatalytic HCR biocircuit. Source: Adapted with permission. Copyright 2020, Wiley-VCH.

in real samples. Furthermore, Lv et al. [54] designed a three-layer cascaded hyperbranched HCR to achieve ultrasensitive miRNA detection as low as 1 nM. And due to cascade amplification, the three-layer HCR performed better than two-layer HCR and one-layer HCR (Figure 6.3).

6.2.2 Single-Nucleotide Variants Detection

SNV refer to single-nucleotide somatic mutations in a specific position of genome sequence located at exons, usually associated with cancerous occurrence [55]. Therefore, SNV can become a potential biomarker for cancer diagnosis [56]. Compared with cancerous

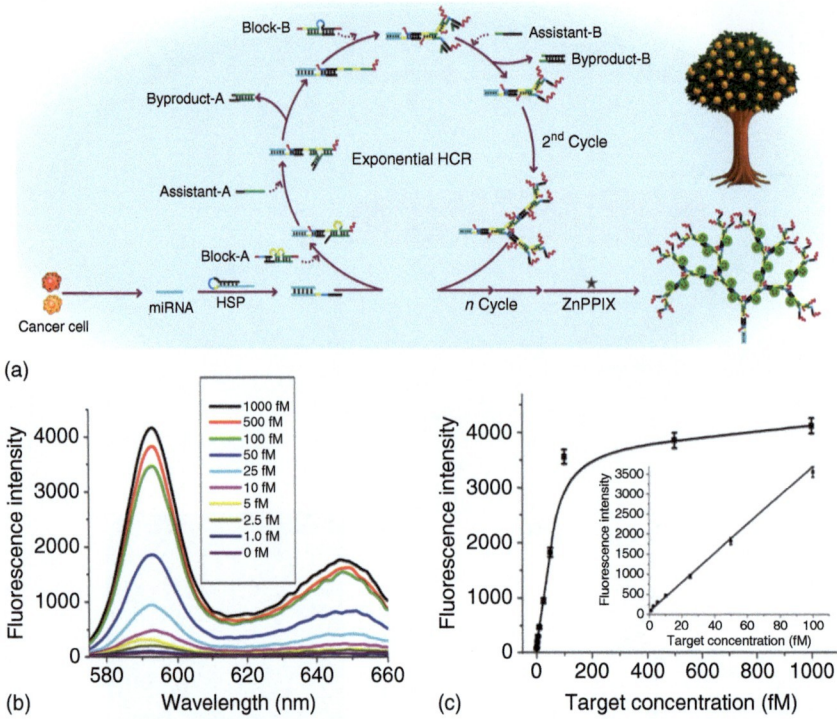

Figure 6.3 Dendritic HCR system and its application in detecting miR-21. (a) The description of dendritic HCR system. The dendritic HCR product could form multiple G-quadruplex, where Zn^{II}PPIX could be intercalated into, and release fluorescence. Source: Reproduced with permission. Copyright 2018, American Chemical Society. (b) The fluorescence spectrum of dendritic HCR system in analyzing miR-21 with different concentrations. Source: Reproduced with permission. Copyright 2018, American Chemical Society. (c) The fluorescence curve of dendritic HCR system in analyzing miR-21 with different concentrations. Source: Reproduced with permission. Copyright 2018, American Chemical Society.

miRNA and circulating DNA, SNV challenges HCR-based detection due to abundant inferences from normal nucleotide mutations and nucleic acid steric structures. The molecular beacon can cause more intense undesirable background signals due to off-target recognition, thus making it challenging to discriminate between a mutation-type and a wild-type gene.

Some scholars have utilized recombinase polymerase amplification (RPA) to acquire short-length sequences. The RPA could serve as a pre-amplification step and produce short strands. The RPA products could be bound with capture probe on the chip. The HCR could be activated by the linker, which is hybridized with RPA products, amplifying the detection signaling and improving sensitivity. HCR-based detection strategy reduced genomic mutant abundance to 0.2%, fivefold lower than the PCR-based method [57]. Compared to DNA mutation detection, the weaker affinity between RNA target and DNA probe constrains SNV detection sensitivity. Jiang et al. designed two kinds of probes to bind with cDNA from wild or mutated types. With the help of DNA ligase, complementary probes could be ligated with an anchor probe. The tagged initiator on wild or mutated-type probe could induce branched HCR and amplify fluorescence. The highly sensitive method could achieve single-mRNA mutation imaging at single-cell level [58].

Nevertheless, this allele mutation detection method depends on enzyme assistance during RPA or DNA ligation progress. Compared to two-step detection, Marras et al. [59] designed a pair of arm-donating hairpins and arm-acceptor hairpins. The former bound to target mRNA with loop structure, and the donating hairpin arm was disassociated to hybridize with acceptor hairpin arm via TMSD. The dissociated acceptor's initial sequence could trigger HCR of two fluorophore-labeled hairpins. After removing extra acceptor hairpins, the various mutation positions could be identified by designing different donating hairpins with the mutated sequence in the loop [59]. Therefore, the HCR-based SNV detection can identify different SNV genes and specific positions by designing different donating hairpins. Due to its portability, sensitivity, and selectivity, the HCR-based SNV method can be applied in POC detection, which could help to identify mutated transcripts.

6.3 Protein Detection

During cancer progression, some changes can occur in protein function and expression [60]. These proteins are derived from the genes associated with cellular cycle, proliferation, and adhesion [61]. Mutated genes can lead to abnormal expression and exert a harmful effect on the physiological state. Besides genome mutation, epigenetic changes such as phosphorylation and methylation can also alter protein function [62]. Recently, the widely-used tumor-related antigens include human epidermal growth factor receptor 2 (HER2), epithelial cell adhesion molecule, carcinoembryonic antigen, DNA methyltransferase, and so on [63]. Nevertheless, due to the low specificity of these antigens, the protein can only act as the indicator of cancerous risk, severity, and prognosis instead of direct diagnosis. HCR can also play a vital role in monitoring the expression and function of these tumor antigens.

6.3.1 Antibody-Based HCR System

Recently, tumor cell membrane antigen detection has been based on the fluorophore-labeled secondary antibody, which increases handling procedures and acquires weak signaling. Choi et al. [64] conjugated initial sequences of HCR to capture antibodies to recognize target proteins. After removing the extra antibody, the initial oligonucleotide attached to the antibody could spur HCR involving a pair of complementary hairpins. Because of signaling amplification, an antibody-based HCR detection strategy can be utilized to identify cancer cell membrane antigens. Rafiee et al. [65] designed a HER2 antibody conjugated with an HCR initiator to detect breast cancer cells within 30 minutes. And the method showed high specificity and could distinguish HER2$^+$ cancer cells in peripheral blood [65]. Besides, antibody-based HCR may become a novel strategy to amplify ELISA detection signaling. However, the detection strategy was limited by background noises caused by the remaining antibodies in the reaction buffer. And the direct conjugation between antibody and oligonucleotide is rather difficult (Figure 6.4).

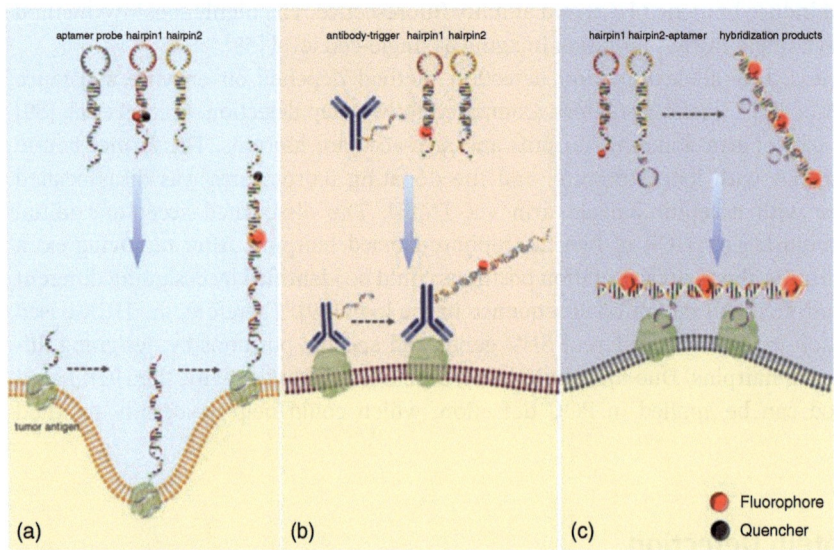

Figure 6.4 Tumor antigen detection based on HCR. (a) The aptamer probe contains a locked aptamer sequence. When encountering tumor antigen, the aptamer probe will be unlocked and expose HCR trigger sequence, which could initiate HCR subsequently. The fluorescence could recover due to the far distance from quenchers in long-nicked DNA duplexes. (b) HCR triggers are conjugated to antibodies. The trigger could initiate hybridization between hairpin1 and hairpin2 after removing extra antibodies. (c) Aptamer sequences are bound with hairpin2. After HCR occurred, the HCR products carried multiple aptamers, which could bind with tumor antigens.

6.3.2 Aptamer-Based HCR System

Compared with traditional antibody detection, in addition to high affinity and specificity, aptamers also have the advantages of low cost, stable chemical properties, and low immunogenicity. Some scholars appended initiator of HCR to aptamer sequence. The initiator could be anchored in the surface of the cell membrane when aptamer binds to the target membrane protein. The fluorophore-labeled complementary hairpins hybridized and formed an extended DNA duplex. However, the isolation of unreacted hairpins could cause high background noise. To lower background noise, Dong et al. [66] loaded hairpins onto Fe_3O_4@DOP nanoparticles, and the nanoparticles could quench the fluorescence of hairpins. When HCR occurs, hairpins will be detached from nanoparticles and release fluorescence. And the unreacted hairpins could be cleared by magnetic separation. Besides, because the fluorophore and quencher were attached to the identical hairpins, the fluorescence would be quenched due to FRET, which also lowers background noises [66]. To strengthen tumor cell capture efficiency, Li et al. [67] attached an aptamer to the hairpin's sticking end. The hybridization between two hairpins could form long-nicked HCR products labeled with multiple aptamers and quantum dots (QDs). QDs own excellent optical properties compared with common fluorophores. The HCR product could bind with protein tyrosine kinase (PTK) 7 on the surface of cancer cells via multivalent binding by the aptamers. The cancer cells could be discriminated via surface QD expression (Figure 6.4).

However, the above two strategies both depend on removing unbinding aptamer, which increases detection instability. Wang et al. [66] locked aptamers and initiators into a metastable hairpin structure. The aptamer hairpin and two complementary hairpins remained stable until the target membrane protein occurred. The aptamer-hairpin could free up the blocked HCR initiator when aptamer binds with PTK7, and the initiator triggered hairpin hybridization. Due to the quenching effect of BHQ3, the fluorescence could recover after forming HCR products due to the separation between fluorophore and quencher. Due to PTK7 being associated with cell carcinogenesis, thus HCR-based system could serve as one of the real-time cancer detection methods. Moreover, due to signaling amplification by HCR, a minimum of 16 PTK7-expressing cells could be detected by flow cytometry (Figure 6.4). Moreover, Li et al. [68] also utilize a similar blocked-aptamer hairpin design, and the exposed initial sequences were amplified by cascade rolling circle amplification. The former representative initiator could activate hyperbranched HCR, and the adjacent hairpins could constitute multiple G-quadruplex structures, enhancing N-methy mesoporphyrin IX fluorescence signal release. The hyperbranched HCR could improve detection sensitivity compared to linear HCR, with detection limits down to 10 cells.

To solve repetitive wash and unspecific signaling, besides locking aptamer into the hairpin structure, Li et al. [69] split aptamer into two parts, which were conjugated with split initiators, respectively. When encountering the target protein, the split aptamer could bind with the target protein and form an intact and nicked initiator, which triggered HCR subsequently. The split aptamer-induced HCR detection achieved highly sensitive cancer cell counting with a detection limit of 18 cells in a 180 µl buffer. Besides, the split aptamer-based detection system could be applied to complex microenvironments such as serum and mixed cell samples due to its specificity. Chen et al. [70] utilized similar methods to identify tumor-derived exosomes. The sensitive and specific detection system could achieve trace exosome detection down to 2.08×10^5 particles per ml and discriminate exosomes from different donor sources.

6.4 Multiple Target Detection

6.4.1 Combined HCR-Based Probe

The multiplexed and simultaneous target detection, including nucleic acid and protein, facilitates accurate cancer occurrence and type diagnosis. Zada et al. [71] fabricated three pairs of HCR-based probes with different fluorophores to achieve three kinds of miRNAs imaging in living cells. The multiple miRNA detection systems could facilitate the sensitive detection of three miRNAs with low detection limits down to the femtomolar level. The cellular miRNA was produced from precursor miRNAs by RNase Dicer shearing, but molecular beacon cannot discriminate miRNA from precursor miRNAs. Yang et al. [72] designed two pairs of programmed hairpins to identify precursor miRNA and mature miRNA. The bulge loop structure of precursor miRNA could be recognized by hairpins, hybridized DNA long-nicked duplex could be formed, and the fluorescence could be recovered. In contrast, the mature miRNA was locked in stem regions of precursor miRNA, which cannot trigger HCR. Therefore, multiple miRNA detection could effectively discriminate between mature and precursor miRNA. Furthermore, due to the long luminescence lifetime of lanthanide, the FRET-caused sensitized time of fluorophore would be delayed to microseconds. Guo et al. [24] designed hairpins with different FRET distances between lanthanide and Cy5.5 to detect miR-20a and miR-21, respectively in different periods. The method could effectively distinguish two kinds of miRNA and reduce background noise. Besides multiplex miRNA detection, Zhang et al. [28] utilized two hairpins to achieve membrane nucleolin and intracellular miR-21 ultrasensitive detection, which are the representative biomarker for cancer cell identification.

Microfluidic technology, which can capture target analyte from complicated solutions, can be applied in multiplex HCR-based detection. Kim et al. [73] immobilized capture probe into hydrogel microparticles. The captured miRNA could be conjugated with biotinylated universal adapter under T4 DNA ligase. The biotinylated initiator conjugated with the adapter with neutravidin, which could spur HCR. The neutravidin could bind with three biotinylated initiators, amplifying detection signaling. However, the capture effect could be affected due to the heterogeneous distribution and steric hindrance of the immobilized probe. Qu et al. [74] integrated capture probe into DNA tetrahedron, which could maintain probe orientation and density. The captured miRNA could hybridize with complementary hairpins, and the fluorescence could occur in the microfluidic platform after removing extra hairpins. The simultaneous multiplex miRNA could facilitate accurate and sensitive gastric cancer diagnosis (Figure 6.5).

6.4.2 HCR-Based Logic Gate

Adenosine triphosphate (ATP) maintains a high concentration in the tumor microenvironment and tumor cells, which plays a crucial role in tumor cell metastasis and immune regulation [75]. Therefore, the local high-expressing ATP can become vital clues for cancer diagnosis and prognosis. Meng et al. [76] developed a kind of ATP-responsive HCR-based logic gate. The ATP aptamer locked the toehold of one hairpin. When ATP is bound with the aptamer, the exposed toehold could recognize miR-155 and subsequently trigger HCR. The AND logic gate could release fluorescence only when the co-existence of intracellular ATP and miR-155. And the significant fluorescence could be observed in tumor cells rather than normal cells. Besides ATP increasing, increased glycolysis increases lactic acid production, which lowers tumor microenvironment pH value. Chen et al. [77] integrated pH-responsive sequences into partial complementary base pairing duplex, TaTb. When PTK7 was absent, the PTK7 aptamer was locked into the hairpin structure. PTK7 could cause aptamer-hairpin structure switching, exposed invading strand to induce strand displacement of TaTb. When pH is low, down to 6.3, the pH-responsive sequence located in TaTb could form i-motif tetraplex structures. Only when low pH and

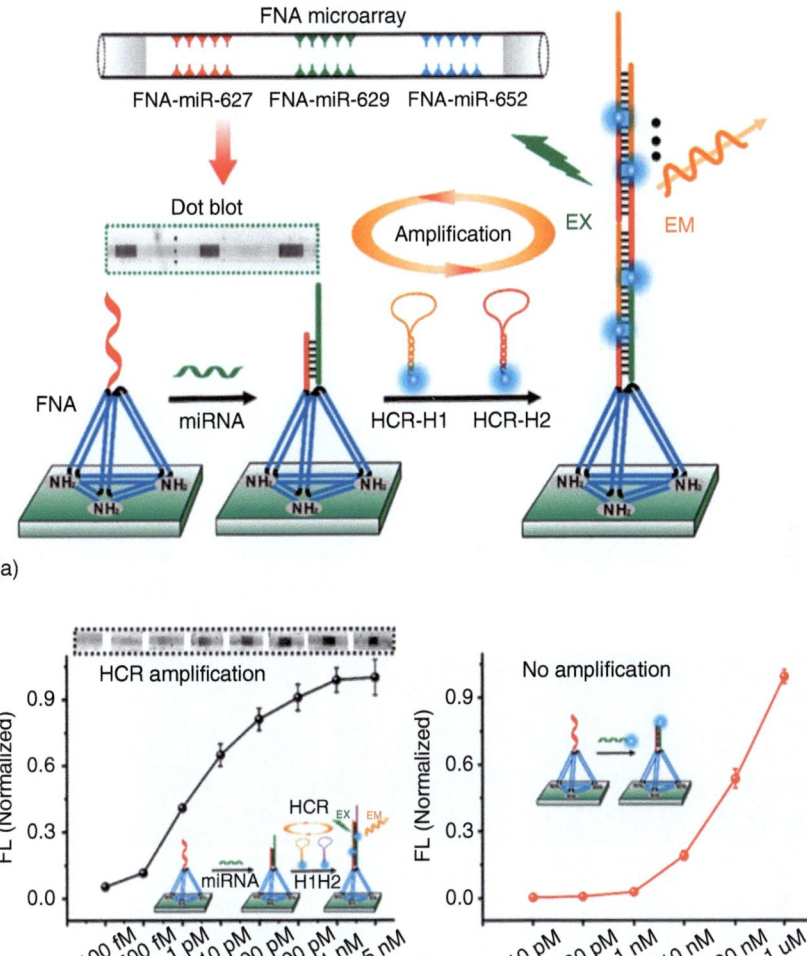

Figure 6.5 The multiplex miRNA detection based on combined HCR probes. (a) The microfluidic platform loads multiple HCR probes, which could detect different miRNAs. The different miRNAs could be analyzed in corresponding positions. Source: Reproduced with permission. Copyright 2018, American Chemical Society. (b, c) The fluorescence intensity of HCR and molecular beacon in detecting miRNA with different concentrations. Source: Reproduced with permission. Copyright 2018, American Chemical Society.

PTK7 exist simultaneously can the i-motif structure form an intact and nicked trigger to activate HCR. Ma et al. also utilized a pH-responsive i-motif switch to initiate HCR. The pH-responsive sequences were located in dendrimers. When pH decreased, the formed i-motif would be disassociated from dendrimers and exposed to the bipedal initiator of HCR. These pH-responsive or ATP-responsive logic gates could detect tumor extracellular acid microenvironment and intracellular miRNA (Figure 6.6). Besides, Quan et al. [78] developed a dual miRNA detection system based on DNAzyme and HCR. The miR-122

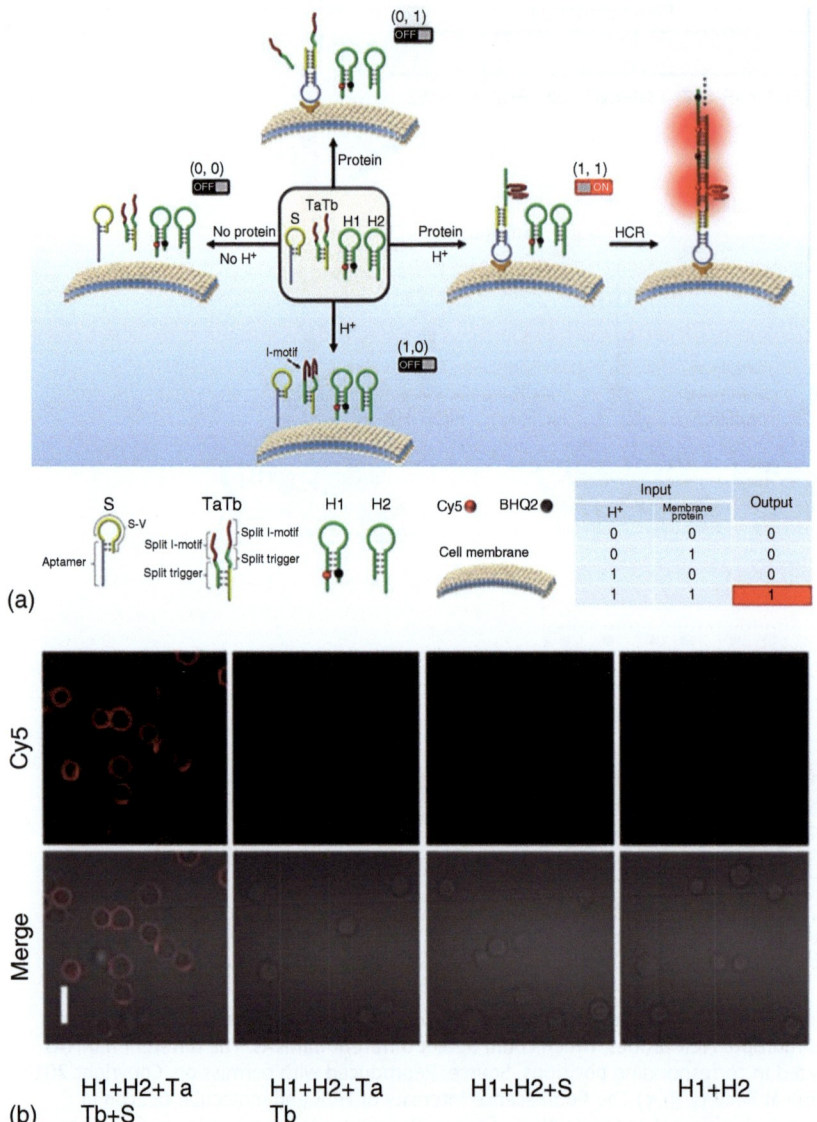

Figure 6.6 DNA logic gate-based HCR and its application (a) The tumor antigen could induce S probe structure changing, exposing invading strands, and causing toehold-mediated strand displacement with TaTb. The acid condition could induce i-motif formation and form intact nicked HCR triggers. The fluorescence could release only when tumor antigens and acid conditions coexist. Source: Reproduced with permission. Copyright 2022, American Chemical Society. (b) The results of confocal laser microscope. Source: Reproduced with permission. Copyright 2022, American Chemical Society.

could trigger metal-dependent DNAzyme unit formation. The hairpin would be formed after cutting the ribonucleotide position. The formed hairpin hybridized with another hairpin when miR-21 existed. Due to the adjacent position, the FRET caused TRMA intensity to increase. The AND logic gate could release positive signaling only when the detection target exists simultaneously, facilitating accurate cancer diagnosis.

6.5 HCR-Based Assembly Nanoplatforms

The high loading and endocytosis efficiency can facilitate intracellular sensitive target detection [79]. And the metal-DNA hybridization can provide metal ions for DNAzyme enzyme reaction [50].

Besides increasing endocytosis, some nanomaterials, such as gold, graphene oxide (GO), and MoS_2, have unique fluorescence quenching effects. And these nanomaterials have a stronger affinity for single hairpins but a weaker affinity for double DNA strands. Therefore, when the target is absent, the hairpin labeled with fluorophore is quenched by the nanoplatforms. However, when the target is present, long HCR products are formed, which are detached from the nanoplatform and recover fluorescence. Some scholars utilized GO- or MoS_2-loaded hairpins labeled with fluorophores to detect mutated mRNA and exosome-derived miRNA [80, 81].

Some scholars assembled HCR-based nanoplatforms with mediators and effectors to detect tumor-related enzyme activity. The mediator is the detached initiator owing to enzyme catalysis, and the effector exhibits detection signaling via HCR. Some scholars fixed the mediator strand and telomerase primer into gold nanoparticles via base-complementary pairing with immobilized template strands. The telomerase elongated primer and free-up mediator, which could promote the hybridization of hairpins absorbed onto gold particles [82, 83]. Besides, Zheng et al. [84] designed sub-peptide-modified silica nanoparticles, which could absorb mediator DNA strands. When protein kinase A (PKA) phosphorylates sub-peptide, the negative repulsion effect could free up the mediator. The mediator activated down-streaming HCR and recovered the fluorescence. The mediator–effector nanoplatform could enrich reaction molecules and amplify detection signaling. And these strategies could evaluate tumor-related enzyme activity such as telomerase and PKA (Figure 6.7).

Besides, in the reaction buffer, two complementary hairpins encounter via molecular collision, which affects the reaction rate due to the far distance between them. Therefore, some scholars applied DNA tetrahedrons to load two reactive hairpins, which pulled the two hairpins close significantly and increased local hairpin concentration. It was reported that tetrahedron-loaded hairpins could greatly promote reaction speed than nude hairpins. Furthermore, DNA tetrahedrons have excellent endocytosis and penetration advantages over other nucleic acid nanomaterials due to their four vertex structures. The increased endocytosis could facilitate intracellular target detection. The DNA tetrahedron can strengthen hairpin stability under serum conditions. Zhang et al. [28] utilized tetrahedron-loaded hairpins to achieve real-time tumor tissue imaging in living mice.

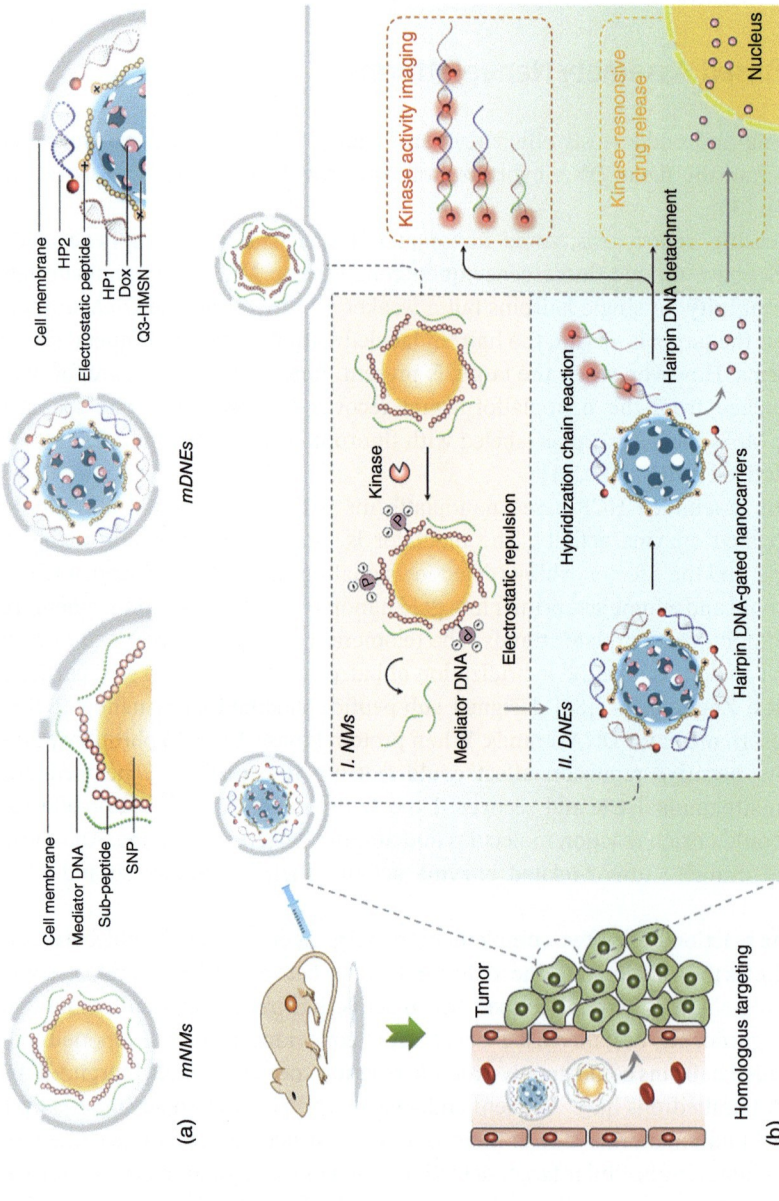

Figure 6.7 Mediator–effector-assembled nanoplatform based on HCR for PKA detection. (a) The fabrication of mediator and effector. Source: Reproduced with permission. Copyright 2022, American Chemical Society. (b) The PKA phosphates sub-peptide and increased negative charges could free up the mediator. The mediator induced HCR on the surface of nanocarriers. The increased charge repulsion freed up HCR products and released fluorescence. Source: Reproduced with permission. Copyright 2021, Wiley VCH.

6.6 HCR-Based Tumor Biotherapy

6.6.1 Chemotherapy

Tan et al. [85] constructed a kind of aptamer-trigger chimeras to bind with sgc8 antigen on the surface of tumor cells. After removing extra unreacted chimeras, the trigger sequences induced HCR. HCR products could enrich the antitumor drug payload. Therefore, the aptamer-tethered nano-trains based on HCR could achieve target delivery of anticancer drugs with higher efficiency. HCR products could be internalized into the lysosome, and the loaded drug could be released after enzyme digestion. To further weaken the nonspecific cytotoxicity of normal cells, Wang et al. [66] conjugated cisplatin prodrug into one of the hairpins. Lysosomes contain a higher concentration of thiols than the extracellular environment. The active cisplatin could be released from the HCR duplex via thiols. In contrast, nude hairpin with cisplatin is rather difficult to be taken in by cells. Therefore, the prodrug-loaded hairpins could minimize nonspecific cytotoxicity and increase endocytosis (Figure 6.8).

6.6.2 Photodynamic Therapy

Some scholars utilized HCR to recognize tumor cells and amplify photosensitizers to strengthen tumor treatment effects. Xiong et al. [80] conjugated Ce6 to hairpins, which could be activated to produce $^1O_2$. Due to the quenching effect of GO, the Ce6 cannot be activated in the absence of mutant mRNA in the normal cells. When HCR occurred, the amplified Ce6 could be activated under irradiation, and $^1O_2$ could accurately kill tumor cells (Figure 6.8). Besides, some scholars developed a label-free HCR-based PTD method. Multiple G-quadruplex units could be formed in the long duplex when HCR occurred triggered by miRNA or antigen. Zn^{II}PPIX could be intercalated into G-quadruplex units and generate reactive oxygen species under light irradiation. Therefore, the label-free

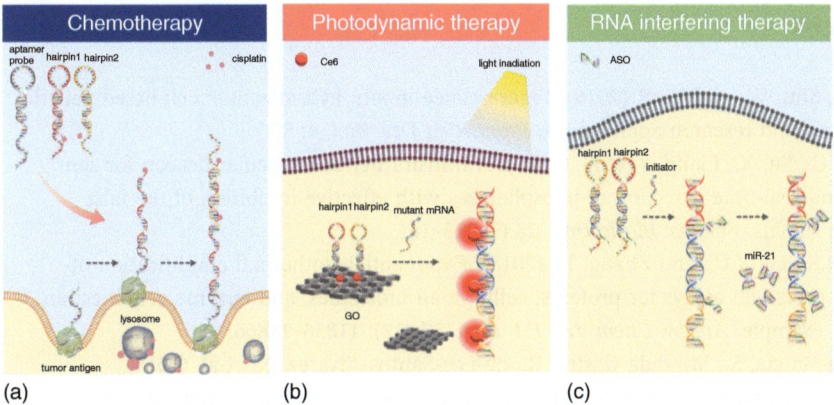

Figure 6.8 The application of HCR in tumor biotherapy. (a) HCR can enrich anticancer drugs and increase endocytosis. (b) The GO-loaded hairpins with photosensitizers can recognize mutant mRNA and release $^1O_2$ under light irradiation. (c) The HCR products can carry multiple antisense oligonucleotides, which can silence miR-21.

method can accurately lead to tumor cell death only in the presence of cancer-related miRNA and antigens.

6.6.3 RNA Interfering Therapy

Currently, the delivery of oligonucleotide drugs mainly depends on liposomes and viruses, which can be limited by tumor targeting and biocompatibility. Therefore, some scholars utilized HCR to achieve the amplification and targeted delivery of oligonucleotide drugs. Due to the small size of ASO and siRNA, they could be conjugated to hairpins. Initiated by tumor-related biomarkers, HCR products can accelerate cellular uptake of ASO and siRNA because of stronger endocytosis. Due to programmable assembly, ATP- or pH-responsive release of oligonucleotide drugs from HCR duplexes can be achieved [86]. Therefore, an HCR-based oligonucleotide drug delivery system can improve antitumor effects by enhancing stability, target delivery, cellular uptake, and controllable release (Figure 6.8).

6.7 Conclusion

HCR-based methods provide excellent sensitivity, specificity, and portability for POC real-time detection and good efficacy for antitumor treatment. Besides, due to superior programmability, a variety of modified HCR methods have been proposed to increase detection efficiency, including nuclease-assisted HCR systems, autocatalytic HCR systems, and nonlinear HCR systems. In summary, the assembly with microelectronics and microfluidic technology, DNA computers, and DNA chips based on HCR is expected to achieve a simple and sensitive "sample-answer" setting, which significantly promotes the clinical application of POC detection in the field. And the diagnosis and therapy of tumors based on HCR can facilitate the fabrication of integrated tumor treatment platforms.

References

1 Cui, C., Shu, W., and Li, P. (2016). Fluorescence in situ hybridization: cell-based genetic diagnostic and research applications. *Front. Cell Dev. Biol.* 4: 89.
2 Zhang, C., Su, X., Liang, Y. et al. (2011). A transformer of molecular beacon for sensitive and real-time detection of phosphatases with effective inhibition of the false positive signals. *Biosens. Bioelectron.* 28 (1): 13–16.
3 Reid, M.S., Le, X.C., and Zhang, H. (2018). Exponential isothermal amplification of nucleic acids and assays for proteins, cells, small molecules, and enzyme activities: an EXPAR example. *Angew. Chem. Int. Ed. Engl.* 57 (37): 11856–11866.
4 Barreda-Garcia, S., Miranda-Castro, R., de-Los-Santos-Alvarez, N. et al. (2018). Helicase-dependent isothermal amplification: a novel tool in the development of molecular-based analytical systems for rapid pathogen detection. *Anal. Bioanal.Chem.* 410 (3): 679–693.
5 Mohsen, M.G. and Kool, E.T. (2016). The discovery of rolling circle amplification and rolling circle transcription. *Acc. Chem. Res.* 49 (11): 2540–2550.

6 Zhou, Y., Yang, S., Xiao, Y. et al. (2019). Cytoplasmic protein-powered in situ fluorescence amplification for intracellular assay of low-abundance analyte. *Anal. Chem.* 91 (23): 15179–15186.

7 Dirks, R.M. and Pierce, N.A. (2004). Triggered amplification by hybridization chain reaction. *PNAS* 101 (43): 15275–15278.

8 Wong, Z.W., Ng, J.F., and New, S.Y. (2021). Ratiometric detection of microRNA using hybridization chain reaction and fluorogenic silver nanoclusters. *Chem. Asian J.* 16 (24): 4081–4086.

9 Chen, H.J., Hu, Y., Yao, P. et al. (2021). Accurate and efficient lipoprotein detection based on the HCR-DNAzyme platform. *Anal. Chem.* 93 (15): 6128–6134.

10 Wang, Y., Song, W., Zhao, H. et al. (2021). DNA walker-assisted aptasensor for highly sensitive determination of Ochratoxin A. *Biosens. Bioelectron.* 182: 113171.

11 Kaewarsa, P., Vilaivan, T., and Laiwattanapaisal, W. (2021). An origami paper-based peptide nucleic acid device coupled with label-free DNAzyme probe hybridization chain reaction for prostate cancer molecular screening test. *Anal. Chim. Acta* 1186: 339130.

12 Kong, Q., Cheng, S., Hu, X. et al. (2022). Ultrasensitive detection of tumor-derived small extracellular vesicles based on nonlinear hybridization chain reaction fluorescence signal amplification and immunomagnetic separation. *Analyst*.

13 Jiang, B., Wei, Y., Xu, J. et al. (2017). Coupling hybridization chain reaction with DNAzyme recycling for enzyme-free and dual amplified sensitive fluorescent detection of methyltransferase activity. *Anal. Chim. Acta* 949: 83–88.

14 Jiang, Y., Guo, Z., Wang, M. et al. (2022). Construction of fluorescence logic gates responding to telomerase and miRNA based on DNA-templated silver nanoclusters and the hybridization chain reaction. *Nanoscale* 14 (3): 612–616.

15 Li, C., Li, H., and Jie, G. (2020). Click chemistry reaction-triggered DNA walker amplification coupled with hyperbranched DNA nanostructure for versatile fluorescence detection and drug delivery to cancer cells. *Mikrochim. Acta* 187 (11): 625.

16 Wang, X., Yuan, Y., Wu, Z., and Jiang, J.H. (2020). Self-tracking multifunctional nanotheranostics for sensitive miRNA imaging guided photodynamic therapy. *ACS Appl. Bio Mater.* 3 (5): 2597–2603.

17 Shang, J., Li, C., Li, F. et al. (2021). Construction of an enzyme-free initiator-replicated hybridization chain reaction circuit for amplified methyltransferase evaluation and inhibitor assay. *Anal. Chem.* 93 (4): 2403–2410.

18 Figg, C.A., Winegar, P.H., Hayes, O.G., and Mirkin, C.A. (2020). Controlling the DNA hybridization chain reaction. *JACS* 142 (19): 8596–8601.

19 Zhang, W., Hao, W., Liu, X. et al. (2020). Visual detection of miRNAs using enzyme-free amplification reactions and ratiometric fluorescent probes. *Talanta* 219: 121332.

20 Ang, Y.S. and Yung, L.Y. (2016). Rational design of hybridization chain reaction monomers for robust signal amplification. *Chem. Commun. (Camb)* 52 (22): 4219–4222.

21 Oe, M., Miki, K., Ueda, Y. et al. (2021). Deep-red/near-infrared turn-on fluorescence probes for aldehyde dehydrogenase 1A1 in cancer stem cells. *ACS Sens.* 6 (9): 3320–3329.

22 Liu, H., Liu, S., Xiao, Y. et al. (2021). A pH-reversible fluorescent probe for in situ imaging of extracellular vesicles and their secretion from living cells. *Nano Lett.* 21 (21): 9224–9232.

23 Zhang, F., Wang, S., Feng, J. et al. (2019). MoS2-loaded G-quadruplex molecular beacon probes for versatile detection of MicroRNA through hybridization chain reaction signal amplification. *Talanta* 202: 342–348.

24 Guo, J., Mingoes, C., Qiu, X., and Hildebrandt, N. (2019). Simple, amplified, and multiplexed detection of MicroRNAs using time-gated FRET and hybridization chain reaction. *Anal. Chem.* 91 (4): 3101–3109.

25 Wang, J., Wang, D.X., Ma, J.Y. et al. (2019). Three-dimensional DNA nanostructures to improve the hyperbranched hybridization chain reaction. *Chem. Sci.* 10 (42): 9758–9767.

26 Yang, C.H., Wu, T.H., Chang, C.C. et al. (2021). Biosensing amplification by hybridization chain reaction on phase-sensitive surface plasmon resonance. *Biosens.-Basel* 11 (3).

27 Glineburg, M.R., Zhang, Y., Krans, A. et al. (2021). Enhanced detection of expanded repeat mRNA foci with hybridization chain reaction. *Acta Neuropathol. Commun.* 9 (1): 73.

28 Zhang, B., Tian, T., Xiao, D. et al. (2022). Facilitating in situ tumor imaging with a tetrahedral DNA framework-enhanced hybridization chain reaction probe. *Adv. Funct. Mater.*

29 Li, H., Zhang, B., He, X. et al. (2021). Intracellular CircRNA imaging and signal amplification strategy based on the graphene oxide-DNA system. *Anal. Chim. Acta* 1183: 338966.

30 Kimura, Y., Soma, T., Kasahara, N. et al. (2016). Edesign: primer and enhanced internal probe design tool for quantitative PCR experiments and genotyping assays. *PLoS One* 11 (2): e0146950.

31 Borum, R.M., Moore, C., Chan, S.K. et al. (2021). A photoacoustic contrast agent for miR-21 via NIR fluorescent hybridization chain reaction. *Bioconjugate Chem.*

32 Liu, G., Chai, H., Tang, Y., and Miao, P. (2020). Bright carbon nanodots for miRNA diagnostics coupled with concatenated hybridization chain reaction. *Chem. Commun. (Camb)* 56 (8): 1175–1178.

33 Yang, L., Ma, P., Chen, X. et al. (2022). High-sensitivity fluorescence detection for lung cancer CYFRA21-1 DNA based on accumulative hybridization of quantum dots. *J. Mater. Chem. B* 10 (9): 1386–1392.

34 Zhao, Y., Feng, Y., Zhang, Y. et al. (2020). Combining competitive sequestration with nonlinear hybridization chain reaction amplification: an ultra-specific and highly sensitive sensing strategy for single-nucleotide variants. *Anal. Chim. Acta* 1130: 107–116.

35 Jin, X., Chen, Y., Chen, H. et al. (2017). Evaluation of tumor-derived exosomal miRNA as potential diagnostic biomarkers for early-stage non-small cell lung cancer using next-generation sequencing. *Clin. Cancer Res.* 23 (17): 5311–5319.

36 Selvi, A., Devi, K., Manimekalai, R. et al. (2021). High-throughput miRNA deep sequencing in response to drought stress in sugarcane. *3 Biotech* 11 (7): 312.

37 Forero, D.A., Gonzalez-Giraldo, Y., Castro-Vega, L.J., and Barreto, G.E. (2019). qPCR-based methods for expression analysis of miRNAs. *Biotechniques* 67 (4): 192–199.

38 Rupaimoole, R. and Slack, F.J. (2017). MicroRNA therapeutics: towards a new era for the management of cancer and other diseases. *Nat. Rev. Drug Discovery* 16 (3): 203–222.

39 Kasai, A., Kakihara, S., Miura, H. et al. (2016). Double in situ hybridization for MicroRNAs and mRNAs in brain tissues. *Front. Mol. Neurosci.* 9: 126.

40 Ogata, M., Hayashi, G., Ichiu, A., and Okamoto, A. (2020). L-DNA-tagged fluorescence in situ hybridization for highly sensitive imaging of RNAs in single cells. *Org. Biomol. Chem.* 18 (40): 8084–8088.

41 Reilly, S.K., Gosai, S.J., Gutierrez, A. et al. (2021). Direct characterization of cis-regulatory elements and functional dissection of complex genetic associations using HCR-FlowFISH. *Nat. Genet.* 53 (8): 1166–1176.

42 Hosseinzadeh, E., Ravan, H., Mohammadi, A., and Pourghadamyari, H. (2020). Colorimetric detection of miRNA-21 by DNAzyme-coupled branched DNA constructs. *Talanta* 216: 120913.

43 Pu, J., Liu, M., Li, H. et al. (2021). One-step enzyme-free detection of the miRNA let-7a via twin-stage signal amplification. *Talanta* 230: 122158.

44 Ji, P., Han, G., Huang, Y. et al. (2021). Ultrasensitive ratiometric detection of Pb(2+) using DNA tetrahedron-mediated hyperbranched hybridization chain reaction. *Anal. Chim. Acta* 1147: 170–177.

45 Moon, W.J., Huang, P.J., and Liu, J. (2021). Probing metal-dependent phosphate binding for the catalysis of the 17E DNAzyme. *Biochemistry* 60 (24): 1909–1918.

46 Wang, S., Ding, J., and Zhou, W. (2019). An aptamer-tethered, DNAzyme-embedded molecular beacon for simultaneous detection and regulation of tumor-related genes in living cells. *Analyst* 144 (17): 5098–5107.

47 Wu, Q., Wang, H., Gong, K. et al. (2019). Construction of an autonomous nonlinear hybridization chain reaction for extracellular vesicles-associated MicroRNAs discrimination. *Anal. Chem.* 91 (15): 10172–10179.

48 Gong, K., Wu, Q., Wang, H. et al. (2020). Autocatalytic DNAzyme assembly for amplified intracellular imaging. *Chem. Commun. (Camb)* 56 (77): 11410–11413.

49 Liu, F., Li, X.L., and Zhou, H. (2020). Biodegradable MnO_2 nanosheet based DNAzyme-recycling amplification towards sensitive detection of intracellular MicroRNAs. *Talanta* 206: 120199.

50 Wei, J., Wang, H., Wu, Q. et al. (2020). A smart, autocatalytic, DNAzyme biocircuit for in vivo, amplified, MicroRNA imaging. *Angew. Chem. Int. Ed. Engl.* 59 (15): 5965–5971.

51 Bi, S., Yue, S., and Zhang, S. (2017). Hybridization chain reaction: a versatile molecular tool for biosensing, bioimaging, and biomedicine. *Chem. Soc. Rev.* 46 (14): 4281–4298.

52 Li, P., Zhang, H., Wang, D. et al. (2018). An efficient nonlinear hybridization chain reaction-based sensitive fluorescent assay for in situ estimation of calcium channel protein expression on bone marrow cells. *Anal. Chim. Acta* 1041: 25–32.

53 Xue, Q., Liu, C., Li, X. et al. (2018). Label-free fluorescent DNA dendrimers for microRNA detection based on nonlinear hybridization chain reaction-mediated multiple G-quadruplex with low background signal. *Bioconjugate Chem.* 29 (4): 1399–1405.

54 Lv, W.Y., Li, C.H., Li, Y.F. et al. (2021). Hierarchical hybridization chain reaction for amplified signal output and cascade DNA logic circuits. *Anal. Chem.* 93 (7): 3411–3417.

55 Prashant, N.M., Alomran, N., Chen, Y. et al. (2021). SCReadCounts: estimation of cell-level SNVs expression from scRNA-seq data. *BMC Genomics* 22 (1): 689.

56 Winter, H., Kaisaki, P.J., Harvey, J. et al. (2019). Identification of circulating genomic and metabolic biomarkers in intrahepatic cholangiocarcinoma. *Cancers (Basel)* 11 (12).

57 Lazaro, A., Maquieira, A., and Tortajada-Genaro, L.A. (2022). Discrimination of single-nucleotide variants based on an allele-specific hybridization chain reaction and smartphone detection. *ACS Sens.*

58 Tang, Y., Zhang, X.L., Tang, L.J. et al. (2017). In situ imaging of individual mRNA mutation in single cells using ligation-mediated branched hybridization chain reaction (ligation-bHCR). *Anal. Chem.* 89 (6): 3445–3451.

59 Marras, S.A.E., Bushkin, Y., and Tyagi, S. (2019). High-fidelity amplified FISH for the detection and allelic discrimination of single mRNA molecules. *PNAS* 116 (28): 13921–13926.

60 Murdocca, M., De Masi, C., Pucci, S. et al. (2021). LOX-1 and cancer: an indissoluble liaison. *Cancer Gene Ther.* 28 (10-11): 1088–1098.

61 Montalto, F.I. and De Amicis, F. (2020). Cyclin D1 in cancer: a molecular connection for cell cycle control, adhesion and invasion in tumor and stroma. *Cells* 9 (12).

62 Kasuga, A., Semba, T., Sato, R. et al. (2021). Oncogenic KRAS-expressing organoids with biliary epithelial stem cell properties give rise to biliary tract cancer in mice. *Cancer Sci.* 112 (5): 1822–1838.

63 Jayanthi, V., Das, A.B., and Saxena, U. (2017). Recent advances in biosensor development for the detection of cancer biomarkers. *Biosens. Bioelectron.* 91: 15–23.

64 Choi, J., Love, K.R., Gong, Y. et al. (2011). Immuno-hybridization chain reaction for enhancing detection of individual cytokine-secreting human peripheral mononuclear cells. *Anal. Chem.* 83 (17): 6890–6895.

65 Rafiee, S.D., Kocabey, S., Mayer, M. et al. (2020). Detection of HER2(+) breast cancer cells using bioinspired DNA-based signal amplification. *ChemMedChem* 15 (8): 661–666.

66 Wang, Y.M., Wu, Z., Liu, S.J., and Chu, X. (2015). Structure-switching aptamer triggering hybridization chain reaction on the cell surface for activatable theranostics. *Anal. Chem.* 87 (13): 6470–6474.

67 Li, Z., He, X., Luo, X. et al. (2016). DNA-programmed quantum dot polymerization for ultrasensitive molecular imaging of cancer cells. *Anal. Chem.* 88 (19): 9355–9358.

68 Li, W., Wang, L., Wang, Y., and Jiang, W. (2018). Binding-induced nicking site reconstruction strategy for quantitative detection of membrane protein on living cell. *Talanta* 189: 383–388.

69 Li, L., Jiang, H., Meng, X. et al. (2021). Highly sensitive detection of cancer cells via split aptamer mediated proximity-induced hybridization chain reaction. *Talanta* 223 (Pt 1): 121724.

70 Chen, J., Tang, J., Meng, H.M. et al. (2020). Recognition triggered assembly of split aptamers to initiate a hybridization chain reaction for wash-free and amplified detection of exosomes. *Chem. Commun. (Camb)* 56 (63): 9024–9027.

71 Zada, S., Lu, H., Dai, W. et al. (2022). Multiple amplified microRNAs monitoring in living cells based on fluorescence quenching of Mo2B and hybridization chain reaction. *Biosens. Bioelectron.* 197: 113815.

72 Yang, F., Cheng, Y., Cao, Y. et al. (2019). Sensitively distinguishing intracellular precursor and mature microRNA abundance. *Chem. Sci.* 10 (6): 1709–1715.

73 Kim, J., Shim, J.S., Han, B.H. et al. (2021). Hydrogel-based hybridization chain reaction (HCR) for detection of urinary exosomal miRNAs as a diagnostic tool of prostate cancer. *Biosens. Bioelectron.* 192: 113504.

74 Qu, X., Xiao, M., Li, F. et al. (2018). Framework nucleic acid-mediated pull-down MicroRNA detection with hybridization chain reaction amplification. *ACS Appl. Bio Mater.* 1 (3): 859–864.
75 Vultaggio-Poma, V., Sarti, A.C., and Di Virgilio, F. (2020). Extracellular ATP: a feasible target for cancer therapy. *Cells* 9 (11).
76 Meng, X., Wang, H., Yang, M. et al. (2021). Target-cell-specific bioorthogonal and endogenous ATP control of signal amplification for intracellular MicroRNA imaging. *Anal. Chem.* 93 (3): 1693–1701.
77 Chen, B., Ma, W., Long, X. et al. (2022). Membrane protein and extracellular acid heterogeneity-driven amplified DNA logic gate enables accurate and sensitive identification of cancer cells. *Anal. Chem.* 94 (5): 2502–2509.
78 Quan, K., Li, J., Wang, J. et al. (2019). Dual-microRNA-controlled double-amplified cascaded logic DNA circuits for accurate discrimination of cell subtypes. *Chem. Sci.* 10 (5): 1442–1449.
79 Jia, Y., Shen, X., Sun, F. et al. (2020). Metal-DNA coordination based bioinspired hybrid nanospheres for in situ amplification and sensing of microRNA. *J. Mater. Chem. B* 8 (48): 11074–11081.
80 Xiong, M., Rong, Q., Kong, G. et al. (2019). Hybridization chain reaction-based nanoprobe for cancer cell recognition and amplified photodynamic therapy. *Chem. Commun. (Camb)* 55 (21): 3065–3068.
81 Wei, J., He, S., Mao, Y. et al. (2021). A simple "signal off-on" fluorescence nanoplatform for the label-free quantification of exosome-derived microRNA-21 in lung cancer plasma. *Mikrochim. Acta* 188 (11): 397.
82 Wang, X., Yang, D., Liu, M. et al. (2019). Highly sensitive fluorescence biosensor for intracellular telomerase detection based on a single patchy gold/carbon nanosphere via the combination of nanoflare and hybridization chain reaction. *Biosens. Bioelectron.* 137: 110–116.
83 Hong, M., Xu, L., Xue, Q. et al. (2016). fluorescence imaging of intracellular telomerase activity using enzyme-free signal amplification. *Anal. Chem.* 88 (24): 12177–12182.
84 Zheng, F., Meng, T., Jiang, D. et al. (2021). Nanomediator-effector cascade systems for amplified protein kinase activity imaging and phosphorylation-induced drug release in vivo. *Angew. Chem. Int. Ed. Engl.* 60 (39): 21565–21574.
85 Zhu, G., Zheng, J., Song, E. et al. (2013). Self-assembled, aptamer-tethered DNA nanotrains for targeted transport of molecular drugs in cancer theranostics. *PNAS* 110 (20): 7998–8003.
86 Li, F., Yu, W., Zhang, J. et al. (2021). Spatiotemporally programmable cascade hybridization of hairpin DNA in polymeric nanoframework for precise siRNA delivery. *Nat. Commun.* 12 (1): 1138.

7

Application and Prospects of Framework Nucleic Acid-Based Nanomaterials in Tumor Therapy

Tianyu Chen and Xiaoxiao Cai

Sichuan University, State Key Laboratory of Oral Diseases, West China Hospital of Stomatology, No.14, 3Rd Section of Ren Min Nan Rd., Chengdu 610041, PR China

In recent years, cancer has become a significant threat to human health because of its extremely high fatality rate. According to a report by the International Agency for Research on Cancer (IARC), approximately 10 million people succumbed to cancer in 2020, and the number of cancer patients will increase by 60% globally in the next 20 years [1–3]. Current clinical treatments for cancer mainly include radiotherapy, chemotherapy, and surgery; however, these treatments have shortcomings. Surgical treatment is only suitable for the early stages of cancer, and the recurrence rate of cancer after surgery remains high. Radiotherapy and chemotherapy have toxic side effects on normal tissues and are prone to drug resistance. Therefore, the development of new tumor treatment regimens has become a research hotspot in recent years. Compared with traditional antitumor drugs, nucleic acid nanomaterials possess excellent programmability, narrow particle size distribution, low cytotoxicity, and stable spatial structures, and these nanomaterials can help improve the stability, delivery efficiency, and targeting ability of traditional chemotherapy drugs [4, 5].

DNA is an important biological molecule that carries genetic information from the body. Because of the stability of complementary base pairing, DNA is regarded as a bionanomaterial with extremely wide application prospects [6, 7]. In recent years, a variety of nucleic acid nanomaterials have been synthesized and widely used in the field of targeted delivery of antitumor drugs, such as DNA origami structures, tetrahedral framework nucleic acids (tFNAs), and dynamic DNA nanostructures. Because of their good biocompatibility and modified specific targeting of nucleic acid nanomaterials, they can effectively transport drugs to the tumor site, thereby improving the efficacy of drug delivery systems [4].

Many nucleic acid nanomaterials have presented outstanding achievements in chemotherapy, photodynamic therapy (PDT), gene therapy, and immunotherapy for tumors, and they are considered promising potential treatment methods [8–10]. In this review, we outline the application and prospects of nucleic acid nanomaterials in tumor therapy and demonstrate the advantages and disadvantages of nucleic acid nanomaterials in tumor therapy by introducing their functions and properties.

Nucleic Acid-Based Nanomaterials: Stabilities and Applications, First Edition.
Edited by Yunfeng Lin and Shaojingya Gao.
© 2024 WILEY-VCH GmbH. Published 2024 by WILEY-VCH GmbH.

7.1 Development of Nucleic Acid Nanomaterials

Nucleic acid nanotechnology was first proposed by Nadrian Seeman in 1982, who first used DNA sequences as building blocks to build structures, thereby creating a new field of research [11]. Over the past 40 years, nucleic acid nanotechnology has developed rapidly, producing various spatial nanostructures in one to three dimensions. In 2006, Paul W. K. Rothemund first reported DNA origami technology, which can achieve precise regulation of DNA nanostructures and assembly of nucleic acid nanomaterials of various shapes [12]. Over the past ten years, advances in technology have led to the development of software to assist in the design of DNA nanostructures, such as caDNAno, oxDNA, and CanDo, thereby enabling DNA origami technology to design more complex and delicate structures [13–15]. For example, Han et al. designed a 3D DNA nanostructure with complex curvatures by tuning the specific positions and patterns of the crossovers between adjacent DNA double helices [16]. In 2018, Zhang et al. used DNA origami blocks to construct a "tensegrity triangle" with the assistance of caDNAno and further assembled it into a 3D rhombohedral crystalline structure [17].

In addition to its structural diversity, DNA origami technology can precisely modify DNA nanostructures based on its ability to locate each base of the nanostructure. In 2015, Qiao enhanced the antitumor effect of DNA origami nanomaterials by modifying the surface of gold nanorods, thereby providing a promising candidate material for cancer diagnosis and treatment [8].

In 2018, Fan Chunhai et al. first proposed the concept of framework nucleic acids (FNAs), which are a new type of nucleic acid nanomaterial with unique physicochemical and biological properties [18]. Among the various FNA structures, tFNAs are widely used because of their simple synthesis, high yield, and good economic performance. tFNA is a nucleic acid nanomaterial with excellent anti-inflammatory and antioxidant capacities; thus, it plays an important role in various disease models [19–21]. For example, Qi discovered that tFNAs can exhibit protective effects by inhibiting apoptosis and alleviating oxidative stress, thereby achieving the effect of treating acute kidney injury [22]. Junyao et al. reported that tFNAs can alleviate cellular inflammation by activating the AKT signaling pathway [23]. On the other hand, as a drug delivery vehicle, tFNAs can enhance the efficacy of chemotherapy drugs; Figure 7.1 shows several representative tetrahedral FNA drug delivery systems for cancer treatment [24–28].

In recent years, dynamic DNA nanostructures have been the focus of nucleic acid nanomaterial development. For example, Juul et al. designed a special three-dimensional DNA nanostructure that can realize reversible encapsulation and release of enzyme cargo without any form of covalent or noncovalent linkage for heat-sensitive drug delivery [29]. In 2021, Tian Taoran designed a melittin-loaded DNA nanostructure for targeted therapy, and it undergoes a conformational change to release the drug after binding to the target protein. All the aforementioned points highlight the broad potential of dynamic DNA structures in biomedicine [30].

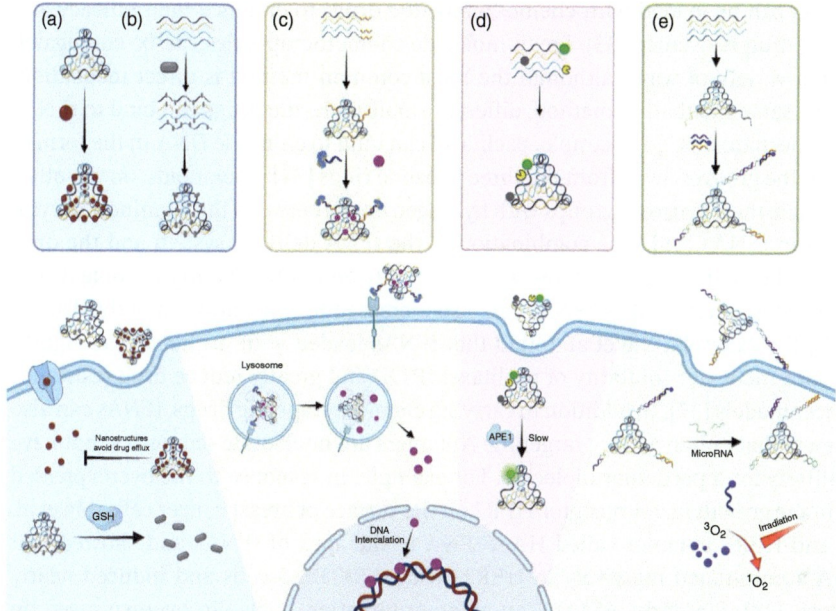

Figure 7.1 tFNA nanostructure functionalization for tumor therapy. (a) Self-assembled DNA tetrahedron for the treatment of multidrug-resistant cancer cells. (b) Camptothecin-grafted DNA tetrahedron for the inhibition of tumor growth. (c) Cell targeting was achieved by locating multi-functional components via intercalation or insertion at one end of the tFNA. (d) Through intracellular ape1 binding, an apurinic/apyrimidinic endonuclease (ape1)-site-modified tFNA was employed to interfere with enzymatic catalysis, increasing tumor cell sensitivity to chemotherapeutic treatments. (e) A DNA-tetrahedron-based logic device was utilized to recognize microRNA for specific cell selection, after which a photosensitizer was activated for synergetic chemotherapy and photodynamic effects in the treatment of solid tumor tissue and metastasis. Source: (a) Adapted with permission from [24]. Copyright 2013, Royal Society of Chemistry. (b) Adapted with permission from [25]. Copyright 2019, Wiley-VCH. (c) Adapted with permission from [26]. Copyright 2018, Royal Society of Chemistry. (d) Adapted under the terms of the CC-BY Creative Commons Attribution 3.0 Unported license (https://creativecommons.org/licenses/by/3.0) from [27]. Copyright 2019, Royal Society of Chemistry. (e) Adapted with permission from [28]. Copyright 2021, Royal Society of Chemistry.

7.2 Properties and Applications of Nucleic Acid Nanomaterials

7.2.1 tFNAs

tFNAs are highly editable, and small-molecule drugs, polypeptides, oligonucleotides, and antibodies can bind to tFNAs through covalent bonding or charge attraction [31, 32]. Moreover, these functionally modified tFNAs can still carry cargo into the cell. The strong carrying capacity and editability of tFNAs have enabled their wide involvement in multiple aspects of tumor therapy.

First, tFNAs can be loaded with chemotherapeutic drugs to enhance their efficacy and reduce tumor drug resistance [33]. Small-molecule chemotherapeutics can be conjugated to tFNAs in a variety of ways, although the most common method is direct incubation. Even with the same incubation method, different small-molecule drugs can bind to tFNAs via different mechanisms. For example, paclitaxel can bind to chimeric DNA in the form of monomers in the grooves away from the three benzene rings [34]. Flavonoids, on the other hand, intercalate their hydroxyl groups with hydrogen bonds between the guanine and cytosine residues in tFNAs [35]. The combination of the tFNA delivery system and the drug can overcome the characteristics of low water solubility and difficult entry of some drugs, and the endocytosis mediated by tFNAs can also increase the concentration of the drug in the cell [36]. For example, Xie et al. found that tFNAs loaded with paclitaxel can significantly enhance the water solubility of paclitaxel (PTX) and greatly reduce drug resistance in lung cancer models [37]. In addition to carrying chemotherapeutic drugs, tFNAs can also carry aptamers that enhance drug targeting. Aptamers are nucleotide sequences that have a unique affinity for a particular molecule. For example, in response to the overexpressed transmembrane growth factor receptor HER2 on the surface of breast cancer cells, Ma et al. linked an anti-HER2 aptamer called HApt-tFNA to the apex of tFNAs and showed that HApt-tFNA accumulated massively in HER2-positive SK-BR-3 cells and induced nearly 50% apoptosis [38]. Similarly, AS1411, an aptamer targeting nucleolin overexpressed on the surface of tumor cells, can help tFNAs to rapidly target tumor cells, and the tFNA complex linked to AS1411 and the anticancer drug 5-fluorouracil (5-FU) can simultaneously enhance the targeting of 5-FU and promote the apoptosis of nucleolin-positive MCF-7 cells [39].

In addition, tFNAs can be used for gene therapy because they can link oligonucleotides to regulate tumor-related genes. The connection between tFNAs and oligonucleotides is relatively conservative. Oligonucleotides can be hydrogen-bonded with sticky ends extending from the apex of tFNAs or directly covalently bonded to one end of the single strand as an extended cantilever. In addition to the aptamers described above, oligonucleotides have been studied extensively and include microRNAs (miRNAs) and small interfering RNAs (siRNAs). miRNA is a small nucleotide with a length of approximately 20–25 bp, and it is mainly produced by cells to regulate various physiological activities. siRNA is a synthetic double-stranded RNA that can interfere with the transcription of mRNA to regulate the physiological processes of cells. It is difficult for both types of oligonucleotides to cross cell membranes without carriers because of their negative charges. tFNAs can enhance the transmembrane capacity of oligonucleotides, such as miRNAs and siRNAs, and protect them from degradation. Li et al. found that tFNAs help miRNA214-3p enter A549 cells to silence the SURVIVIN gene and promote apoptosis [40]. Similarly, tFNAs can also be loaded with siRNA and aptamers to target and interfere with the normal physiological processes of tumor cells to promote their apoptosis. For example, Xiao et al. linked tFNAs with AS1411 to siBraf, which interferes with Braf gene expression in melanoma; moreover, the complex showed strong cleavage of Braf mRNA and promoted apoptosis in melanoma cells [10, 41].

In addition to directly acting on tumor cells, tFNAs can be involved in indirect tumor therapy in various ways. Tumor immunotherapy refers to enhancing the immunogenicity of tumor cells or the killing effect of the immune system in various ways to achieve antitumor activity. Cytosine-phosphate-guanine oligodeoxynucleotides (CpG ODNs) are commonly used immunostimulants that stimulate dendritic cells and macrophages to secrete pro-inflammatory factors, such as tumor necrosis factor-α and interleukin-1, resulting in enhanced innate immune strength. Liu et al. co-linked antigens and CpGs on tFNAs because CpGs can help the immune system generate stronger and longer-lasting immune effects than tFNAs linked to antigens alone [42]. In addition, tFNAs linked to miRNAs can achieve similar effects. Qin et al. demonstrated that tFNAs linked to microRNA-155 can promote the polarization of macrophages to a proinflammatory type and enhance the strength of innate immunity [43]. In PDT, specific wavelengths of light, photosensitizers, and oxygen are used to generate reactive oxygen species during treatment, during which tumor cells are killed through reactive oxygen species-induced apoptosis and necrosis. The intensity of PDT action was positively correlated with intracellular photosensitizer and oxygen concentrations. However, typical photosensitizers are hydrophobic and difficult to transport into the cells. Therefore, similar to small-molecule drugs, tFNAs enhance photosensitizer hydrophilicity, help photosensitizers enter cells, and increase photosensitizer concentrations, thereby enhancing PDT efficacy. Jin et al. combined methylene blue with tFNAs to enhance light-induced cytotoxicity [44]. Wang et al. combined IR780, a photosensitizer with photothermal and photosensitizing effects, with tFNAs and showed that tFNAs not only helped IR780 achieve a stronger PDT killing effect but also significantly increased the accumulation of IR780 in tumor sites; moreover, under infrared light irradiation, IR780 loaded with tFNAs exhibited stronger tumor-imaging effects than IR780 alone [45].

Interestingly, tFNAs have four vertices; that is, the mixing of multiple therapeutic strategies will be the future development direction of tFNAs in the field of tumor therapy. For example, when chemotherapy drugs are combined with immunotherapy, reports have indicated that tFNAs can be linked to doxorubicin (DOX) and CpG to achieve better efficacy [46]. Ren et al. reported tFNAs with multiple functions of targeting aptamers, e.g. DOX and DNase, which further expanded the therapeutic scope of tFNAs for tumors [47].

Therefore, in the future, multiple functional modifications of tFNAs will make tFNA-based treatments more comprehensive and complex, which can further improve the efficiency of treatment.

7.2.2 DNA Origami

In 2006, Rothemund first proposed a new DNA self-assembly method called DNA origami technology [12], which applies the principle of complementary base pairing to fold specific regions on long-chain DNA (scaffold) and then fix the short-chain DNA (staples) to construct the expected structure. Because the experimental conditions for DNA origami technology are relatively simple and the assembly is highly efficient, nanomaterials or molecules assembled using DNA origami technology can be used to prepare nanodevices or drug carriers with certain optoelectronic properties. Under the guidance of DNA origami technology, nanoarchitectures of various shapes and sizes have been synthesized, such as diverse two-dimensional nanostructures (rectangles, triangles, pentacles, smiling faces, etc.) [12], nanotubes [48], polyhedral nanostructures [16, 49], and other complicated DNA origami nanostructures (DONs) [50–52].

In 1986, Matsumura et al. [53] reported the enhanced permeability and retention (EPR) effects of these nanostructures. Since then, the development and research of macromolecular anticancer drugs have increased, and various nanomaterials have performed well in tumor targeting based on EPR [54, 55]. In addition to their targeting properties, DONs have good stability and biocompatibility, excellent cellular membrane penetration, and multiple modification sites for a series of biomolecules. Hence, DONs are promising platforms for creating multifunctional nanomaterials for cancer therapy [56].

To serve as a drug delivery system, DONs must exhibit excellent stability in circulation conditions. Many studies have shown that DNA nanostructures are more stable than single- or double-stranded DNA. The complex structure of DNA origami may induce good stability by hindering the function of nucleases [57]. Moreover, encapsulation modifications have been performed to improve the stability of DNA nanostructures. Perrault et al. [58] utilized a lipid bilayer to encapsulate DONs, and this updated form led to longer blood circulation times compared with the plain form. Other molecules, such as PEG, polypeptides, and proteins, have been used for the same purpose [59, 60].

DOX is an antitumor micromolecule that can inhibit the synthesis of RNA and DNA, and it has been applied as a cell cycle-nonspecific drug with a broad antitumor spectrum. However, DOX might produce toxic effects, such as leukopenia, thrombocytopenia, cardiotoxicity, nausea, and loss of appetite. Considering its intercalation into DNA double strands, DOX serves as a classic drug that can be modified into DONs for cancer therapy, which presents higher tumor-targeting efficiency and fewer toxic effects. Zhao et al. [61] encapsulated DOX in screwy DNA nanotubes and studied their release rates, and the results indicated that DNA nanotubes could improve the cellular uptake of DOX to kill cancer cells. Jiang et al. [62] also reported that DNA nanocarriers exhibit an enrichment effect and reduce drug resistance to a certain extent (Figure 7.2), and they also used triangular DNA

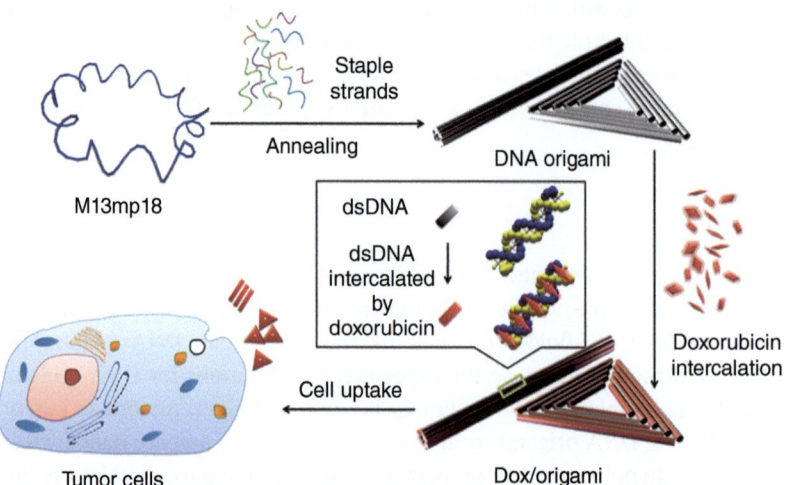

Figure 7.2 DNA origami and doxorubicin origami delivery system assembly. The long single-strand M13mp18 genomic DNA scaffold strand (blue) is folded into the triangle and tube structures through the hybridization of rationally designed staple strands. Watson–Crick base pairs in the double helices serve as docking sites for doxorubicin intercalation. After incubation with doxorubicin, the drug-loaded DNA nanostructure delivery vessels were administered to MCF 7 cells, and the effects were investigated. Source: Reproduced from Jiang et al. [62]. Copyright 2012, American Chemical Society.

nanocarriers to deliver DOX [63]. The research revealed that DOX-loaded DONs showed better antitumor effects and lower systemic toxicity in mouse models compared with free DOX molecules.

Currently, an increasing number of researchers are focusing on gene therapy, which applies exogenous nucleic acid strands, such as siRNA, miRNA, and CpG [64], to inhibit cancer cells. However, these therapeutic sequences are unstable in a complicated *in vivo* environment; thus, it is urgent to find suitable delivery systems for these sequences. DONs can easily extend the length of DNA sequences to ensure a connection with exogenous nucleic acid strands, thereby protecting nucleic acid strands and targeting tissues. Schuller et al. [65] utilized DONs to deliver CpG sequences, resulting in a strong immune response and fewer side effects. Rahman et al. [66] reported DNA bricks modified with siRNA that could knock down the antiapoptotic protein Bcl-2. These results indicated that DNA bricks could stabilize siRNA and effectively inhibit the growth of lung tumors.

Aptamers are short, simple-stranded DNA or RNA oligonucleotides that specifically bind to target molecules [67]. Aptamers tend to fold and form stable three-dimensional structures through base pairing, electrostatic interactions, or hydrogen bonding, and they can wrap around target molecules with high affinity. To obtain a high uptake efficiency of DONs, aptamers tend to be extended to certain strands of DNA nanostructures. Song et al. [68] modified DOX and AuNRs on DONs to kill drug-resistant cancer cells, and the tumor-specific aptamer MUC-1 was also used with this multifunctional DNA nanostructure, which significantly enhanced the internalization of the nanostructure by cancer cells. Schaffert et al. [69] synthesized a two-dimensional DNA nanostructure with the modified transport protein transferrin (TF). The results showed that transferrin improved the cellular uptake of the DNA nanostructures, and the internalization efficiency was influenced by the number of modified transferrins.

In recent years, multidrug resistance has become a major obstacle to cancer therapy. To solve this problem, researchers tend to combine multiple chemotherapeutic drugs in the same nano-delivery system, which is called combination therapy. Liu et al. [70] designed a multifunctional DONs to deliver DOX and shRNA to destroy multidrug-resistant breast cancer cells (MCF-7R), and they found that the application of shRNA successfully reversed drug resistance in MCF-7R cells, which led to an enhanced therapeutic effect of DOX. Pan et al. [71] established a versatile DNA origami platform to combine chemotherapy with RNA. In this study, two different antisense oligonucleotides (ASOs) were used to target P-glycoprotein (P-gp) and Bcl-2. Meanwhile, they connected the aptamer MUC-1 to the surface of the DNA platform to target cancer cells. These results demonstrated that this DNA platform increased the therapeutic effect of DOX with the assistance of ASOs and aptamers. These studies indicated that DONs perform well in drug delivery and combination therapy.

In addition to drug delivery, DONs are promising platforms for cancer detection [72]. With the rise in cancer incidence and the development of tumor-targeting therapies, the early diagnosis of cancer has become increasingly crucial. Various DNA origami detectors have been constructed for cellular imaging by modification with fluorescent molecules. For example, Shen et al. [73] incorporated biscyanine into DNA nanotubes to visualize live cells directly, and Samanta et al. [74] combined DNA origami with infrared-emitting QDs for biophotonics applications.

7.2.3 Dynamic DNA Nanostructure

Although tFNAs and DNA origami have demonstrated promising prospects in the fields of drug delivery and cancer therapy, the complicated tumor microenvironment places higher requirements on drug delivery systems. In recent years, dynamic DNA nanostructures have attracted considerable attention from researchers; Table 7.1 shows the timeline of the representative advances in the field of dynamic DNA nanostructures. Based on nucleic acids, proteins, and some physical factors, such as light, temperature, and pH, a range of stimuli-responsive dynamic DNA nanostructures have been designed and applied in various biomedical fields. In addition, dynamic DNA nanostructures can serve as special modifications in other nanomaterials. Kahn et al. [88] combined dynamic DNA nanostructures with polyacrylamide to construct an environmentally sensitive hydrogel.

Considering that single strands of DNA form double-stranded structures through hydrogen bonds, DNA nanostructures are susceptible to changes in pH or ion concentration. Moreover, a wide range of ions have been reported to be media that can be inserted into DNA strands and can induce the formation of complicated structures [89, 90]. In addition, some DNA strands are easily affected by environmental pH. For instance, the i-motif sequence includes abundant cytosine and tends to maintain a linear strand in a neutral environment. However, as the pH decreases, the i-motif folds and forms a secondary structure [91]. Meanwhile, the guanine-rich sequence, which is paired with the i-motif, can construct a G-quadruplex in the same plane [92]. Therefore, the i-motif and G-quadruplex are often used as switch structures. Keum et al. [93] developed a pH-sensitive DNA nanocarrier to control the release of proteins based on the i-motif sequence. Park et al. [94] utilized the i-motif and G-quadruplex to deliver DOX and realize a triggered release. Moreover, they combined photodynamic and photothermal therapy with chemotherapy, which led to enhanced antitumor effects and fewer toxic side effects.

Aptamers bind tightly to target molecules by folding into secondary structures, such as G-quadruplexes and DNA loops, and the target molecules are mainly proteins, including cytokines, kinases, cell adhesion factors, and cell surface receptors. Through their combination with certain proteins, aptamers can perform various biological functions. Researchers believe that proteins cannot approach other biomolecules after binding with aptamers, which inhibits the usual effects of the proteins. AS1411 is a classic G-quadruplex aptamer targeting the surface receptors of cancer cells (nucleolin), and reports have indicated that AS1411 can inhibit the growth of cancer cells, especially blood cancer cells [95]. Considering the multiple functions of aptamers, they have been widely applied in diverse DNA nano-delivery systems [96]. In addition to promoting the target efficacy and therapeutic effect of nanomedicine, aptamers can serve as switches in dynamic DNA nanostructures. Tian et al. [30] constructed a nucleolin-triggered dynamic DNA tetrahedron based on the AS1411 switch (Figure 7.3). This tetrahedron remains intact in a nucleolin-free environment, and when it approaches cancer cells, the AS1411 strand separates and binds to nucleolin, resulting in structural disintegration and drug release. This design significantly reduces the damage to normal tissues caused by chemotherapy drugs.

The basic unit of the DNA double strands is a paired base bound by hydrogen bonds. However, by regulating the affinity of different single DNA strands, competitive binding of these DNA strands is expected, which will lead to dynamic sequence replacement. The replaced

Table 7.1 Timeline of the representative advances in the field of dynamic DNA nanostructures.

Year	Dynamic DNA nanostructures	Year	Dynamic DNA nanostructures
1999	Paired double crossover structure [75]	2015	DNA-based hinges, sliders, and hybrid mechanisms [52]
2001	Toehold-mediated strand displacement (TMSD)-based DNA tweezers [76]	2016	Rotary device made from multiple tightly fitted components [77]
2003	pH-Based DNA nanostructure [78]	2018	Gold nanocrystal-mediated slider [79]
2005	DNA walker [80]	2019	Molecular algorithm that executes logic and outputs literal Arabic numerals [81]
2007	DNA box actuated by TMSD [82]	2019	DNA box actuated by changing pH [83]
2011	DNA beacon, which detects biomolecules [84]	2019	Thermally actuated nanovalve [85]
2012	Logic-gated DNA box, which displays molecular cargo [86]	2019	Self-regulating DNA nanotubes [87]
2015	DNA structure connected with shape complementarity [51]	2021	Robotic nanobee [30]

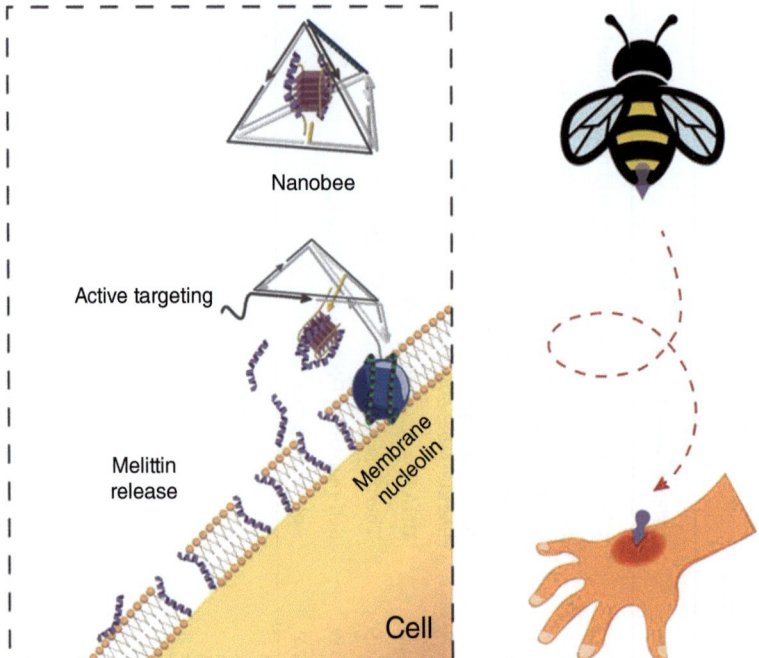

Figure 7.3 A nucleolin-triggered dynamic DNA tetrahedron based on AS1411 to reduce the damage to normal tissues caused by chemotherapy drugs. Source: Adapted with permission from [62]. Copyright 2021, American Chemical Society, Copyright Wiley-VCH.

single strand further induces the replacement of other DNA strands [97]. This theory is based on various DNA reactions, such as hybridization chain reactions (HCRs), catalytic DNA hairpins (CDHs), and catalytic DNA circuits (CDCs). A hairpin-based HCR is a classic example of a DNA replacement reaction. Traditionally, hairpin-based HCRs include two DNA hairpins and an initiator. The initiator induces Hairpin-1 to release a low-affiliative DNA strand, which demonstrates a high affinity to Hairpin-2 and initiates strand displacement of Hairpin-2. Subsequently, the redundant strand from Hairpin-2 further triggers the displacement reaction of Hairpin-1 [98, 99]. Through this cascading reaction, an infinitesimal initiated sequence can be amplified for detection, which builds a theoretical foundation for the use of DNA biosensors to detect proteins, small molecules, and nucleic acids [100]. In addition, dynamic DNA nanostructures demonstrate several merits as biosensors, such as their outstanding biocompatibility and cellular internalization, high structural editability, and diverse detection targets. Based on HCR, Zhang et al. [101] constructed a DNA tetrahedron modified with DNA hairpins and showed that the DNA tetrahedron could rapidly sense tumor cells by amplifying the number of receptors on the cell membrane.

With the development of DNA-editing technology, DNA hydrogels have become a research hotspot in recent years. DNA hydrogels can be formed by covalent interactions between linear or polymeric DNA molecules, which is termed chemical crosslinking. Through chemical crosslinking, DNA hydrogels tend to be irreversible and highly stable in physiological environments [102]. Moreover, physical crosslinking could also promote DNA gelation through hydrogen bonding, electrostatic interactions, or DNA-metal coordination. Compared to chemical crosslinking, physically crosslinked DNA hydrogels are reversible and demonstrate better degradability, biocompatibility, and biosecurity [103]. In addition, by inducing trigger-responsive units, DNA hydrogels have become smart and show prospects in drug delivery, tissue engineering, and gene editing. Ding et al. [104] reported an antitumor DNA hydrogel with embedded siRNA, and it could achieve self-assembly by grafting DNA strands onto polycaprolactone. These results indicated that this hydrogel structure protected siRNA from RNase degradation and did not disturb the function of RNA interference.

7.3 Conclusion

Nucleic acid nanomaterials have become an important research direction for tumor therapy, and various DNA structures have been considered potential drug carriers because of their good biocompatibility, excellent editability, and easy internalization. However, the comprehensive application of nucleic acid nanomaterials for tumor treatment as a drug delivery system still faces huge challenges, including the (i) instability of synthesis cost and high error rate of self-assembly, (ii) various biomedical effects that are not yet clear, and (iii) biosafety concerns arising from the uncertain *in vivo* stability, pharmacokinetic properties, and long-term cytotoxicity. Solving these challenges will greatly enhance the further application of nucleic acid nanomaterials in the field of tumor therapy.

References

1 Ferlay, J., Colombet, M., Soerjomataram, I. et al. (2021). Cancer statistics for the year 2020: an overview. *Int. J. Cancer* 149 (4): 778–789.
2 Siegel, R.L., Miller, K.D., Fuchs, H.E., and Jemal, A. (2021). Cancer statistics, 2021. *CA Cancer J. Clin.* 71 (1): 7–33.
3 Siegel, R.L., Miller, K.D., Fuchs, H.E., and Jemal, A. (2022). Cancer statistics, 2022. *CA Cancer J. Clin.* 72 (1): 7–33.
4 Hu, Q., Li, H., Wang, L. et al. (2019). DNA nanotechnology-enabled drug delivery systems. *Chem. Rev.* 119 (10): 6459–6506.
5 Xu, X., Ho, W., Zhang, X. et al. (2015). Cancer nanomedicine: from targeted delivery to combination therapy. *Trends Mol. Med.* 21 (4): 223–232.
6 Xiao, K., Liu, J., Chen, H. et al. (2017). A label-free and high-efficient GO-based aptasensor for cancer cells based on cyclic enzymatic signal amplification. *Biosens. Bioelectron.* 91: 76–81.
7 Avci-Adali, M., Wilhelm, N., Perle, N. et al. (2013). Absolute quantification of cell-bound DNA aptamers during SELEX. *Nucleic Acid Ther.* 23 (2): 125–130.
8 Jiang, Q., Shi, Y., Zhang, Q. et al. (2015). A self-assembled DNA origami-gold nanorod complex for cancer theranostics. *Small* 11 (38): 5134–5141.
9 Du, Y., Jiang, Q., Beziere, N. et al. (2016). DNA-nanostructure–gold-nanorod hybrids for enhanced in vivo optoacoustic imaging and photothermal therapy. *Adv. Mater.* 28 (45): 10000–10007.
10 Xiao, D., Li, Y., Tian, T. et al. (2021). Tetrahedral framework nucleic acids loaded with aptamer AS1411 for siRNA delivery and gene silencing in malignant melanoma. *ACS Appl. Mater. Interfaces* 13 (5): 6109–6118.

11 Seeman, N.C. (1982). Nucleic acid junctions and lattices. *J. Theor. Biol.* 99 (2): 237–247.
12 Rothemund, P.W.K. (2006). Folding DNA to create nanoscale shapes and patterns. *Nature* 440 (7082): 297–302.
13 Douglas, S.M., Marblestone, A.H., Teerapittayanon, S. et al. (2009). Rapid prototyping of 3D DNA-origami shapes with caDNAno. *Nucleic Acids Res.* 37 (15): 5001–5006.
14 Reshetnikov, R.V., Stolyarova, A.V., Zalevsky, A.O. et al. (2018). A coarse-grained model for DNA origami. *Nucleic Acids Res.* 46 (3): 1102–1112.
15 Linko, V. and Kostiainen, M.A. (2016). Automated design of DNA origami. *Nat. Biotechnol.* 34 (8): 826–827.
16 Dietz, H., Douglas Shawn, M., and Shih William, M. (2009). Folding DNA into twisted and curved nanoscale shapes. *Science* 325 (5941): 725–730.
17 Zhang, T., Hartl, C., Frank, K. et al. (2018). 3D DNA origami crystals. *Adv. Mater.* 30 (28): 1800273.
18 Ge, Z., Gu, H., Li, Q., and Fan, C. (2018). Concept and development of framework nucleic acids. *JACS* 140 (51): 17808–17819.
19 Zhang, T., Tian, T., Zhou, R. et al. (2020). Design, fabrication and applications of tetrahedral DNA nanostructure-based multifunctional complexes in drug delivery and biomedical treatment. *Nat. Protoc.* 15 (8): 2728–2757.
20 Zhang, T., Cui, W., Tian, T. et al. (2020). Progress in biomedical applications of tetrahedral framework nucleic acid-based functional systems. *ACS Appl. Mater. Interfaces* 12 (42): 47115–47126.
21 Li, S., Tian, T., Zhang, T. et al. (2019). Advances in biological applications of self-assembled DNA tetrahedral nanostructures. *Mater. Today* 24: 57–68.
22 Zhang, Q., Lin, S., Wang, L. et al. (2021). Tetrahedral framework nucleic acids act as antioxidants in acute kidney injury treatment. *Chem. Eng. J.* 413: 127426.
23 Zhu, J., Zhang, M., Gao, Y. et al. (2020). Tetrahedral framework nucleic acids promote scarless healing of cutaneous wounds via the AKT-signaling pathway. *Signal Transduction Targeted Ther.* 5 (1): 120.
24 Kim, K.-R., Kim, D.-R., Lee, T. et al. (2013). Drug delivery by a self-assembled DNA tetrahedron for overcoming drug resistance in breast cancer cells. *Chem. Commun.* 49 (20): 2010–2012.
25 Zhang, J., Guo, Y., Ding, F. et al. (2019). A camptothecin-grafted DNA tetrahedron as a precise nanomedicine to inhibit tumor growth. *Angew. Chem. Int. Ed.* 58 (39): 13794–13798.
26 Jorge, A.F., Aviñó, A., Pais, A.A.C.C. et al. (2018). DNA-based nanoscaffolds as vehicles for 5-fluoro-2′-deoxyuridine oligomers in colorectal cancer therapy. *Nanoscale* 10 (15): 7238–7249.
27 Zhang, Y., Deng, Y., Wang, C. et al. (2019). Probing and regulating the activity of cellular enzymes by using DNA tetrahedron nanostructures. *Chem. Sci.* 10 (23): 5959–5966.
28 Lin, Q., Cai, S., Zhou, B. et al. (2021). Dual-MicroRNA-regulation of singlet oxygen generation by a DNA-tetrahedron-based molecular logic device. *Chem. Commun.* 57 (32): 3873–3876.
29 Juul, S., Iacovelli, F., Falconi, M. et al. (2013). Temperature-controlled encapsulation and release of an active enzyme in the cavity of a self-assembled DNA nanocage. *ACS Nano* 7 (11): 9724–9734.

30 Tian, T.R., Xiao, D.X., Zhang, T. et al. (2021). A framework nucleic acid based robotic nanobee for active targeting therapy. *Adv. Funct. Mater.* 31 (5).

31 Fu, W., Ma, L., Ju, Y. et al. (2021). Therapeutic siCCR2 loaded by tetrahedral framework DNA nanorobotics in therapy for intracranial hemorrhage. *Adv. Funct. Mater.* 31 (33): 2101435.

32 Cui, W., Yang, X., Chen, X. et al. (2021). Treating LRRK2-related Parkinson's disease by inhibiting the mTOR signaling pathway to restore autophagy. *Adv. Funct. Mater.* 31 (38): 2105152.

33 Kim, K.-R., Kim, H.Y., Lee, Y.-D. et al. (2016). Self-assembled mirror DNA nanostructures for tumor-specific delivery of anticancer drugs. *J. Controlled Release* 243: 121–131.

34 Krishna, A.G., Kumar, D.V., Khan, B.M. et al. (1998). Taxol–DNA interactions: fluorescence and CD studies of DNA groove binding properties of taxol. *Biochim. Biophys. Acta, Gen. Subj.* 1381 (1): 104–112.

35 Rusak, G., Piantanida, I., Mašić, L. et al. (2010). Spectrophotometric analysis of flavonoid-DNA interactions and DNA damaging/protecting and cytotoxic potential of flavonoids in human peripheral blood lymphocytes. *Chem. Biol. Interact.* 188 (1): 181–189.

36 Zhang, T., Tian, T., and Lin, Y. (2021). Functionalizing framework nucleic-acid-based nanostructures for biomedical application. *Adv. Mater.* 2107820.

37 Xie, X., Shao, X., Ma, W. et al. (2018). Overcoming drug-resistant lung cancer by paclitaxel loaded tetrahedral DNA nanostructures. *Nanoscale* 10 (12): 5457–5465.

38 Ma, W., Zhan, Y., Zhang, Y. et al. (2019). An intelligent DNA nanorobot with in vitro enhanced protein lysosomal degradation of HER2. *Nano Lett.* 19 (7): 4505–4517.

39 Zhan, Y., Ma, W., Zhang, Y. et al. (2019). DNA-based nanomedicine with targeting and enhancement of therapeutic efficacy of breast cancer cells. *ACS Appl. Mater. Interfaces* 11 (17): 15354–15365.

40 Li, S., Sun, Y., Tian, T. et al. (2020). MicroRNA-214-3p modified tetrahedral framework nucleic acids target survivin to induce tumour cell apoptosis. *Cell Proliferation* 53 (1): e12708.

41 Lee, H., Lytton-Jean, A.K.R., Chen, Y. et al. (2012). Molecularly self-assembled nucleic acid nanoparticles for targeted in vivo siRNA delivery. *Nat. Nanotechnol.* 7 (6): 389–393.

42 Liu, X., Xu, Y., Yu, T. et al. (2012). A DNA nanostructure platform for directed assembly of synthetic vaccines. *Nano Lett.* 12 (8): 4254–4259.

43 Qin, X., Xiao, L., Li, N. et al. (2022). Tetrahedral framework nucleic acids-based delivery of microRNA-155 inhibits choroidal neovascularization by regulating the polarization of macrophages. *Bioact. Mater.* 14: 134–144.

44 Kim, K.-R., Bang, D., and Ahn, D.-R. (2016). Nano-formulation of a photosensitizer using a DNA tetrahedron and its potential for in vivo photodynamic therapy. *Biomater. Sci.* 4 (4): 605–609.

45 Yan, J., Zhan, X., Zhang, Z. et al. (2021). Tetrahedral DNA nanostructures for effective treatment of cancer: advances and prospects. *J. Nanobiotechnol.* 19 (1): 412.

46 Ren, T., Deng, Z., Liu, H. et al. (2019). Co-delivery of DNAzyme and a chemotherapy drug using a DNA tetrahedron for enhanced anticancer therapy through synergistic effects. *New J. Chem.* 43 (35): 14020–14027.

47 Wang, S., Liu, Z., Tong, Y. et al. (2021). Improved cancer phototheranostic efficacy of hydrophobic IR780 via parenteral route by association with tetrahedral nanostructured DNA. *J. Controlled Release* 330: 483–492.

48 Douglas, S.M., Chou, J.J., and Shih, W.M. (2007). DNA-nanotube-induced alignment of membrane proteins for NMR structure determination. *PNAS* 104 (16): 6644–6648.

49 Iinuma, R., Ke, Y., Jungmann, R. et al. (2014). Polyhedra self-assembled from DNA tripods and characterized with 3D DNA-PAINT. *Science* 344 (6179): 65–69.

50 Benson, E., Mohammed, A., Gardell, J. et al. (2015). DNA rendering of polyhedral meshes at the nanoscale. *Nature* 523 (7561): 441–444.

51 Gerling, T., Wagenbauer Klaus, F., Neuner Andrea, M., and Dietz, H. (2015). Dynamic DNA devices and assemblies formed by shape-complementary, non–base pairing 3D components. *Science* 347 (6229): 1446–1452.

52 Marras Alexander, E., Zhou, L., Su, H.-J., and Castro Carlos, E. (2015). Programmable motion of DNA origami mechanisms. *Proc. Natl. Acad. Sci.* 112 (3): 713–718.

53 Matsumura, Y. and Maeda, H. (1986). A new concept for macromolecular therapeutics in cancer chemotherapy: mechanism of tumoritropic accumulation of proteins and the antitumor agent smancs. *Cancer Res.* 46 (12 Pt 1): 6387–6392.

54 Brigger, I., Dubernet, C., and Couvreur, P. (2002). Nanoparticles in cancer therapy and diagnosis. *Adv. Drug Delivery Rev.* 54 (5): 631–651.

55 Acharya, S. and Sahoo, S.K. (2011). PLGA nanoparticles containing various anticancer agents and tumour delivery by EPR effect. *Adv. Drug Delivery Rev.* 63 (3): 170–183.

56 Mishra, S., Feng, Y., Endo, M., and Sugiyama, H. (2020). Advances in DNA origami-cell interfaces. *ChemBioChem* 21 (1-2): 33–44.

57 Mei, Q., Wei, X., Su, F. et al. (2011). Stability of DNA origami nanoarrays in cell lysate. *Nano Lett.* 11 (4): 1477–1482.

58 Perrault, S.D. and Shih, W.M. (2014). Virus-inspired membrane encapsulation of DNA nanostructures to achieve in vivo stability. *ACS Nano* 8 (5): 5132–5140.

59 Chopra, A., Krishnan, S., and Simmel, F.C. (2016). Electrotransfection of polyamine folded DNA origami structures. *Nano Lett.* 16 (10): 6683–6690.

60 Ponnuswamy, N., Bastings, M.M.C., Nathwani, B. et al. (2017). Oligolysine-based coating protects DNA nanostructures from low-salt denaturation and nuclease degradation. *Nat. Commun.* 8: 15654.

61 Zhao, Y.-X., Shaw, A., Zeng, X. et al. (2012). DNA origami delivery system for cancer therapy with tunable release properties. *ACS Nano* 6 (10): 8684–8691.

62 Jiang, Q., Song, C., Nangreave, J. et al. (2012). DNA origami as a carrier for circumvention of drug resistance. *JACS* 134 (32): 13396–13403.

63 Zhang, Q., Jiang, Q., Li, N. et al. (2014). DNA origami as an in vivo drug delivery vehicle for cancer therapy. *ACS Nano* 8 (7): 6633–6643.

64 Vollmer, J. and Krieg, A.M. (2009). Immunotherapeutic applications of CpG oligodeoxynucleotide TLR9 agonists. *Adv. Drug Delivery Rev.* 61 (3): 195–204.

65 Schüller, V.J., Heidegger, S., Sandholzer, N. et al. (2011). Cellular immunostimulation by CpG-sequence-coated DNA origami structures. *ACS Nano* 5 (12): 9696–9702.

66 Rahman, M.A., Wang, P., Zhao, Z. et al. (2017). Systemic delivery of Bcl2-targeting siRNA by DNA nanoparticles suppresses cancer cell growth. *Angew. Chem. Int. Ed. Engl.* 56 (50): 16023–16027.

67 Jayasena, S.D. (1999). Aptamers: an emerging class of molecules that rival antibodies in diagnostics. *Clin. Chem.* 45 (9): 1628–1650.

68 Song, L., Jiang, Q., Liu, J. et al. (2017). DNA origami/gold nanorod hybrid nanostructures for the circumvention of drug resistance. *Nanoscale* 9 (23): 7750–7754.

69 Schaffert, D.H., Okholm, A.H., Sørensen, R.S. et al. (2016). Intracellular delivery of a planar DNA origami structure by the transferrin-receptor internalization pathway. *Small* 12 (19): 2634–2640.

70 Liu, J., Song, L., Liu, S. et al. (2018). A tailored DNA nanoplatform for synergistic RNAi-/chemotherapy of multidrug-resistant tumors. *Angew. Chem. Int. Ed. Engl.* 57 (47): 15486–15490.

71 Pan, Q., Nie, C., Hu, Y. et al. (2020). Aptamer-functionalized DNA origami for targeted codelivery of antisense oligonucleotides and doxorubicin to enhance therapy in drug-resistant cancer cells. *ACS Appl. Mater. Interfaces* 12 (1): 400–409.

72 Li, C. (2014). A targeted approach to cancer imaging and therapy. *Nat. Mater.* 13 (2): 110–115.

73 Shen, X., Jiang, Q., Wang, J. et al. (2012). Visualization of the intracellular location and stability of DNA origami with a label-free fluorescent probe. *Chem. Commun. (Camb.)* 48 (92): 11301–11303.

74 Samanta, A., Deng, Z., and Liu, Y. (2014). Infrared emitting quantum dots: DNA conjugation and DNA origami directed self-assembly. *Nanoscale* 6 (9): 4486–4490.

75 Mao, C., Sun, W., Shen, Z., and Seeman, N.C. (1999). A nanomechanical device based on the B–Z transition of DNA. *Nature* 397 (6715): 144–146.

76 Yurke, B., Turberfield, A.J., Mills, A.P. et al. (2000). A DNA-fuelled molecular machine made of DNA. *Nature* 406 (6796): 605–608.

77 Ketterer, P., Willner Elena, M., and Dietz, H. Nanoscale rotary apparatus formed from tight-fitting 3D DNA components. *Sci. Adv.* 2 (2): e1501209.

78 Liu, D. and Balasubramanian, S. (2003). A proton-fuelled DNA nanomachine. *Angew. Chem. Int. Ed.* 42 (46): 5734–5736.

79 Urban, M.J., Both, S., Zhou, C. et al. (2018). Gold nanocrystal-mediated sliding of doublet DNA origami filaments. *Nat. Commun.* 9 (1): 1454.

80 Shin, J.-S. and Pierce, N.A. (2004). A Synthetic DNA walker for molecular transport. *J. Am. Chem. Soc.* 126 (35): 10834–10835.

81 Woods, D., Doty, D., Myhrvold, C. et al. (2019). Diverse and robust molecular algorithms using reprogrammable DNA self-assembly. *Nature* 567 (7748): 366–372.

82 Andersen, E.S., Dong, M., Nielsen, M.M. et al. (2009). Self-assembly of a nanoscale DNA box with a controllable lid. *Nature* 459 (7243): 73–76.

83 Ijäs, H., Hakaste, I., Shen, B. et al. (2019). Reconfigurable DNA origami nanocapsule for pH-controlled encapsulation and display of cargo. *ACS Nano* 13 (5): 5959–5967.

84 Kuzuya, A., Sakai, Y., Yamazaki, T. et al. (2011). Nanomechanical DNA origami 'single-molecule beacons' directly imaged by atomic force microscopy. *Nat. Commun.* 2 (1): 449.

85 Arnott, P.M. and Howorka, S. (2019). A temperature-gated nanovalve self-assembled from DNA to control molecular transport across membranes. *ACS Nano* 13 (3): 3334–3340.

86 Douglas Shawn, M., Bachelet, I., and Church George, M. (2012). A logic-gated nanorobot for targeted transport of molecular payloads. *Science* 335 (6070): 831–834.

87 Green, L.N., Subramanian, H.K.K., Mardanlou, V. et al. (2019). Autonomous dynamic control of DNA nanostructure self-assembly. *Nat. Chem.* 11 (6): 510–520.

88 Kahn, J.S., Hu, Y., and Willner, I. (2017). Stimuli-responsive DNA-based hydrogels: from basic principles to applications. *Acc. Chem. Res.* 50 (4): 680–690.

89 Kohwi, Y. and Kohwi-Shigematsu, T. (1988). Magnesium ion-dependent triple-helix structure formed by homopurine-homopyrimidine sequences in supercoiled plasmid DNA. *Proc. Natl. Acad. Sci. U.S.A.* 85 (11): 3781–3785.

90 Wu, Y.-Y., Zhang, Z.-L., Zhang, J.-S. et al. (2015). Multivalent ion-mediated nucleic acid helix-helix interactions: RNA versus DNA. *Nucleic Acids Res.* 43 (12): 6156–6165.

91 Abou Assi, H., Garavís, M., González, C., and Damha, M.J. (2018). i-Motif DNA: structural features and significance to cell biology. *Nucleic Acids Res.* 46 (16): 8038–8056.

92 Chu, B., Zhang, D., and Paukstelis, P.J. (2019). A DNA G-quadruplex/i-motif hybrid. *Nucleic Acids Res.* 47 (22): 11921–11930.

93 Keum, J.-W. and Bermudez, H. (2012). DNA-based delivery vehicles: pH-controlled disassembly and cargo release. *Chem. Commun. (Camb.)* 48 (99): 12118–12120.

94 Park, H., Kim, J., Jung, S., and Kim, W.J. (2018). DNA-Au nanomachine equipped with i-Motif and G-quadruplex for triple combinatorial anti-tumor therapy. *Adv. Funct. Mater.* 28 (5): 1705416.

95 Keefe, A.D., Pai, S., and Ellington, A. (2010). Aptamers as therapeutics. *Nat. Rev. Drug Discovery* 9 (7): 537–550.

96 Bunka, D.H.J., Platonova, O., and Stockley, P.G. (2010). Development of aptamer therapeutics. *Curr. Opin. Pharmacol.* 10 (5): 557–562.

97 Qian, L. and Winfree, E. (2011). Scaling up digital circuit computation with DNA strand displacement cascades. *Science* 332 (6034): 1196–1201.

98 Liu, L., Rong, Q., Ke, G. et al. (2019). Efficient and reliable microRNA imaging in living cells via a FRET-based localized hairpin-DNA cascade amplifier. *Anal. Chem.* 91 (5): 3675–3680.

99 Dirks, R.M. and Pierce, N.A. (2004). Triggered amplification by hybridization chain reaction. *PNAS* 101 (43): 15275–15278.

100 Tang, J., Lei, Y., He, X. et al. (2020). Recognition-driven remodeling of dual-split aptamer triggering in situ hybridization chain reaction for activatable and autonomous identification of cancer cells. *Anal. Chem.* 92 (15): 10839–10846.

101 Zhang, B.W., Tian, T.R., Xiao, D.X. et al. Facilitating in situ tumor imaging with a tetrahedral DNA framework-enhanced hybridization chain reaction probe. *Adv. Funct. Mater.*

102 Shahbazi, M.-A., Bauleth-Ramos, T., and Santos, H.A. (2018). DNA hydrogel assemblies: bridging synthesis principles to biomedical applications. *Adv. Ther.* 1 (4): 1800042.

103 Hu, W.K., Wang, Z.J., Xiao, Y. et al. (2019). Advances in crosslinking strategies of biomedical hydrogels. *Biomater. Sci.* 7 (3): 843–855.

104 Ding, F., Mou, Q.B., Ma, Y. et al. (2018). A crosslinked nucleic acid nanogel for effective siRNA delivery and antitumor therapy. *Angew. Chem. Int. Ed.* 57 (12): 3064–3068.

8

Application of Framework Nucleic Acid-Based Nanomaterials in the Treatment of Endocrine and Metabolic Diseases

Jingang Xiao

Southwest Medical University, Stomatological Hospital of Southwest Medical University, Zhongshan Road, Luzhou 646000, PR China

Since the pioneering days of nucleic acid nanotechnology, the artificial encoding and production of nucleic acid materials have become increasingly sophisticated and have shown great potential in various biological fields [1]. From simple single-stranded nucleic acid structures to nucleic acid nanostructures with complex spatial configurations, nucleic acid nanomaterials exhibit strikingly different physical–chemical properties [2, 3]. Currently, nucleic acid nanomaterials have been used in numerous fields such as biosensing, bioimaging, targeted delivery, and disease treatment to promote human health [4–6]. The intractability and complexity of endocrine and metabolic diseases is a challenge that has plagued many medical researchers. Therefore, researchers have introduced nucleic acid nanomaterials into this field and have obtained numerous research results. In this paper, we focus on the progress of research on nucleic acid nanomaterials in common endocrine and metabolic diseases such as diabetes, osteoporosis (OP), obesity, and nonalcoholic fatty liver disease in this paper (Figure 8.1).

8.1 Endocrine and Metabolic Diseases

Metabolic diseases commonly refer to the disorders of one or more links of metabolism, and metabolic disorders mainly caused by primary organ diseases are included in the category of organ diseases (such as endocrine gland diseases) [7]. From the perspective of molecular etiology, there is no clear boundary between some metabolic diseases and endocrine diseases [8]. Obesity, for example, is an energy metabolic illness caused by a disturbance of fat metabolism; nevertheless, the reasons of obesity are nearly entirely connected to the action of hormones that regulate substance metabolism and their receptors, such as insulin, epinephrine, and leptin [9]. Lesions of most endocrine organs lead to disorders of the metabolic system, which in turn lead to the development of disease. The metabolism involving a variety of substances such as sugars, fats, and proteins is both regulated by the relevant hormones and shows different phenotypic differences in response to environmental factors.

Nucleic Acid-Based Nanomaterials: Stabilities and Applications, First Edition.
Edited by Yunfeng Lin and Shaojingya Gao.
© 2024 WILEY-VCH GmbH. Published 2024 by WILEY-VCH GmbH.

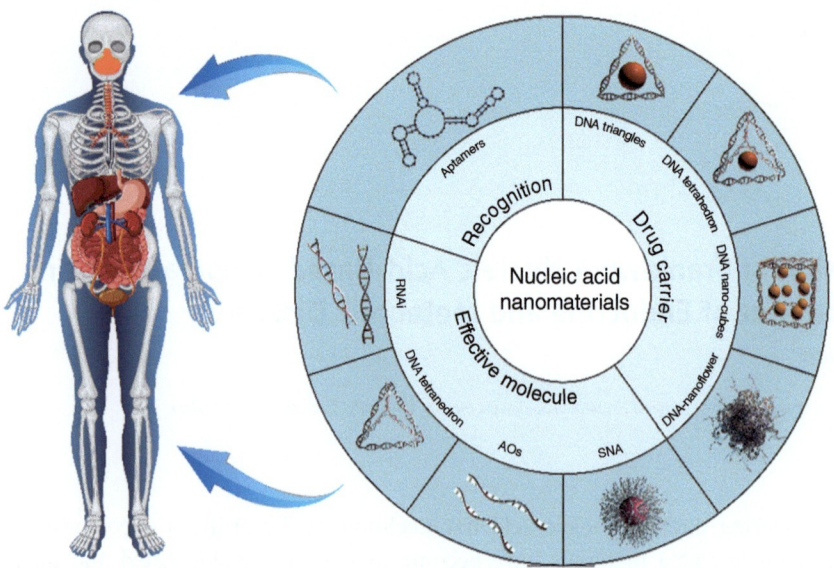

Figure 8.1 Nucleic acid nanomaterials with different functional components in drugs.

In recent decades, endocrine and metabolic diseases such as obesity, diabetes, and OP have become progressively more prevalent and pose a serious threat to human health [10]. Endocrine and metabolic illnesses can create serious complications and terrible pain for individuals if they are not treated swiftly and efficiently. For instance, diabetes causes many complications that can damage almost all vital organs (kidneys, heart, brain, and lungs) as well as the eyes, peripheral nerves, and feet [11, 12]. The dynamic balance of bone formation and bone resorption is maintained after adulthood. If there are factors that cause a disruption of this balance, such as estrogen deficiency and elevated glucocorticoid levels, it can lead to bone loss and the development of OP [13]. OP is a common systemic skeletal disease characterized by damage to the microstructure of bone tissue and a loss of bone mass, which can increase the risk of bone fragility and fracture [14]. Obesity is a risk factor for chronic noncommunicable diseases such as hypertension, diabetes, cardiovascular disease, and tumors, as well as a pathological underpinning for them [15]. In addition, Nonalcoholic fatty liver disease is becoming a major chronic liver disease worldwide and can cause end-stage liver diseases such as cirrhosis and liver cancer. Due to its complex pathogenesis, there are no effective therapeutic drugs available. The prevalence of NAFLD in adults in China is also increasing year by year and was close to 30% in 2018 [16]. NAFLD-related advanced liver disease and its mortality along with its overall disease burden are expected to increase substantially [17]. Meanwhile, metabolic syndrome (Met S), which includes obesity, insulin resistance, glucose intolerance, hypertension, and dyslipidemia, has emerged as a serious health hazard in the modern world [18]. Currently, pharmacological therapy is the primary therapeutic option for endocrine and metabolic illnesses, with virtually all patients requiring long-term medication or injections [19, 20]. Although a variety of medications are applied in the therapeutic treatment of endocrine and metabolic illnesses, there are still numerous obstacles to overcome. As a consequence, discovering new therapeutic agents and carriers is essential for the treatment of endocrine and metabolic illnesses.

8.2 Nucleic Acid Nanomaterials

Nucleic acids, as the most significant component of life, carry the most vital genetic information in nature and are essential to an organism's growth, development, and reproduction. The discovery of the principle of complementary base pairing and the double helix structure of DNA ushered in the era of molecular biology, with the result that the study of DNA has expanded from genetic function to the molecular level [1]. On the basis of this, continued improvements and innovations in the design of nucleic acid nanomaterials have facilitated the emergence of numerous novel nanostructures [2]. As scientific research on the subject of nucleic acids has progressed, the variety and complexity of their functions have been astonishing, with a wide range of applications in biosensing [21, 22], bioimaging [23], targeted delivery [24], and disease therapy [25].

Since the emergence of the family of nongenetic nucleic acids, the concept of functional nucleic acids (FNAs) has evolved. FNAs, such as DNA enzymes, aptamers, DNA tiles, and DNA origami [26], are a type of nucleic acid molecules and analogs that have distinct structural functions and can perform unique biological nongenetic functions (Figure 8.1). With the development of nucleic acid nanotechnology, scientists can create nanostructures with well-defined shapes and sizes that are programmable, addressable, and biocompatible [27, 28]. Simple nucleic acid nanostructures such as cubes, stars, and three-dimensional geometries were initially able to be synthesized [29]. With further exploration of these potential nanostructures, nucleic acid nanomaterials have given rise to multi-component nanocomplexes, including flexible linkages and combinations with therapeutic molecules [30–33]. Artificial DNA nanostructures exhibit high structural potential for engineering intracellular and *in vivo* biological processes [34]. Nucleic acid nanoparticles' three-dimensional space makes them ideal for encapsulating various therapeutic or diagnosis agents as well as adding functional agents on surfaces [35].

In the field of endocrine and metabolic diseases, there have been many studies on the applications of aptamers and DNA polyhedrons. The laboratory results demonstrate the high identification and drug delivery capability of nucleic acid nanostructures in endocrine and metabolic disorders, demonstrating the possibilities and research prospects of nucleic acid nanomaterials. The potential of DNA nanostructures for drug delivery and targeted therapeutics has become apparent in recent decades. As drug carriers, DNA polyhedrons possess the following typical properties: excellent biocompatibility and biostability, feasibility and convenience of cargo loading, targeting specificity, ability to penetrate physical barriers, high cellular uptake efficiency, and low side effects [36].

8.3 Nucleic Acid and Drugs

Nucleic acid-based gene therapy is a hot spot in the biomedical field. Antisense, RNA interference (RNAi)-based therapeutic systems have been extensively discussed, and they correspond to a variety of nucleic acid drugs, including antisense oligonucleotides (AOs) [37, 38], small interfering RNA (siRNA) [39], and small hairpin RNA (shRNA) [40]. The therapeutic potential of siRNAs has been convincingly realized in preclinical models of various diseases. Therapeutic oligonucleotides can reduce the production of target proteins by inhibiting gene expression, thus avoiding a direct response to undruggable proteins.

Also, oligonucleotides selectively and specifically bind to mRNA transcripts, thus avoiding the inhibition of nontarget protein isoforms and reducing off-target effects [41]. Treatment of diabetic mice with lipid nanoparticles (LNPs) and leptin-encapsulated glucagon-siRNA (Gcgr-siRNA) is effective in reducing blood glucose levels [42]. Targeted silencing of the lipoprotein packaging enzymes ApoB and PSCK9 has been effective in animal studies of hyperlipidemia, significantly reducing serum LDL and total cholesterol levels [43, 44]. The use of siRNA to target and silence diacylglycerol O-acyltransferase 2 (DGAT2), a key enzyme in triglyceride synthesis, can prevent and reverse the accumulation of triglycerides, thereby significantly improving the fatty liver phenotype [45].

Natural small RNA is rapidly degraded by nucleases in the serum and is rapidly cleared by the kidneys and liver. Furthermore, biological membranes are impermeable to small RNAs, and, as a result, small RNA cannot readily enter the cell [46]. Even when injected directly into local sites, such as the vitreous humor of the eye, siRNA must escape nuclease activity and localize to the cytoplasm of the cell in order to function [47]. Systemic drug delivery of gene-related therapeutics remains a major challenge due to poor stability and limited cellular absorption. Gene drugs based on viral vectors have remarkably high transfection efficiency. However, several fundamental limitations of viral vectors, including carcinogenesis, immunogenicity, and mutagenesis, still need to be considered [48]. Nonviral vectors are expected to address these limitations, but improvements in pharmacodynamics and enhanced stability under physiological conditions are still required [49].

According to the most recent research, the DNA tetrahedral architecture conferred higher resistance to RNase and serum on the cargo siRNA and ensured its tight integration with the carrier during delivery, providing a new approach for siRNA transport [50]. By systemic administration, siRNA preferentially accumulates in the liver when delivered via DNA tetrahedron, suggesting that DNA tetrahedrons are a promising vehicle for liver-targeted delivery of therapeutic nucleic acids [51]. At present, a variety of nucleic acid drugs for endocrine and metabolic diseases have been researched and entered into clinical trials [52]. Simultaneously, nucleic acid nanomaterials have become an essential area of drug research and development and participate in the treatment of diseases.

Aptamers are oligonucleotides (DNA or RNA) that can be acquired by phylogenetic evolution of ligands *in vitro* by exponential enrichment (SELEX) [53]. Production begins with a large number of single-chained nucleic acid molecules, followed by iterative *in vitro* selection by challenging specific targets (Figure 8.1) [54]. After subsequent amplification by polymerase chain reaction, aptamers are obtained with high affinity and specificity for their targets. Aptamers help guide therapeutic molecule to identify specific organs or cells, culminating in drug enrichment. The pharmacokinetic profile of aptamers could be altered by chemical modification. Polyethylene glycol (PEG) or cholesterol can be added to the ends of the aptamers (the 2′-hydroxyl position on pyrimidines) and increase the half-life to several hours [55]. Aptamers can specifically bind targets ranging from small molecules to complex structures, making them suitable for diagnostic and therapeutic applications in many diseases. Due to the special 3D structure of aptamer, it is able to bind to target epitopes with high specificity, which is not possible with larger antibodies [56]. Therefore, aptamers have been used in the treatment of various diseases, such as cancer [57], blood diseases [58], and metabolic diseases [59]. From the results of the present study,

Table 8.1 Recent advances of nucleic acid nanomaterials in endocrine and metabolic diseases.

Type	Drugs	Diseases	Receptors	Target	Model	References
DNA nano-cubes	Vildagliptin	Diabetes Mellitus	N/A	Dipeptidyl-peptidase-4 (DPP4)	Db/Db-mice	[81]
DNA nano-cubes	Vildagliptin	Diabetes Mellitus	N/A	Dipeptidyl-peptidase-4 (DPP4)	N/A	[82]
DNA triangles	Vildagliptin	Diabetes Mellitus	N/A	Dipeptidyl-peptidase-4 (DPP4)	Db/Db-mice	[83]
DNA tetrahedron	N/A	Diabetes Mellitus	N/A	PI3K/Akt Pathway	HFD	[84]
DNA tetrahedron	Resveratrol	Diabetes Mellitus	N/A	Immune tolerance	HFD	[85]
DNA tetrahedron	N/A	Diabetes Mellitus	N/A	Immune tolerance	NOD	[86]
Aptamers	N/A	Diabetes Mellitus	N/A	CCL2/CCL2 receptor axis	N/A	[87]
DNA-RNA-Spiegelmer	N/A	Diabetes Mellitus	N/A	Glucagon	STZ	[88]
Aptamers	Plekho1 siRNA	Osteoporosis	Osteoblasts	Plekho1 (also known as casein kinase-2 interacting protein-1 (CKIP-1))	OVX rats	[89]
Aptamers	Vitamin D	Osteoporosis	Osteoblasts	N/A	Female rats	[90]
DNA cross-linking	Rho-inhibiting C3 toxin	Osteoporosis	N/A	Rho-mediated signal-transduction	N/A	[91]
DNA cross-linking	Dexamethasone	Osteoporosis	N/A	N/A	Female rats	[92]
Aptamers	N/A	Obesity	APMAP (Adipocyte Plasma Membrane Associated Protein)	N/A	HFD	[93]
Aptamers' DNA-nanoflower	Allicin	Obesity	APMAP (Adipocyte Plasma Membrane Associated Protein)	N/A	HFD	[94]
Aptamers	N/A	NAFLD	Membrane-bound CD36	Membrane-bound CD36	HFD	[95]
Spherical nucleic acids	TLR antagonist oligonucleotides	NAFLD	N/A	Immunoregulatory	STAM	[96]

HFD: high-fat diet mice; NOD: non-obese diabetic mice; STZ: Injection of the toxin streptozotocin; OVX: ovariectomy; STAM: high-fat diet after injection of the toxin streptozotoci.

it appears that therapeutic aptamers are progressing more slowly than their applications in the analytical and diagnostic fields.

DNA polyhedrons, particularly DNA tetrahedrons [60], have been reported to have a therapeutic impact on illnesses and may also be employed as a drug carrier to increase the permeability and retention rate of pharmaceuticals [61]. With DNA origami techniques were developed, resulting in the formation of multicomponent nanocomplexes with flexible connections and combinations of medicinal compounds [62, 63]. For example, by loading paclitaxel with DNA tetrahedron, anticancer medications could be transported into the cytoplasm and released, contributing to increased drug accumulation and relatively high efficacy while ameliorating drug resistance [61, 64]. In comparison to conventional DNA molecules, DNA nanostructures provide a number of unique features, including (i) a simple synthesis process [65]; (ii) excellent editability [2]; (iii) remarkable biocompatibility and biostability [66, 67]; (iv) significant internalization through endocytosis [68]; and (v) DNA polyhedron can be connected in a different way to tiny molecules, peptides, and nucleic acids [69, 70]. Thus, DNA polyhedrons possess a wide range of potential applications as nanocarriers in illness therapy and diagnosis [71, 72]. Additionally, nucleic acid nanomaterials have therapeutic properties of their own. For example, after the DNA tetrahedron therapy, blood levels of pancreatitis-related indicators dramatically decreased, alleviating the disease's subsequent multiorgan damage in mice [72]. Meanwhile, tetrahedral framework nucleic acids (tFNAs) show an excellent ability to the cross blood–brain barrier and have antiepileptic effects [73], showing potential therapeutic applications for neurological illnesses [74–76]. DNA tetrahedron also has positive effects on hepatocyte proliferation [77], skin injury healing [78], and promotion of angiogenesis [79, 80]. In summary, nucleic acid nanomaterials offer a wide range of possible applications, whether as a medicine or a carrier (Table 8.1).

8.4 Nucleic Acid Nanomaterials for Endocrine and Metabolic Diseases

8.4.1 Diabetes Mellitus

Diabetes mellitus (DM) is a chronic metabolic disorder characterized by hyperglycemia caused by defective insulin secretion, resistance to insulin action, or a combination of both [97]. So far, the primary treatment goal for diabetes is still the control of blood glucose levels [98]. Several oral hypoglycemic agents are commonly used, such as metformin, biguanides, glitazones, sulfonylureas, a-glucosidase inhibitors, glucagon-like peptide-1 (GLP1) receptor agonists, and dipeptidyl peptidase-4 inhibitors [99–101]. However, these drugs do not stop the progression of T2D due to insulin resistance, and at the same time, oral hypoglycemic drugs can cause adverse effects and affect patient compliance [102].

According to recent research from Lin et al. [84], tFNAs may play a potential therapeutic role in type 2 diabetes mellitus (T2DM) by increasing glucose synthetic glycogen and reducing insulin resistance (IR). Through the establishment of the mouse model of T2DM, they confirmed that tFNAs can increase glucose uptake, reduce blood glucose levels, improve insulin resistance, and alleviate type 2 diabetes (Figure 8.1). In glucosamine-stimulated

HepG2 cells, they employed PI3K inhibitors to further clarify that tFNAs reduced IR through PI3K/Akt pathway. A hallmark of type 1 diabetes is the immune system's assault and death of insulin-producingβ-cells [11]. In another of their research studies, treatment with tFNAs prevented the onset of diabetes in nonobese diabetic mice with prediabetes. Immune tolerance was achieved with the use of tFNAs, which retained insulin-producing β-cell mass and function and inhibited autoreactive T cells [86]. Similarly, tFNAs containing resveratrol to reduce inflammation increased insulin sensitivity and reduced insulin resistance in high-fat diet-fed (HFD) rats, as shown in another research [85].

Vildagliptin is an inhibitor of plasma DPP-IV activity. DPP-IV inhibitor drugs increase blood incretin levels while simultaneously inhibiting glucagon levels (Figure 8.2a) [83, 103]. As a core scaffold material to coat the oral drug formed with alginate and poly-L-lysine polymer, DNA nanocubes loaded with vildagliptin are used to provide effective resistance to gastric acid and increased drug access to the intestine, which is vildagliptin's site of action, resulting in improved control of blood glucose levels (Figure 8.2d) [81, 82].

Although DM therapies differ depending on subtype, the main objective is to regulate blood glucose levels. The proinflammatory chemokine C–C motif-ligand 2 (CCL2) is implicated in the development of insulin resistance, macrophage infiltration, and inflammation [104]. Emapticap pegol (NOX-E36) is a modified aptamer that binds and suppresses CCL2 with high affinity and specificity. Subcutaneous injection of NOX-E36 in patients with diabetic nephropathy decreased fasting blood glucose levels and glycosylated hemoglobin levels in individuals with diabetic nephropathy [87]. NOX-G15, a developed mirror DNA aptamer that inhibits glucagon activity through direct interaction, significantly improves glucose tolerance in streptozotocin-induced diabetic mice and type 2 diabetes mouse models [88].

AOs are single-stranded nucleic acids with 15–30 nucleotides that can regulate the expression of a target gene by specifically binding to the target RNA via Watson–Crick

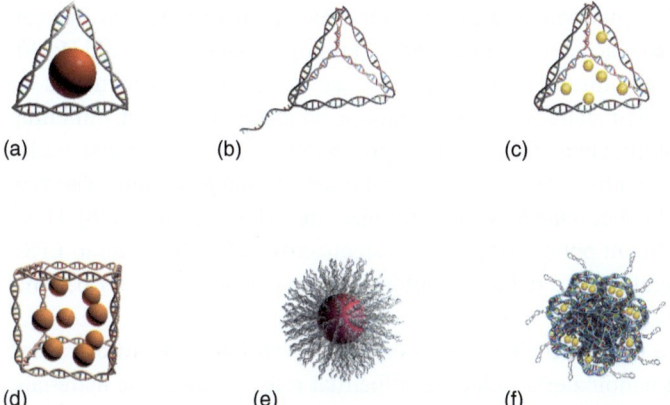

Figure 8.2 A schematic diagram depicting various types of nucleic acid nanomaterials loaded with different small molecules. (a) DNA triangles loaded with vildagliptin. (b) Oligonucleotide-hanged DNA tetrahedral. (c) DNA Tetrahedral loaded with small-molecular-weight drugs. (d) DNA cubes loaded with drugs. (e) Spherical nanonucleic acid formed by organizing nucleic acids radially around a nanoparticle core. (f) DNA nanoflowers loaded with allicin and functionalized adipo-8 aptamer.

base-pairing [105]. Research in the development of AOs-based therapeutics has focused on blocking the production of insulin resistance and hyperglycemia-associated proteins. These include sodium glucose cotransporter-2 (SGLT2) [106–108], protein tyrosine phosphatase-1B (PTP1B) [109, 110], glucocorticoid receptor (GCCR) [111–113], diacylglycerol acyltransferase-2 (DGAT2) [114, 115], and apolipoprotein C-III (APOC3) [116, 117].

SGLT2 plays a crucial role in renal glucose reabsorption, and SGLT2 inhibition increases urinary glucose excretion and lowers blood glucose levels [118]. Zanardi et al. treated mice and crab monkeys with SGLT2-targeted AO and reduced renal SGLT2 mRNA production in a dose-dependent manner compared to saline controls after 13 weeks of treatment [119].

PTP1B negatively regulates insulin signaling by dephosphorylating the insulin receptor (IR) and insulin receptor substrate (IRS) [120, 121]. Amelioration of hyperglycemia and hyperinsulinemia due to reduced PTP1B is also associated with elevated lipocalin, which enhances insulin signaling primarily through activation of the IRS [122, 123].

Glucagon is a hormone produced by pancreatic α-cells, and its elevated levels in the body can lead to hyperglycemia [124]. Treatment of rodent models of diabetes with GCGR-targeted AOs resulted in reduced production of enzymes associated with gluconeogenesis and glycogenolysis and improved pancreatic β-cell function to normalize blood glucose [125]. Some of these AOs are already in clinical trials, and antisense-based antidiabetic drugs for clinical use will become a reality in the near future.

8.4.2 Osteoporosis

Bone formation and repair are dynamic processes co-regulated by coordinated interactions between osteoblasts and osteoclasts [126]. An imbalance in this complex interaction can lead to an imbalance between bone formation and resorption, which can lead to serious bone diseases, including OP. Surveys have shown that there are 90 million OP patients worldwide, or 7.1% of the population, and that the incidence is increasing year by year, with the number expected to reach 221 million in 2050. It is now listed by the World Health Organization as one of the top three diseases of middle age [127]. Current treatments for OP focus on pharmacological modalities such as promoting bone mineralization, inhibiting bone resorption, and promoting bone formation. Bisphosphonate (BPS) is the first choice of anti-OP drug recommended by several domestic and international guidelines. The goal of treatment is to prevent further bone loss and minimize the risk of fracture [128]. However, there is no ideal treatment option for possible osteonecrosis of the jaw due to BPSs, but mainly conservative treatment to reduce symptoms; surgery may be ideal for patients with extensive involvement [129].

Several researchers have focused their research on the prevention of fractures caused by OP, where nucleic acid nanomaterials play an influential role. As injectable materials, DNA-based nanocomposite hydrogels have a wide range of biomedical uses, including controlled medication delivery and bone tissue regeneration [130]. A biodegradable and biocompatible protein–DNA hybrid hydrogel containing the Rho-suppressing C3 toxin was constructed for the purpose of specifically inhibiting osteoclast development and activity. The hydrogel is cross-linked by DNA hybridization without the use of reactive organic reagents or catalysts. Owing to mild and efficient DNA hybridization hydrogel formation procedure, protein molecule activity can be maintained. The toxin was released in a time- and location-specific manner as a result of the hydrogel being degraded by DNase I. It is expected to be used prophylactically in osteoporotic patients with high fracture risk [91]. Another DNA skeleton hydrogel material combined with silicate has also been developed as a carrier for bone regeneration and sustained release of therapeutic drugs, and the osteogenic drug dexamethasone was successfully carried [92].

Notch receptors play a critical role in the regulation of skeletal development and bone remodeling [131, 132]. In a mouse model of OP constructed with a NOTCH2 gain-of-function mutation, systemic administration of Notch2 AOs reduced the induction of TNF superfamily member 11 (Tnfsf11, encoding the osteoclast protein RANKL) mRNA levels in cultured mouse osteoblasts and decreased RANKL and osteoclast production, thereby improving their osteopenic phenotype [133].

Furthermore, to improve the targeted delivery of siRNA, aptamer-functionalized LNPs were designed. The researchers screened the osteoblast-targeting aptamer CH6 and combined it with LNPs to allow osteoblasts to specifically uptake the siRNA-loaded LNPs. At the same time, Plekho1, an intracellular negative regulator of bone formation, was selected as the target gene of siRNA. An aptamer-functionalized LNP delivery system enhances the targeted delivery of siRNA to the cellular level. Through the above system it promotes bone formation while reducing side effects [89]. Based on this research, some researchers used CH6 aptamer and the fourth-generation polyamidoamine dendrimer modified with peptide C11 to construct a drug delivery carrier. The C11 peptide is derived from 11 amino acids at the C-terminus of amelogenin and has a strong interaction with hydroxyapatite, which is an important site for regulating the formation of hydroxyapatite. The nanocarrier can accumulate in target cells, mineralized areas, and tissues, delivering drugs to the active site of osteoblasts [90].

8.4.3 Obesity

Lifestyle and behavioral interventions are the first-line treatment options for obesity. For patients with obesity combined with complications such as hyperglycemia, hypertension, and dyslipidemia, pharmacological treatment is recommended based on lifestyle and behavioral interventions [134, 135].

Weight gain in obese patients is associated with altered methionine metabolism. Methionine adenosyltransferase (MAT) catalyzes the first reaction of the methionine cycle and plays an essential role in the regulation of lipid metabolism [136]. Prevention and reversal of obesity and obesity-related insulin resistance and hepatotoxicity by increasing hepatocyte energy expenditure in diet-induced obese or genetically obese mice using AO inhibition of Mat1a, a regulatory process dependent on fibroblast growth factor 21 (FGF21) in hepatocytes [137].

Increasing energy expenditure is a major strategy in obesity treatment. The thyroid hormone T3 is extremely powerful in increasing basal metabolic rate and inducing browning of white adipose tissue [138, 139]. However, thyroid hormone is prohibited for obesity treatment because of its effects on the brain, heart, and muscles, leading to anxiety, heart failure, and muscle atrophy [140]. However, it is possible to avoid the above problems by combining T3 with AO by certain technical means. Using the liver-targeting effect of AOs, sulfosuccinimidyl-4-(N-maleimid-omethyl)cyclohexane-1-carboxylate (Sulfo-SMCC) was used to couple T3 and AO, which successfully resulted in reduced body weight and fat in mice [141]. However, the coupling of small molecules with AO may affect the stability and activity of AO, while the payload and delivery efficiency of T3 need to be improved.

Various studies have extensively demonstrated that nutritional overload can lead to IκB kinase (IKK)β activation *in vivo* and *in vitro* [142, 143]. The activation of IKKβ in the hypothalamic and hepatic tissues leads to obesity and insulin resistance and is a key molecular link between obesity and metabolic disorders [144–146]. Inhibition of IKKβ using AO *in vivo* ameliorated obesity symptoms and metabolic disorders in a diet-induced mouse model. Surprisingly, IKKβ AO also inhibited HFD-induced adipocyte differentiation and suppressed the growth of adipose tissue [147].

Clinical studies have found elevated HIF1α expression in adipose tissue of obese patients and reduced HIF1α expression after weight loss, suggesting that HIF1α itself may provide a new avenue for the treatment of obesity [148–150]. HIF1α and GCGR have also been chosen as targets for AO to treat obesity and improve hyperglycemia and hyperlipidemia, both validated *in vivo* and *in vitro* experiments [151].

Adipose tissue (mainly white adipose tissue) is widely distributed and is the main place for human fat storage. Its hypertrophy and hyperplasia lead to obesity and play an important role in fat metabolism. Adipo-8 is an aptamer that specifically recognizes adipocytes, and it interacts with the recognition target adipocyte plasma membrane-associated protein (APMAP) to improve fat deposition *in vitro* and *in vivo* [93]. Allicin, a phytochemical that induces browning of adipose tissue, is considered a potential ideal drug for obesity treatment, and targeting it to release fat cells is a key point. Recently, some researchers have used DNA nanoflower structure to load allicin and adipo-8 aptamer, avoiding the adsorption of allicin by lysosomes, successfully encapsulating, transporting, and releasing drugs in white adipose tissue, effectively promoting the browning of adipocytes (Figure 8.2f) [94].

8.4.4 Nonalcoholic Fatty Liver Disease

Nonalcoholic fatty liver disease (NAFLD) is a clinicopathological syndrome associated with systemic metabolism and characterized by hepatic steatosis. The age of onset of NAFLD is getting younger, and the prevalence is rising rapidly [152]. As the pathogenesis of NAFLD becomes clearer, there is a diversity of drug targets under development [152, 153]. Most of these drugs target metabolic disorders, oxidative stress, and inflammation [154], where metabolic regulation includes bile acid metabolism, lipid metabolism, and glucose metabolism [155].

Tan used the SELEX system to generate a NAFLD cell-specific aptamer and named it NAFLD01. The aptamer can specifically recognize membrane-bound CD36, and NAFLD01 can improve fatty acid degradation in NAFLD cells, increase the expression of peroxisome proliferator-activated receptor-α (PPARα), and reduce triglyceride levels, while *in vivo* experimental results show that NAFLD01 exerts the above functions by interacting with CD36 [95].

Nucleic acids can stimulate or modulate immunity by binding to endosomal toll-like receptors (TLRs), a finding that researchers have used to develop immunotherapies for NAFLD [156]. They designed spherical nucleic acids (SNAs) using metal nanoparticles as a core surrounded by an oligonucleotide shell (Figure 8.2e). Immunoregulatory-SNAs can significantly mitigate liver fibrosis in NAFLD model mice by enhancing immunomodulatory function. Compared with unformulated free oligonucleotides, SNAs can enter cells without other delivery vehicles, and after being taken up by cells, they can resist the degradation of intracellular nucleases and prolong the lifespan of oligonucleotides in cells [96].

DGAT2 is an enzyme involved in triglyceride synthesis. AO therapy targeting DGAT2 significantly reduced hepatic diacylglycerol and triglycerides and decreased the expression of adipogenic genes while increasing the expression of genes involved in lipid oxidation [115]. It was also found that treatment targeting DGAT2 significantly improved insulin sensitivity, which was validated in an obesity model experiment in mice [114].

Similarly, serine/threonine protein kinase (STK)25, a member of the sterile 20-kinase superfamily, is a key regulator of systemic energy homeostasis [157, 158]. STK 25 AOs effectively reversed the HFD-induced systemic hyperglycemia, improved insulin sensitivity, and suppressed hepatic fatty accumulation and NASH features such as alanine aminotransferase (ALT) secretion and oxidative damage in obese mice [159, 160].

Silencing apolipoprotein B (ApoB), a crucial component of LDL particles [161], has proven effective at lowering serum LDL cholesterol levels in animal studies [44, 162, 163]. DNA tetrahedrons are used as vectors for liver-specific delivery of siRNA targeting ApoB1 mRNA. The siRNA preferentially accumulates and downregulates the ApoB1 protein in the liver, thereby lowering blood cholesterol levels [51]. Notably, the trial delivered the drug by systemic administration and successfully targeted the liver site.

The TE20-type kinase MST3, which is primarily localized to intracellular lipid droplets, is a key regulator of ectopic fat accumulation in human hepatocytes, and MST3 protein levels were found to be positively correlated with the severity of NAFLD [164, 165]. MST3 AOs inhibit acetyl coenzyme and carboxylase protein abundance, as well as adipogenic gene expression, and substantially reduce oxidative stress and endoplasmic reticulum stress in the liver of obese mice. Administration of MST3-targeted ASOs in mice with NAFLD induced by a HFD effectively ameliorated hepatic steatosis, inflammation, fibrosis, and hepatocyte injury [165].

8.5 Conclusion and Outlook

At present, drug therapy is an important way to treat endocrine and metabolic diseases. Previous antisense, RNAi-based therapeutic systems have been extensively studied and used in clinical trials [166]. Researchers began to use nucleic acid nanostructures (DNA polyhedron) as carriers for siRNA and AO [167, 168]. Compared with previous viral vectors, the carriers of nucleic acid nanomaterials had reduced cytotoxicity. Another strategy is to make siRNA and AO self-assemble into special configurations or form SNA-like structures, pass through the cell membrane through endocytosis, and improve the stability of oligonucleotides to exert their biological functions [169]. At the same time, some special nucleic acid nanostructures are used for drug target recognition and delivery. Nucleic acid-based drugs have a strong competitive advantage over small-molecule drugs and protein inhibitor drugs, including (i) AOs bind selectively and specifically to mRNA transcripts of target proteins, thereby avoiding off-targeting and effects on nontargeted proteins; (ii) for some proteins that cannot be directly acted upon by drugs, therapeutic oligonucleotides reduce their production by inhibiting target gene expression (transcription/translation); (iii) The diversity of AOs antisense mechanisms and the possibility of binding to different small molecules provide sufficient potential options and tools to optimize AO drug candidates to improve their efficacy, stability, and safety.

Aptamer, as a new "antibody," can be synthesized and produced in batches, and its specific recognition function is helpful for the targeting of drugs. Although aptamers are particularly well suited for diagnostic purposes, including molecular probes, biosensors, and aptamer-based immunoassays, their role in drug targeting cannot be ignored. Aptamers can improve their biological stability through special modifications on nucleotides, providing a new strategy for drug development [170, 171]. Furthermore, aptamers have been reported to modify cellular properties by constructing artificial aptamer-lipid receptors [172, 173], and these applications provide the basis for the development of aptamers for the treatment of endocrine and metabolic diseases and their complications. In addition, although a variety of aptamer-based drugs have entered clinical trials, unexplained adverse reactions have also been reported, and their pharmacological behaviors should be continuously studied to avoid potential adverse reactions [174].

DNA nanostructures can be modified by suspending functional domains or adding small molecular weight drugs during synthesis to achieve different therapeutic purposes [175, 176]. DNA tetrahedra have been successfully applied in various fields, such as stem cells, biosensors, and tumor therapy. Its own immunomodulatory function and distinct spatial configuration hold great promise for applications in disease treatment [177]. Nucleic acid nanomaterials have natural biocompatibility and satisfactory cell membrane permeability, together with good editability and relative biological stability under complex conditions. Nucleic acid nanomaterials could enter cells via cell membrane endocytosis without the requirement for transfection (Figure 8.3). Although nucleic acid nanomaterials have many

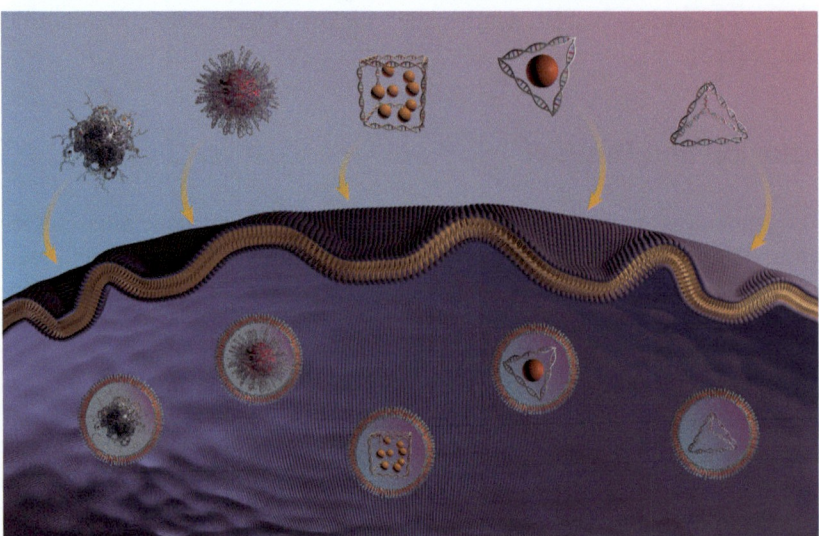

Figure 8.3 Nucleic acid nanomaterials are internalized into various cells via caveolin-mediated endocytosis.

advantages and great application potential, there are still some problems in their clinical application. For example, nucleic acid nanomaterials have a considerable degree of stability, but still cannot meet the requirements of long-term application *in vivo*. At the same time, nucleic acid nanomaterials are still limited in the size of carrying bioactive molecules, and still cannot be combined and transported for macromolecules. Moreover, while multiple cargo types have been successfully incorporated into DNA nanostructures, the exact mechanism of interaction is unclear, and the mechanisms of the various relevant cellular responses remain unknown. This limits the application of nucleic acid nanomaterials to complex physiological or pathological conditions.

Apart from being utilized to cure diseases, nucleic acid nanomaterials can benefit the advancement of multiple fields of biological study. Probes, immunological compounds, and other binders may be employed to provide nucleic acid nanomaterials with extra-specific features, and these composite nanostructures will enhance technological advancements in detection and sensing [178].

Overall, with continuous research and exploration in the field of nucleic acid nanomaterials, it will occupy an irreplaceable position in various biological fields, including disease treatment. Nucleic acid nanomaterials may serve as a conventional drug loading and delivery platform, allowing the targeted recognition and *in vivo* delivery of a variety of biomolecules and demonstrating a larger application value.

References

1 Seeman, N.C. (1982). Nucleic acid junctions and lattices. *J. Theor. Biol.* 99 (2): 237–247.
2 Zhang, T., Tian, T., and Lin, Y. (2022). Functionalizing framework nucleic-acid-based nanostructures for biomedical application. *Adv. Mater.* 34 (46): e2107820.
3 Nummelin, S. et al. (2018). Evolution of structural DNA nanotechnology. *Adv. Mater.* 30 (24): e1703721.
4 Peng, P. et al. (2019). Reconfigurable bioinspired framework nucleic acid nanoplatform dynamically manipulated in living cells for subcellular imaging. *Angew. Chem. Int. Ed. Engl.* 58 (6): 1648–1653.
5 Sundah, N.R. et al. (2019). Barcoded DNA nanostructures for the multiplexed profiling of subcellular protein distribution. *Nat. Biomed. Eng.* 3 (9): 684–694.
6 Wiraja, C. et al. (2019). Framework nucleic acids as programmable carrier for transdermal drug delivery. *Nat. Commun.* 10 (1): 1147.
7 Hotamisligil, G.S. (2006). Inflammation and metabolic disorders. *Nature* 444 (7121): 860–867.
8 Diemieszczyk, I. et al. (2021). Metabolic syndrome etiology and pathogenesis. *Wiad. Lek.* 74 (10 pt 1): 2510–2515.
9 Engin, A. (2017). The definition and prevalence of obesity and metabolic syndrome. *Adv. Exp. Med. Biol.* 960: 1–17.
10 Collaborators, G.B.D.O. et al. (2017). Health effects of overweight and obesity in 195 countries over 25 years. *N Engl. J. Med.* 377 (1): 13–27.
11 DiMeglio, L.A., Evans-Molina, C., and Oram, R.A. (2018). Type 1 diabetes. *Lancet* 391 (10138): 2449–2462.

12 Kahn, S.E., Cooper, M.E., and Del Prato, S. (2014). Pathophysiology and treatment of type 2 diabetes: perspectives on the past, present, and future. *Lancet* 383 (9922): 1068–1083.

13 Qaseem, A. et al. (2017). Treatment of low bone density or osteoporosis to prevent fractures in men and women: a clinical practice guideline update from the American College of Physicians. *Ann. Intern. Med.* 166 (11): 818–839.

14 Ensrud, K.E. and Crandall, C.J. (2017). Osteoporosis. *Ann. Intern. Med.* 167 (3): ITC17–ITC32.

15 Caballero, B. (2019). Humans against obesity: who will win? *Adv. Nutr.* 10 (suppl_1): S4–S9.

16 Zhou, J. et al. (2020). Epidemiological features of NAFLD from 1999 to 2018 in China. *Hepatology* 71 (5): 1851–1864.

17 Estes, C. et al. (2018). Modeling NAFLD disease burden in China, France, Germany, Italy, Japan, Spain, United Kingdom, and United States for the period 2016-2030. *J. Hepatol.* 69 (4): 896–904.

18 Kassi, E. Pervanidou, P., Kaltsas, G., and Chrousos, G. (2011). Metabolic syndrome definitions and controversies. *BMC Med.* 9: 48.

19 Persaud, S.J. and Bowe, J.E. (2018). Editorial overview: endocrine and metabolic diseases druggable diabetes: identification of therapeutic opportunities. *Curr. Opin. Pharmacol.* 43: iii–v.

20 Bewick, G.A. (2017). Editorial overview: endocrine and metabolic diseases: Busting BMI: new strategies for the treatment of obesity and metabolic disease. *Curr. Opin. Pharmacol.* 37: ix–xii.

21 Xiao, M. et al. (2019). Rationally engineered nucleic acid architectures for biosensing applications. *Chem. Rev.* 119 (22): 11631–11717.

22 Lei, J. and Ju, H. (2012). Signal amplification using functional nanomaterials for biosensing. *Chem. Soc. Rev.* 41 (6): 2122–2134.

23 Chakraborty, K. et al. (2016). Nucleic acid-based nanodevices in biological imaging. *Annu. Rev. Biochem.* 85: 349–373.

24 Hu, Q. et al. (2019). DNA nanotechnology-enabled drug delivery systems. *Chem. Rev.* 119 (10): 6459–6506.

25 Wu, X. et al. (2020). Gene therapy based on nucleic acid nanostructure. *Adv. Healthcare Mater.* 9 (19): e2001046.

26 Douglas, S.M. et al. (2009). Rapid prototyping of 3D DNA-origami shapes with caDNAno. *Nucleic Acids Res.* 37 (15): 5001–5006.

27 Zhang, T. et al. (2020). Progress in biomedical applications of tetrahedral framework nucleic acid-based functional systems. *ACS Appl. Mater. Interfaces* 12 (42): 47115–47126.

28 Lin, C., Rinker, S., Wang, X. et al. (2008). In vivo cloning of artificial DNA nanostructures. *Proc. Natl. Acad. Sci. U.S.A.* 105 (46): 17626–17631.

29 Zhang, C., Su, M., He, Y. et al. (2008). Conformational flexibility facilitates self-assembly of complex DNA nanostructures. *Proc. Natl. Acad. Sci. U. S. A.* 105 (31): 10665–10669.

30 Lin, C. et al. (2008). In vivo cloning of artificial DNA nanostructures. *Proc. Natl. Acad. Sci. U. S. A.* 105 (46): 17626–17631.

31 Jiang, D., England, C.G., and Cai, W. (2016). DNA nanomaterials for preclinical imaging and drug delivery. *J. Controlled Release* 239: 27–38.

32 Peng, R. et al. (2017). Facile assembly/disassembly of DNA nanostructures anchored on cell-mimicking giant vesicles. *J. Am. Chem. Soc.* 139 (36): 12410–12413.

33 Goodman, R.P. et al. (2005). Rapid chiral assembly of rigid DNA building blocks for molecular nanofabrication. *Science* 310 (5754): 1661–1665.

34 Zhang, T. et al. (2020). Design, fabrication and applications of tetrahedral DNA nanostructure-based multifunctional complexes in drug delivery and biomedical treatment. *Nat. Protoc.* 15 (8): 2728–2757.

35 Liu, Y. et al. (2020). Tetrahedral framework nucleic acids deliver antimicrobial peptides with improved effects and less susceptibility to bacterial degradation. *Nano Lett.* 20 (5): 3602–3610.

36 Madhanagopal, B.R. et al. (2018). DNA nanocarriers: programmed to deliver. *Trends Biochem. Sci* 43 (12): 997–1013.

37 Mou, Q. et al. (2019). Two-in-one chemogene assembled from drug-integrated antisense oligonucleotides to reverse chemoresistance. *J. Am. Chem. Soc.* 141 (17): 6955–6966.

38 Tan, X. et al. (2016). Blurring the role of oligonucleotides: spherical nucleic acids as a drug delivery vehicle. *JACS* 138 (34): 10834–10837.

39 Alterman, J.F. et al. (2019). A divalent siRNA chemical scaffold for potent and sustained modulation of gene expression throughout the central nervous system. *Nat. Biotechnol.* 37 (8): 884–894.

40 Vasher, M.K., Yamankurt, G., and Mirkin, C.A. (2022). Hairpin-like siRNA-based spherical nucleic acids. *JACS* 144 (7): 3174–3181.

41 Sud, R., Geller, E.T., and Schellenberg, G.D. (2014). Antisense-mediated exon skipping decreases tau protein expression: a potential therapy for tauopathies. *Mol. Ther. Nucleic Acids* 3 (7): e180.

42 Shende, P. and Patel, C. (2018). siRNA: an alternative treatment for diabetes and associated conditions. *J. Drug Targeting* 27 (2): 174–182.

43 Horton, J.D., Cohen, J.C., and Hobbs, H.H. (2007). Molecular biology of PCSK9: its role in LDL metabolism. *Trends Biochem. Sci* 32 (2): 71–77.

44 Semple, S.C. et al. (2010). Rational design of cationic lipids for siRNA delivery. *Nat. Biotechnol.* 28 (2): 172–176.

45 Yenilmez, B. et al. (2022). An RNAi therapeutic targeting hepatic DGAT2 in a genetically obese mouse model of nonalcoholic steatohepatitis. *Mol. Ther.* 30 (3): 1329–1342.

46 Bumcrot, D. et al. (2006). RNAi therapeutics: a potential new class of pharmaceutical drugs. *Nat. Chem. Biol.* 2 (12): 711–719.

47 Amadio, M. et al. (2016). Nanosystems based on siRNA silencing HuR expression counteract diabetic retinopathy in rat. *Pharmacol. Res.* 111: 713–720.

48 Li, C. and Samulski, R.J. (2020). Engineering adeno-associated virus vectors for gene therapy. *Nat. Rev. Genet.* 21 (4): 255–272.

49 Allen, T.M. and Cullis, P.R. (2013). Liposomal drug delivery systems: from concept to clinical applications. *Adv. Drug Delivery Rev.* 65 (1): 36–48.

50 Gao, Y. et al. (2022). A lysosome-activated tetrahedral nanobox for encapsulated siRNA delivery. *Adv. Mater.* e2201731.

51 Kim, K.R. et al. (2020). A self-assembled DNA tetrahedron as a carrier for in vivo liver-specific delivery of siRNA. *Biomater. Sci.* 8 (2): 586–590.
52 Chen, S., Sbuh, N., and Veedu, R.N. (2021). Antisense oligonucleotides as potential therapeutics for type 2 diabetes. *Nucleic Acid Ther.* 31 (1): 39–57.
53 Yang, J. and Bowser, M.T. (2013). Capillary electrophoresis-SELEX selection of catalytic DNA aptamers for a small-molecule porphyrin target. *Anal. Chem.* 85 (3): 1525–1530.
54 Bunka, D.H. and Stockley, P.G. (2006). Aptamers come of age - at last. *Nat. Rev. Microbiol.* 4 (8): 588–596.
55 Gupta, S. et al. (2017). Pharmacokinetic properties of DNA aptamers with base modifications. *Nucleic Acid Ther.* 27 (6): 345–353.
56 Wu, L. et al. (2021). Aptamer-based detection of circulating targets for precision medicine. *Chem. Rev.* 121 (19): 12035–12105.
57 Zhan, Y. et al. (2019). Diversity of DNA nanostructures and applications in oncotherapy. *Biotechnol. J.* 15 (1).
58 Ng, E.W. et al. (2006). Pegaptanib, a targeted anti-VEGF aptamer for ocular vascular disease. *Nat. Rev. Drug Discovery* 5 (2): 123–132.
59 Hu, J., Ye, M., and Zhou, Z. (2016). Aptamers: novel diagnostic and therapeutic tools for diabetes mellitus and metabolic diseases. *J. Mol. Med.* 95 (3): 249–256.
60 He, Y. et al. (2008). Hierarchical self-assembly of DNA into symmetric supramolecular polyhedra. *Nature* 452 (7184): 198–201.
61 Zhang, M. et al. (2022). Anti-inflammatory activity of curcumin-loaded tetrahedral framework nucleic acids on acute gouty arthritis. *Bioact. Mater.* 8: 368–380.
62 Zhang, Q., Jiang, Q., Li, N. et al. (2014). DNA origami as an in vivo drug delivery vehicle for cancer therapy. *Am. Chem. Soc. Nano.*
63 Qin, X. et al. (2022). Tetrahedral framework nucleic acids-based delivery of microRNA-155 inhibits choroidal neovascularization by regulating the polarization of macrophages. *Bioact. Mater.* 14: 134–144.
64 Xie, X. et al. (2018). Overcoming drug-resistant lung cancer by paclitaxel loaded tetrahedral DNA nanostructures. *Nanoscale* 10 (12): 5457–5465.
65 Li, S. et al. (2019). Advances in biological applications of self-assembled DNA tetrahedral nanostructures. *Mater. Today* 24: 57–68.
66 Stopar, A. et al. (2018). Binary control of enzymatic cleavage of DNA origami by structural antideterminants. *Nucleic Acids Res.* 46 (2): 995–1006.
67 Zagorovsky, K., Chou, L.Y., and Chan, W.C. (2016). Controlling DNA-nanoparticle serum interactions. *PNAS* 113 (48): 13600–13605.
68 Li, J. et al. (2011). Self-assembled multivalent DNA nanostructures for noninvasive intracellular delivery of immunostimulatory CpG oligonucleotides. *ACS Nano* 5 (11): 8783–8789.
69 Ma, W. et al. (2022). Biomimetic nanoerythrosome-coated aptamer-DNA tetrahedron/maytansine conjugates: pH-responsive and targeted cytotoxicity for HER2-positive breast cancer. *Adv. Mater.* e2109609.
70 Li, J. et al. (2022). Repair of infected bone defect with clindamycin-tetrahedral DNA nanostructure complex-loaded 3D bioprinted hybrid scaffold. *Chem. Eng. J.* 435: 134855.

71 Zhang, B. et al. (2022). Facilitating in situ tumor imaging with a tetrahedral DNA framework-enhanced hybridization chain reaction probe. *Adv. Funct. Mater.* 32 (16).

72 Wang, Y. et al. (2022). Tetrahedral framework nucleic acids can alleviate taurocholate-induced severe acute pancreatitis and its subsequent multiorgan injury in mice. *Nano Lett.* 22 (4): 1759–1768.

73 Zhu, J. et al. (2022). Antiepilepticus effects of tetrahedral framework nucleic acid via inhibition of gliosis-induced downregulation of glutamine synthetase and increased AMPAR internalization in the postsynaptic membrane. *Nano Lett.* 22 (6): 2381–2390.

74 Chen, X. et al. (2022). Positive neuroplastic effect of DNA framework nucleic acids on neuropsychiatric diseases. *ACS Mater. Lett.* 4 (4): 665–674.

75 Li, J. et al. (2022). Modulation of the crosstalk between schwann cells and macrophages for nerve regeneration: a therapeutic strategy based on multifunctional tetrahedral framework nucleic acids system. *Adv. Mater.* e2202513.

76 Fu, W. et al. (2021). Therapeutic siCCR2 loaded by tetrahedral framework DNA nanorobotics in therapy for intracranial hemorrhage. *Adv. Funct. Mater.* 31 (33).

77 Chen, Y. et al. (2022). Therapeutic effects of self-assembled tetrahedral framework nucleic acids on liver regeneration in acute liver failure. *ACS Appl. Mater. Interfaces* 14 (11): 13136–13146.

78 Jiang, Y. et al. (2022). Tetrahedral framework nucleic acids inhibit skin fibrosis via the pyroptosis pathway. *ACS Appl. Mater. Interfaces* 14 (13): 15069–15079.

79 Chen, R. et al. (2022). Treatment effect of DNA framework nucleic acids on diffuse microvascular endothelial cell injury after subarachnoid hemorrhage. *Cell Proliferation* 55 (4): e13206.

80 Zhao, D. et al. (2022). Tetrahedral framework nucleic acid carrying angiogenic peptide prevents bisphosphonate-related osteonecrosis of the jaw by promoting angiogenesis. *Int. J. Oral Sci.* 14 (1): 23.

81 Baig, M.M.F.A. et al. (2019). DNA scaffold nanoparticles coated with HPMC/EC for oral delivery. *Int. J. Pharm.* 562: 321–332.

82 Baig, M.M.F.A. et al. (2019). Design, synthesis and evaluation of DNA nano-cubes as a core material protected by the alginate coating for oral administration of anti-diabetic drug. *J. Food Drug Anal.* 27 (3): 805–814.

83 Baig, M.M.F.A. et al. (2018). Vildagliptin loaded triangular DNA nanospheres coated with eudragit for oral delivery and better glycemic control in type 2 diabetes mellitus. *Biomed. Pharmacother.* 97: 1250–1258.

84 Li, Y. et al. (2021). Tetrahedral framework nucleic acids ameliorate insulin resistance in type 2 diabetes mellitus via the PI3K/Akt pathway. *ACS Appl. Mater. Interfaces* 13 (34): 40354–40364.

85 Li, Y. et al. (2021). Tetrahedral framework nucleic acid-based delivery of resveratrol alleviates insulin resistance: from innate to adaptive immunity. *Nano Micro Lett.* 13 (1).

86 Gao, S. et al. (2021). Tetrahedral framework nucleic acids induce immune tolerance and prevent the onset of type 1 diabetes. *Nano Lett.* 21 (10): 4437–4446.

87 Menne, J. et al. (2016). C-C motif-ligand 2 inhibition with emapticap pegol (NOX-E36) in type 2 diabetic patients with albuminuria. *Nephrol. Dialysis Transplant.* 32 (2): 307–315.

88 Vater, A. et al. (2013). A mixed mirror-image DNA/RNA aptamer inhibits glucagon and acutely improves glucose tolerance in models of type 1 and type 2 diabetes. *J. Biol. Chem.* 288 (29): 21136–21147.

89 Liang, C. et al. (2015). Aptamer-functionalized lipid nanoparticles targeting osteoblasts as a novel RNA interference–based bone anabolic strategy. *Nat. Med.* 21 (3): 288–294.

90 Ren, M. et al. (2021). An oligopeptide/aptamer-conjugated dendrimer-based nanocarrier for dual-targeting delivery to bone. *J. Mater. Chem. B* 9 (12): 2831–2844.

91 Gačanin, J. et al. (2017). Spatiotemporally controlled release of rho-inhibiting C3 toxin from a protein-DNA hybrid hydrogel for targeted inhibition of osteoclast formation and activity. *Adv. Healthcare Mater.* 6 (21).

92 Basu, S. et al. (2018). Harnessing the noncovalent interactions of DNA backbone with 2D silicate nanodisks to fabricate injectable therapeutic hydrogels. *ACS Nano* 12 (10): 9866–9880.

93 Zhong, W. et al. (2020). Adipose specific aptamer adipo-8 recognizes and interacts with APMAP to ameliorates fat deposition in vitro and in vivo. *Life Sci.* 251.

94 Chen, X. et al. (2021). Aptamer-functionalized binary-drug delivery system for synergetic obesity therapy. *ACS Nano*.

95 Pu, Y. et al. (2021). CD36 as a molecular target of functional DNA aptamer NAFLD01 selected against NAFLD cells. *Anal. Chem.* 93 (8): 3951–3958.

96 Radovic-Moreno, A.F. et al. (2015). Immunomodulatory spherical nucleic acids. *Proc. Nat. Acad. Sci.* 112 (13): 3892–3897.

97 American Diabetes, A. (2013). Diagnosis and classification of diabetes mellitus. *Diabetes Care* 36 (Suppl 1): S67–S74.

98 Jia, W. et al. (2019). Standards of medical care for type 2 diabetes in China 2019. *Diabetes Metab. Res. Rev.* 35 (6): e3158.

99 Babiker, A. and Al Dubayee, M. (2017). Anti-diabetic medications: how to make a choice? *Sudan. J. Paediatr.* 17 (2): 11–20.

100 Krentz, A.J. and Bailey, C.J. (2005). Oral antidiabetic agents: current role in type 2 diabetes mellitus. *Drugs* 65 (3): 385–411.

101 Marín-Peñalver, J.J. et al. (2016). Update on the treatment of type 2 diabetes mellitus. *World J. Diabetes* 7 (17): 354–395.

102 Kokil, G.R. et al. (2015). Type 2 diabetes mellitus: limitations of conventional therapies and intervention with nucleic acid-based therapeutics. *Chem. Rev.* 115 (11): 4719–4743.

103 Matthews, D.R. et al. (2019). Glycaemic durability of an early combination therapy with vildagliptin and metformin versus sequential metformin monotherapy in newly diagnosed type 2 diabetes (VERIFY): a 5-year, multicentre, randomised, double-blind trial. *Lancet* 394 (10208): 1519–1529.

104 Kang, Y.S. et al. (2010). CCR2 antagonism improves insulin resistance, lipid metabolism, and diabetic nephropathy in type 2 diabetic mice. *Kidney Int.* 78 (9): 883–894.

105 Goyal, N. and Narayanaswami, P. (2018). Making sense of antisense oligonucleotides: a narrative review. *Muscle Nerv.* 57 (3): 356–370.

106 Liakos, A. et al. (2015). Update on long-term efficacy and safety of dapagliflozin in patients with type 2 diabetes mellitus. *Ther. Adv. Endocrinol. Metab.* 6 (2): 61–67.

107 Jakher, H. et al. (2019). Canagliflozin review - safety and efficacy profile in patients with T2DM. *Diabetes Metab. Syndr. Obes.* 12: 209–215.

108 Markham, A. and Keam, S.J. (2019). Sotagliflozin: first global approval. *Drugs* 79 (9): 1023–1029.

109 Geary, R.S. et al. (2006). Lack of pharmacokinetic interaction for ISIS 113715, a 2'-0-methoxyethyl modified antisense oligonucleotide targeting protein tyrosine phosphatase 1B messenger RNA, with oral antidiabetic compounds metformin, glipizide or rosiglitazone. *Clin. Pharmacokinet.* 45 (8): 789–801.

110 Digenio, A. et al. (2018). Antisense inhibition of protein tyrosine phosphatase 1B with IONIS-PTP-1B(Rx) improves insulin sensitivity and reduces weight in overweight patients with type 2 diabetes. *Diabetes Care* 41 (4): 807–814.

111 van Dongen, M.G. et al. (2015). First proof of pharmacology in humans of a novel glucagon receptor antisense drug. *J. Clin. Pharmacol.* 55 (3): 298–306.

112 Luu, K.T. et al. (2017). Population pharmacokinetics and pharmacodynamics of IONIS-GCGR(Rx), an antisense oligonucleotide for type 2 diabetes mellitus: a red blood cell lifespan model. *J. Pharmacokinet. Pharmacodyn.* 44 (3): 179–191.

113 Morgan, E.S. et al. (2019). Antisense inhibition of glucagon receptor by IONIS-GCGR(rx) improves type 2 diabetes without increase in hepatic glycogen content in patients with type 2 diabetes on stable metformin therapy. *Diabetes Care* 42 (4): 585–593.

114 Yu, X.X. et al. (2005). Antisense oligonucleotide reduction of DGAT2 expression improves hepatic steatosis and hyperlipidemia in obese mice. *Hepatology* 42 (2): 362–371.

115 Choi, C.S. et al. (2007). Suppression of diacylglycerol acyltransferase-2 (DGAT2), but not DGAT1, with antisense oligonucleotides reverses diet-induced hepatic steatosis and insulin resistance. *J. Biol. Chem.* 282 (31): 22678–22688.

116 Graham, M.J. et al. (2013). Antisense oligonucleotide inhibition of apolipoprotein C-III reduces plasma triglycerides in rodents, nonhuman primates, and humans. *Circ. Res.* 112 (11): 1479–1490.

117 Digenio, A. et al. (2016). Antisense-mediated lowering of plasma apolipoprotein C-III by volanesorsen improves dyslipidemia and insulin sensitivity in type 2 diabetes. *Diabetes Care* 39 (8): 1408–1415.

118 Bailey, C.J. (2011). Renal glucose reabsorption inhibitors to treat diabetes. *Trends Pharmacol. Sci.* 32 (2): 63–71.

119 Zanardi, T.A. et al. (2012). Pharmacodynamics and subchronic toxicity in mice and monkeys of ISIS 388626, a second-generation antisense oligonucleotide that targets human sodium glucose cotransporter 2. *J. Pharmacol. Exp. Ther.* 343 (2): 489–496.

120 Kenner, K.A. et al. (1996). Protein-tyrosine phosphatase 1B is a negative regulator of insulin- and insulin-like growth factor-I-stimulated signaling. *J. Biol. Chem.* 271 (33): 19810–19816.

121 Goldstein, B.J. et al. (2000). Tyrosine dephosphorylation and deactivation of insulin receptor substrate-1 by protein-tyrosine phosphatase 1B. Possible facilitation by the formation of a ternary complex with the Grb2 adaptor protein. *J. Biol. Chem.* 275 (6): 4283–4289.

122 Swarbrick, M.M. et al. (2009). Inhibition of protein tyrosine phosphatase-1B with antisense oligonucleotides improves insulin sensitivity and increases adiponectin concentrations in monkeys. *Endocrinology* 150 (4): 1670–1679.

123 Achari, A.E. and Jain, S.K. (2017). Adiponectin, a therapeutic target for obesity, diabetes, and endothelial dysfunction. *Int. J. Mol. Sci.* 18 (6).

124 Dobbs, R. et al. (1975). Glucagon: role in the hyperglycemia of diabetes mellitus. *Science* 187 (4176): 544–547.

125 Sloop, K.W. et al. (2004). Hepatic and glucagon-like peptide-1-mediated reversal of diabetes by glucagon receptor antisense oligonucleotide inhibitors. *J. Clin. Invest.* 113 (11): 1571–1581.

126 Salhotra, A. et al. (2020). Mechanisms of bone development and repair. *Nat. Rev. Mol. Cell Biol.* 21 (11): 696–711.

127 Mukherjee, K. and Chattopadhyay, N. (2016). Pharmacological inhibition of cathepsin K: a promising novel approach for postmenopausal osteoporosis therapy. *Biochem. Pharmacol.* 117: 10–19.

128 Black, D.M. and Rosen, C.J. (2016). Clinical practice. Postmenopausal osteoporosis. *N Engl. J. Med.* 374 (3): 254–262.

129 Janovská, Z. (2012). Bisphosphonate-related osteonecrosis of the jaws. A severe side effect of bisphosphonate therapy. *Acta Med.* 55 (3): 111–115.

130 Basu, S., Pacelli, S., and Paul, A. (2020). Self-healing DNA-based injectable hydrogels with reversible covalent linkages for controlled drug delivery. *Acta Biomater.* 105: 159–169.

131 Zanotti, S. and Canalis, E. (2016). Notch signaling and the skeleton. *Endocr. Rev.* 37 (3): 223–253.

132 Siebel, C. and Lendahl, U. (2017). Notch signaling in development, tissue homeostasis, and disease. *Physiol. Rev.* 97 (4): 1235–1294.

133 Canalis, E. et al. (2020). Antisense oligonucleotides targeting Notch2 ameliorate the osteopenic phenotype in a mouse model of Hajdu-Cheney syndrome. *J. Biol. Chem.* 295 (12): 3952–3964.

134 Semlitsch, T. et al. (2019). Management of overweight and obesity in primary care-A systematic overview of international evidence-based guidelines. *Obes. Rev.* 20 (9): 1218–1230.

135 (2019). Guidelines for primary diagnosis and treatment of obesity. *Chin. J. Gener. Pract.* 2020 (02): 95–101.

136 Alonso, C. et al. (2017). Metabolomic identification of subtypes of nonalcoholic steatohepatitis. *Gastroenterology* 152 (6): 1449–1461.e7.

137 Sáenz de Urturi, D. et al. (2022). Methionine adenosyltransferase 1a antisense oligonucleotides activate the liver-brown adipose tissue axis preventing obesity and associated hepatosteatosis. *Nat. Commun.* 13 (1): 1096.

138 Mullur, R., Liu, Y.Y., and Brent, G.A. (2014). Thyroid hormone regulation of metabolism. *Physiol. Rev.* 94 (2): 355–382.

139 Obregon, M.J. (2014). Adipose tissues and thyroid hormones. *Front. Physiol.* 5: 479.

140 Brent, G.A. (2008). Clinical practice. Graves' disease. *N Engl. J. Med.* 358 (24): 2594–2605.

141 Cao, Y. et al. (2017). Antisense oligonucleotide and thyroid hormone conjugates for obesity treatment. *Sci. Rep.* 7 (1).

142 Jiao, P. et al. (2011). FFA-induced adipocyte inflammation and insulin resistance: involvement of ER stress and IKKβ pathways. *Obesity (Silver Spring)* 19 (3): 483–491.

143 Hotamisligil, G.S. and Erbay, E. (2008). Nutrient sensing and inflammation in metabolic diseases. *Nat. Rev. Immunol.* 8 (12): 923–934.

144 Baker, R.G., Hayden, M.S., and Ghosh, S. (2011). NF-κB, inflammation, and metabolic disease. *Cell Metab.* 13 (1): 11–22.

145 Zhang, X. et al. (2008). Hypothalamic IKKbeta/NF-kappaB and ER stress link overnutrition to energy imbalance and obesity. *Cell* 135 (1): 61–73.

146 Purkayastha, S., Zhang, G., and Cai, D. (2011). Uncoupling the mechanisms of obesity and hypertension by targeting hypothalamic IKK-β and NF-κB. *Nat. Med.* 17 (7): 883–887.

147 Helsley, R.N. et al. (2016). Targeting IκB kinase β in adipocyte lineage cells for treatment of obesity and metabolic dysfunctions. *Stem Cells* 34 (7): 1883–1895.

148 Trayhurn, P. and Wood, I.S. (2004). Adipokines: inflammation and the pleiotropic role of white adipose tissue. *Br. J. Nutr.* 92 (3): 347–355.

149 Lolmède, K. et al. (2003). Effects of hypoxia on the expression of proangiogenic factors in differentiated 3T3-F442A adipocytes. *Int. J. Obes. Relat. Metab. Disord.* 27 (10): 1187–1195.

150 Chen, B. et al. (2006). Hypoxia dysregulates the production of adiponectin and plasminogen activator inhibitor-1 independent of reactive oxygen species in adipocytes. *Biochem. Biophys. Res. Commun.* 341 (2): 549–556.

151 Watts, L.M. et al. (2005). Reduction of hepatic and adipose tissue glucocorticoid receptor expression with antisense oligonucleotides improves hyperglycemia and hyperlipidemia in diabetic rodents without causing systemic glucocorticoid antagonism. *Diabetes* 54 (6): 1846–1853.

152 Friedman, S.L. et al. (2018). Mechanisms of NAFLD development and therapeutic strategies. *Nat. Med.* 24 (7): 908–922.

153 Cotter, T.G. and Rinella, M. (2020). Nonalcoholic fatty liver disease 2020: the state of the disease. *Gastroenterology* 158 (7): 1851–1864.

154 Hong, T. et al. (2021). The role and mechanism of oxidative stress and nuclear receptors in the development of NAFLD. *Oxid. Med. Cell. Longevity* 2021: 6889533.

155 Reimer, K.C. et al. (2020). New drugs for NAFLD: lessons from basic models to the clinic. *Hepatol. Int.* 14 (1): 8–23.

156 Mencin, A., Kluwe, J., and Schwabe, R.F. (2009). Toll-like receptors as targets in chronic liver diseases. *Gut* 58 (5): 704–720.

157 Thompson, B.J. and Sahai, E. (2015). MST kinases in development and disease. *J. Cell Biol.* 210 (6): 871–882.

158 Amrutkar, M. et al. (2015). Genetic disruption of protein kinase STK25 ameliorates metabolic defects in a diet-induced type 2 diabetes model. *Diabetes* 64 (8): 2791–2804.

159 Amrutkar, M. et al. (2016). STK25 is a critical determinant in nonalcoholic steatohepatitis. *FASEB J* 30 (10): 3628–3643.

160 Nuñez-Durán, E. et al. (2018). Serine/threonine protein kinase 25 antisense oligonucleotide treatment reverses glucose intolerance, insulin resistance, and nonalcoholic fatty liver disease in mice. *Hepatol. Commun.* 2 (1): 69–83.

161 Brown, M.S. and Goldstein, J.L. (1986). A receptor-mediated pathway for cholesterol homeostasis. *Science* 232 (4746): 34–47.

162 Love, K.T. et al. (2010). Lipid-like materials for low-dose, in vivo gene silencing. *PNAS* 107 (5): 1864–1869.

163 Akinc, A. et al. (2008). A combinatorial library of lipid-like materials for delivery of RNAi therapeutics. *Nat. Biotechnol.* 26 (5): 561–569.

164 Cansby, E. et al. (2019). Protein kinase MST3 modulates lipid homeostasis in hepatocytes and correlates with nonalcoholic steatohepatitis in humans. *FASEB J.j* 33 (9): 9974–9989.

165 Caputo, M. et al. (2021). Silencing of STE20-type kinase MST3 in mice with antisense oligonucleotide treatment ameliorates diet-induced nonalcoholic fatty liver disease. *FASEB J.j* 35 (5): e21567.

166 Setten, R.L., Rossi, J.J., and Han, S.P. (2019). The current state and future directions of RNAi-based therapeutics. *Nat. Rev. Drug Discovery* 18 (6): 421–446.

167 Xue, H. et al. (2019). DNA tetrahedron-based nanogels for siRNA delivery and gene silencing. *Chem. Commun. (Camb.)* 55 (29): 4222–4225.

168 Yang, J. et al. (2018). Self-assembled double-bundle DNA tetrahedron for efficient antisense delivery. *ACS Appl. Mater. Interfaces* 10 (28): 23693–23699.

169 Yamankurt, G. et al. (2020). The effector mechanism of siRNA spherical nucleic acids. *PNAS* 117 (3): 1312–1320.

170 Nimjee, S.M. et al. (2017). Aptamers as therapeutics. *Annu. Rev. Pharmacol. Toxicol.* 57 (1): 61–79.

171 Sun, Y. et al. (2021). Erythromycin loaded by tetrahedral framework nucleic acids are more antimicrobial sensitive against *Escherichia coli* (*E. coli*). *Bioact. Mater.* 6 (8): 2281–2290.

172 Altman, M.O. et al. (2013). Modifying cellular properties using artificial aptamer-lipid receptors. *Sci. Rep.* 3: 3343.

173 Wu, Y. et al. (2010). DNA aptamer-micelle as an efficient detection/delivery vehicle toward cancer cells. *PNAS* 107 (1): 5–10.

174 Lincoff, A.M. et al. (2016). Effect of the REG1 anticoagulation system versus bivalirudin on outcomes after percutaneous coronary intervention (REGULATE-PCI): a randomised clinical trial. *Lancet* 387 (10016): 349–356.

175 Zhou, M. et al. (2021). A DNA nanostructure-based neuroprotectant against neuronal apoptosis via inhibiting toll-like receptor 2 signaling pathway in acute ischemic stroke. *ACS Nano*.

176 Krissanaprasit, A. et al. (2021). Self-assembling nucleic acid nanostructures functionalized with aptamers. *Chem. Rev.* 121 (22): 13797–13868.

177 Cui, W. et al. (2020). Preventive effect of tetrahedral framework nucleic acids on bisphosphonate-related osteonecrosis of the jaw. *Nanoscale* 12 (33): 17196–17202.

178 Meng, D. et al. (2020). DNA-driven two-layer core-satellite gold nanostructures for ultrasensitive MicroRNA detection in living cells. *Small* 16 (23): e2000003.

9

The Antibacterial Applications of Framework Nucleic Acid-Based Nanomaterials: Current Progress and Further Perspectives

Zhiqiang Liu and Yue Sun

Sichuan University, State Key Laboratory of Oral Diseases, West China Hospital of Stomatology, South Renmin Road, Section 3 No.14, Chengdu 610041, PR China

Due to various reasons, such as the abuse of antibiotics, antibiotic resistance has become an increasingly serious global human health problem [1, 2]. According to clinical statistics, the drug resistance rate of pathogenic bacteria could be as high as 30–50%, and it was increasing by 5% year by year [3–5]. The restricted drug uptake and maintenance of traditional antibiotics owing to low bacterial membrane permeability and induction of efflux pumps lead to drug resistance [5–8]. In addition, several new materials, such as antimicrobial peptides (AMPs) and nucleic acid nanoparticles (NPs), have shown promising potential in reversing antibiotic resistance [9–11]. However, some disadvantages hinder their applications. For example, AMPs and nucleic acid NPs are easily degraded and still lack effective targeted delivery [12–15]. Therefore, effective strategies are urgently needed to promote the delivery of antibacterial materials [16–18].

Since the concept that DNA has potential as a programmable structural material was first proposed by Seaman et al. in the 1980s, DNA nanotechnology has been recognized by the public [19]. Seeman et al. designed the first DNA tetrameric junction mimicking the natural DNA structure Holliday junction in 1983, which marked the beginning of the era of DNA nanostructures [20]. However, DNA nanostructures based on single-stranded DNA were still very simple at the time. The real huge upgrade was achieved by Rothemund, who pioneered the concept of DNA origami [21]. The basic principle of DNA origami was the hybridization of long "scaffold" DNA strands. By designing the staple sequence, scaffold chains could be folded to form various target shapes [22]. As a result, DNA nanostructures were also gradually upgraded from two-dimensional (2D) to three-dimensional (3D). By stacking planar layers or changing the curvature of the structure, the planar shape could be further folded into a 3D geometry structure [23–26].

With more and more research on DNA origami, DNA nanotechnology has flourished in the past decades [27–29]. A variety of DNA nanostructures, including 2D or 3D structure and DNA hydrogel, have emerged and attracted great interest of the public [30–32]. Planar 2D DNA nanostructures containing ribbons [33], 2D crystals [34], triangles [35], quadrilaterals [31], and pseudo-hexagonal arrays [31] initially appeared. Then there were nanoribbons and 2D nanogrids of 4×4 tiles designed by LaBean and colleagues [36].

Nucleic Acid-Based Nanomaterials: Stabilities and Applications, First Edition.
Edited by Yunfeng Lin and Shaojingya Gao.
© 2024 WILEY-VCH GmbH. Published 2024 by WILEY-VCH GmbH.

Afterward, Mao et al. created DNA tetrahedral nanostructure, octahedral nanostructure, dodecahedral nanostructure, and even icosahedral nanostructure [37]. These structures herald the dawn of the era of 3D DNA nanostructure. Moreover, studies on DNA hydrogels have received great attention in the past decade [38, 39]. As a hydrophilic material, DNA could be used as the basic component of hydrogel, using hydrogen bonds of complementary base pairing or interaction with other substances, simple hydrogels or composite hydrogels made of DNA nanostructures were synthesized and studied [40].

Several studies demonstrated that many kinds of DNA nanostructures were widely used in biomedical fields. For example, DNA nanostructures were used in tissue engineering, immune engineering, biosensors, and drug delivery, especially delivery of antibacterial drugs [27]. DNA nanostructures were constructed from oligonucleotide staples [41]. As the constituent elements of DNA nanostructures, these staples could be combined in various ways to carry various antibacterial drugs such as erythromycin, AMPs, lysozyme, and antisense oligonucleotide (ASO) [5, 16, 17]. Based on this principle, a variety of DNA origami structures including 2D DNA nanostructure, 3D DNA nanostructure, and DNA hydrogel were applied to reduce antibiotic resistance. Furthermore, previous studies have shown that DNA nanostructures possessed good compatibility, certain stability, unique editability, and superior drug-loading performance, which played a crucial role in becoming ideal candidates for antimicrobial drug delivery vehicles [42].

Therefore, this review will summarize the progress of various DNA nanostructures including 2D DNA nanostructure, 3D DNA nanostructure, and DNA nanostructure hydrogel in the antibacterial field in recent years, and clarify their future development directions (Figure 9.1).

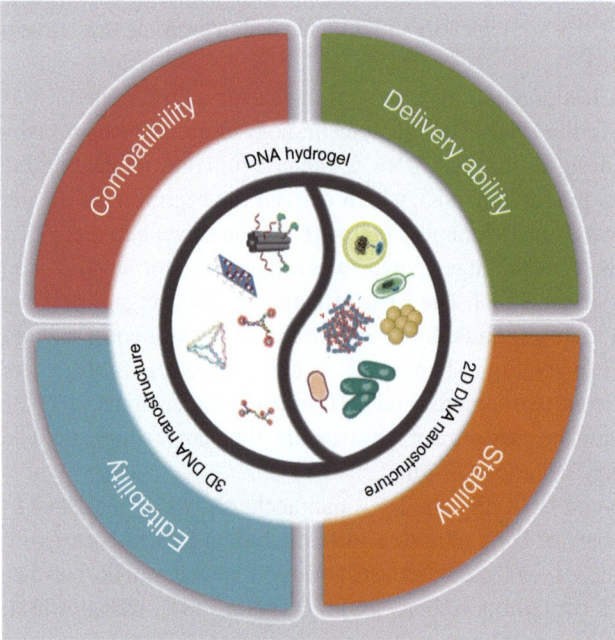

Figure 9.1 The application of DNA nanostructures, including 2D DNA nanostructures, 3D DNA nanostructures, and DNA hydrogels in the antibacterial field has the advantages of biocompatibility, editability, stability, and delivery ability.

9.1 Some Advantages of DNA Nanostructures in the Antibacterial Field

9.1.1 Compatibility of DNA Nanostructures

Several studies have shown that various DNA nanostructures possess good compatibility [5, 43, 44]. As a biological macromolecule, DNA is the basic component of the human body, so DNA nanostructures may have better biocompatibility and low immunogenicity than other nanomaterials. For example, tetrahedral framework DNA loaded with erythromycin showed no obvious cytotoxicity to human umbilical vein endothelial cells (HUVECs) and L929 (mouse fibroblast cells) even at high concentrations [5]. Moreover, there was no significant inhibition on the viability of COS-7 (African green monkey kidney fibroblast-like cells) after treatment with the five "holes" DNA nanostructure loaded with lysozyme [45]. Branched DNA had no inhibitory effect on the culture of various cells including 293T (human embryonic kidney cells), SMCs (vascular smooth muscle cells), and GLC-82 (lung adenocarcinoma cells) [46]. These results suggested that DNA nanostructures might be candidates for drug delivery *in vivo*.

9.1.2 Stability of DNA Nanostructures

Stability is one of the prerequisites for vehicles to deliver drugs. Cell culture media and body fluids are considered as the usual criteria for examining the stability of DNA nanostructures [30]. Yan et al. investigated the stability of naked DNA nanostructures in cell lysates and living cells, respectively. Their research showed that DNA nanostructures were stable in cell lysates for 12 hours and slowly digested in live cells for 72 hours. Compared with the rapid degradation or entanglement of single-stranded and double-stranded DNA in cell lysates, DNA nanostructures possessed greater stability. The possible reason for this was the rigidity of the origami structures. The more compact structure and the larger charge density make DNA nanostructures less prone to degradation and entanglement when treated with cell lysates [47]. However, the stability of DNA nanostructures is still challenging for long-term applications *in vivo* [27, 48].

9.1.3 Editability of DNA Nanostructures

DNA nanostructures are based on oligonucleotide staples, which are loaded with various drugs such as nucleic acids, polypeptides, and traditional antibiotics through different interactions [49]. For example, several oligonucleotide drugs like ASO and aptamers could be linked to DNA nanostructures by phosphodiester bonds [50, 51]; positively charged polypeptides can be linked to negatively charged DNA nanostructures [17, 42]; and antibiotics such as doxorubicin (DOX) could be embedded in DNA nanostructures through intercalation mode [52–54]. The excellent editability of DNA nanostructures was the basis for the drug-loading performance of DNA nanostructures, and it was also the outstanding advantage of DNA nanostructures that distinguished them from other delivery systems. Relying on editability, smarter and fancier DNA nanostructures are expected to be fabricated and applied.

9.1.4 Drug-loading Performance of DNA Nanostructures

The excellent drug delivery ability possessed by various DNA nanostructures enabled them to be applied in various fields [31, 43] [55]. For example, tetrahedral framework DNA could deliver a variety of drugs to treat tumors, eliminate pathogenic bacteria, and inhibit inflammation and reactive oxygen species (ROS) [27, 44]. This superior cellular internalization ability might be related to the small sizes and special spatial conformation of DNA nanostructures [56]. By observing the endocytosis of tetrahedral frame DNA, Liang et al. found that the endocytosis of tetrahedral frame DNA was very rapid. In addition, tetrahedral framework DNA could maintain its structural integrity in cells for a long time [56]. Other DNA nanostructures, such as branched DNA and DNA hydrogel, have shown promising drug delivery capability [46, 57].

9.2 Application of 2D Nanostructures in the Antibacterial Field

9.2.1 Five "Holes" DNA Nanostructure

Dr. Ioanna Mela et al. [45] constructed a 2D DNA nanostructure consisting of a five "holes" origami framework (Figure 9.2a). In this nanostructure, the aptamers were modified at the edges of the planar five "holes" framework to target G^+ bacterial strains *Bacillus subtilis* (*B. subtilis*) and G^- bacterial strains *Escherichia coli* (*E. coli*). Additionally, four fluorophore (Alexa 647) molecules were added to the staples of this nanostructure to facilitate the detection of the nanostructure using fluorescence microscopy. In order to clear pathogens, biotinylated lysozyme, an active antibacterial ingredient, was successfully attached to DNA nanostructure through an efficient binding between biotin and streptavidin. Thus, a novel targeting bacteria platform based on DNA origami nanotechnology was successfully generated.

In this system, a planar framework with five "holes" was used as a vehicle to connect important components, including targeting aptamers, lysozyme, and fluorescent groups. It made the organic combination of each ingredient to achieve the two goals of intelligent targeting and effective killing (Figure 9.1a). At the same time, the shape of the DNA nanostructure with "holes" was chosen because it could help the antibacterial lysozyme to be delivered to the bacterial surface and maximize the chance of the active enzyme contacting with the bacterial surface.

9.2.2 Super Silver Nanoclusters Based on Branched DNA

Yang et al. [46] fabricated super silver nanoclusters (AgNCs) using branched DNA as scaffolds (Figure 9.2b). Through artificial specific regulation, ssDNAs were assembled to form different arm connections and then constructed as branched DNA. Branched DNA based on DNA origami technology had excellent spatial editing. That is to say, different types of DNA scaffolds were constructed by adjusting the length and shape of DNA strands. In this study, three super-AgNCs were synthesized, namely Y-shaped DNA/AgNC (Y-super NC), X-shaped DNA/AgNC (X-super NC), and (Y–X)-shaped DNA/ AgNC ((Y–X)-super NC). The three super-AgNCs possessed different spatial structures and optical properties due to

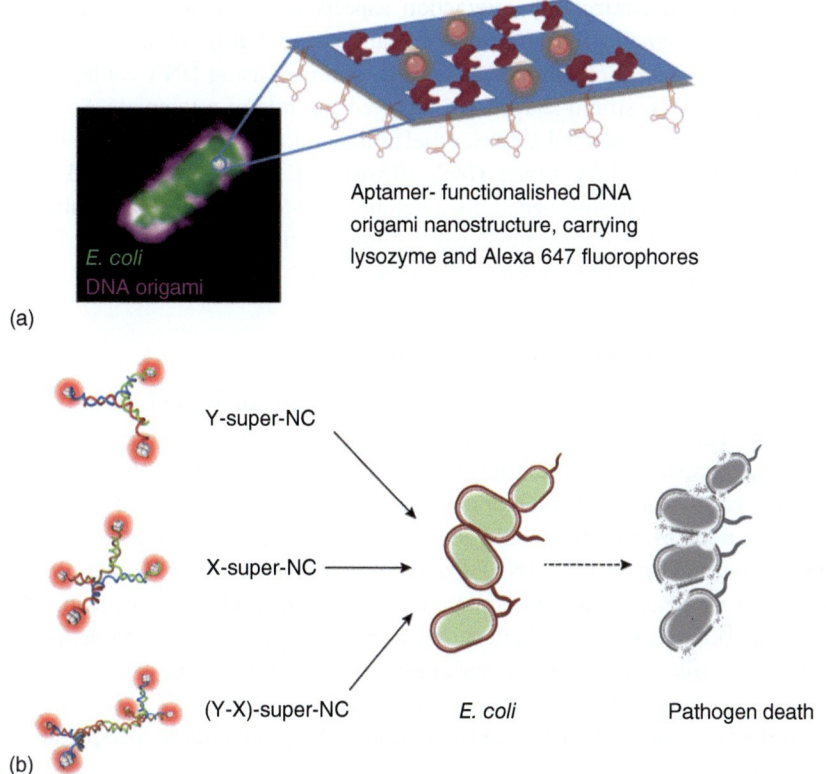

Figure 9.2 Application of 2D DNA nanostructures in the antibacterial field. (a) The planar framework with five "holes" loaded with targeting aptamers, lysozyme, and fluorescent groups, killed and eliminated E. coli. Source: Mela et al. [45]/John Wiley & Sons/CC BY 4.0. (b) Super-AgNCs based on branched DNA inhibited the growth of E. coli. Source: Reproduced from Yang et al. [46]/John Wiley & Sons.

the change in the scaffold shape. Later, the study validated the antibacterial properties of the super-AgNCs.

To be more specific, antibacterial activity was assessed by measuring the growth curve of E. coli after incubation of ssDNA/AgNC, Y-super-NC, X-super-NC, and (Y–X)-super-NC with E. coli. Compared with the control group without antibacterial agent, all three super AgNCs inhibited the growth of E. coli. In addition, the study found that the three AgNCs had a slight inhibitory effect on the growth of the G$^+$ bacteria Staphylococcus aureus (S. aureus). Finally, it was found that the super-AgNCs possessed good biocompatibility with three types of cells, including 293T (human embryonic kidney cells), SMC (vascular smooth muscle cells), and GLC-82 (lung adenocarcinoma cells). This study was the first to discover that branched DNA nanostructure could act as a scaffold to regulate antibacterial activity.

9.2.3 Melamine-DNA-AgNC Complex

DNA conjugation chemistries have also made some progress in the field of antibacterials [22, 58, 59]. In DNA conjugation chemistries, DNA is usually used as programmable "bonds" to control the interaction aspects of the material surface to achieve the purpose of enhancing the performance of antibacterial drugs [22].

Eun et al. [60] constructed a melamine-DNA-AgNC complex using DNA conjugation chemistries. In the complex, strand-shaped DNA played the role of a template. Specifically, hydrogen bonds were generated between melamine and thymine residues in DNA to form a melamine-DNA-AgNC complex (Mel-DNA-AgNC). It has been verified that Mel-DNA-AgNC had excellent fluorescence efficiency and long-term stability. Surprisingly, the bright and stable Mel-DNA-AgNCs obviously exhibited antibacterial activity, and this property was equally applicable to G^+ bacteria *S. aureus* and G^- bacteria *E. coli*.

9.2.4 NET-like Nanogel Based on 2D DNA Networks

Chen et al. [61] extracted kiwifruit genomic DNA as the main component of the DNA nanostructure network. The extracted DNA acquired a large number of hydroxyl groups after HCl treatment to facilitate cross-linking with citric acid. DNA strands with numerous amide bonds were coached into 2D membrane structures of DNA networks. This negatively charged DNA network structure could interact with positively charged ZnO NPs via electrostatic attraction and finally form a NET-like DNA-HCl-ZnO nanogel (NG). The novel DNA-HCl-ZnO NG exerted a good anti-inflammatory effect *in vitro* and *in vivo*. More importantly, it had a remarkable antibacterial effect that could not be ignored. This conclusion was verified in the *E. coli*-induced mouse peritonitis model. Intraperitoneal injection of DNA-HCl-ZnO NG improved survival rates in mice with *E. coli*-induced sepsis. This might be attributed to the fact that DNA-HCl-ZnO NG inhibited the proliferation and spread of *E. coli* in septic mice.

9.2.5 ε-poly-L-lysine-DNA Nanocomplex

The AMP ε-poly-L-lysine (PL) is widely known for its remarkable antibacterial effect and wide variety of antibacterial species [62]. However, PL losed its antibacterial effect due to the strong electrostatic interaction between phosphate and PL. To solve this problem, Jiang et al. [63] created a nanocomplex of ε-poly-L-lysine with DNA. The properties of PL were improved based on various properties of DNA, such as good biocompatibility, low immunogenicity, and hydrophobic groups that increased cell adhesion. The novel DNA-PL nanocomplexes possessed superior antibacterial properties compared to pure PL and had significant effects on a variety of bacteria, including *E. coli*, *S. aureus*, *Bacillus subtilis*, *Baumannii*, and *gas Monospores*. This change might be attributed to the strong adhesion of DNA/PL nanocomplexes. In addition, the small size of the DNA/PL nanocomplex was a guarantee for its entry into bacteria.

9.3 Application of 3D DNA Nanostructures in the Antibacterial Field

9.3.1 Tetrahedral Framework DNA

Tetrahedral framework DNA, as one of the simplest 3D DNA nanostructures, has many advantages, such as rapid synthesis, high yield, good biocompatibility, outstanding drug delivery, and permeability [44, 64–67]. Several studies have shown that tetrahedral

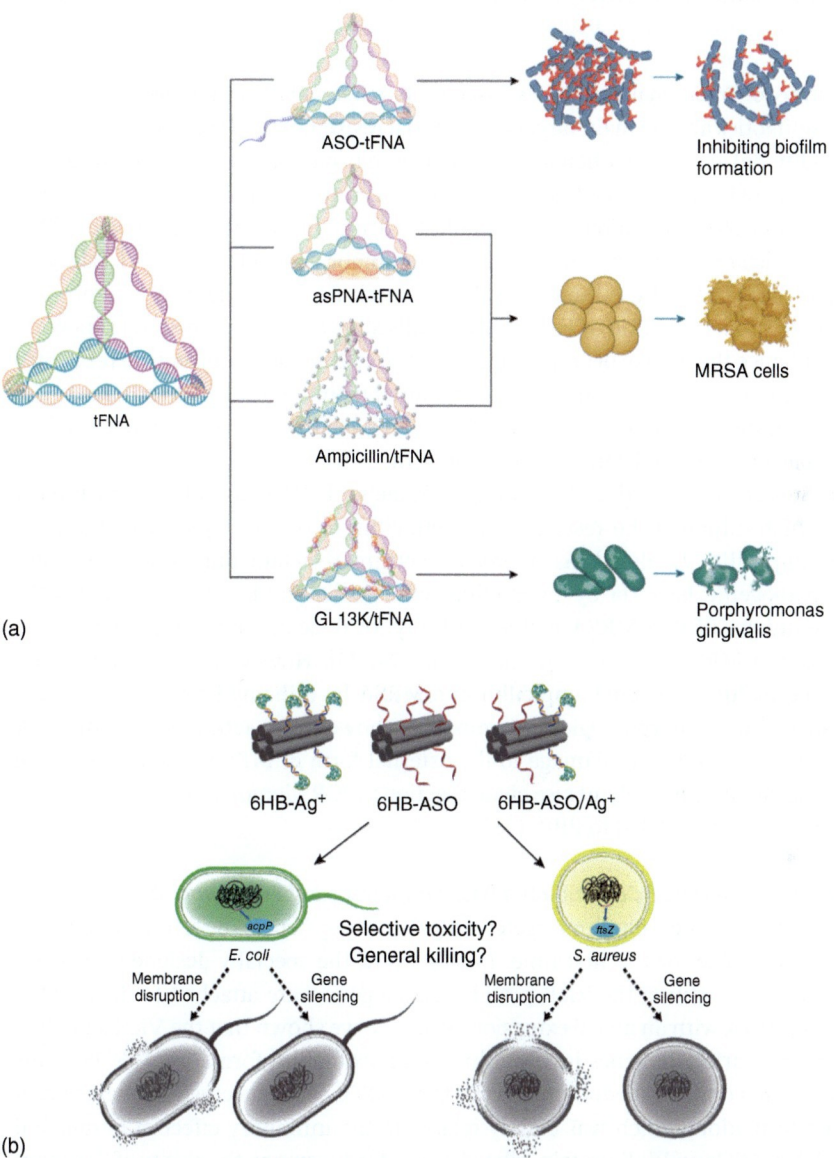

Figure 9.3 Application of 3D DNA nanostructures in the antibacterial field. (a) Tetrahedral framework DNA loaded with traditional antibiotics, nucleic acid antibiotics, or polypeptide antibiotics inhibited the growth of several bacteria and prevented bacterial biofilm formation. Source: Reproduced from Zhang et al. [64]. Copyright 2020, American Chemical Society. (b) DNA 6HB structure loaded with ASOs and Ag were delivered to *E. coli* and *S. aureus*. And 6HB loaded with ASOs and Ag could inhibit the growth of *E. coli* and *S. aureus*. Source: Reproduced from Long et al. [50]. Copyright 2020, American Chemical Society.

framework DNA could successfully deliver three types of antibacterial drugs through various binding methods, effectively solving the practical problem of antibiotic resistance (Figure 9.3a).

9.3.1.1 Delivery of Traditional Antibiotics Based on Tetrahedral Framework DNA

Traditional antibiotics are the main targets of bacterial resistance, and the resistance mechanisms of traditional antibiotics often involve changes in bacterial membrane permeability. This alteration may lead to reduced drug uptake and accumulation by bacteria, leading to drug resistance [68, 69]. In addition, bacterial efflux of drugs through the induction of efflux pumps may also lead to a decrease in intracellular antibiotic concentrations [70]. Based on this principle, Sun et al. [5] used tetrahedral framework DNA as a delivery platform for erythromycin to promote killing of *E. coli*. The results showed that tetrahedral framework DNA could enhance the uptake of erythromycin by *E. coli*, thereby inhibiting the resistance of *E. coli* to erythromycin. The permeability of bacterial membranes was altered due to the interaction of tetrahedral framework DNA with bacterial membranes. This change resulted in increased bacterial internalization of erythromycin.

Similarly, Sun et al. [71] utilized tetrahedral framework DNA to deliver and protect ampicillin, which inhibited the resistance of methicillin-resistant *Staphylococcus aureus* (MRSA) to ampicillin. In detail, as a broad-spectrum β-lactam antibiotic, ampicillin eliminated pathogenic bacteria by interacting with penicillin-binding proteins (PBPs) [72, 73]. The unique PBP of MRSA had low affinity for β-lactam antibiotics, which was the mechanism of MRSA resistance to ampicillin [74, 75]. However, tetrahedral framework DNA successfully delivered ampicillin into MRSA by utilizing its superior delivery capacity. Furthermore, morphological examination showed that tetrahedral framework DNA-ampicillin caused greater damage to the external form of MRSA, compared to free ampicillin. The quantitative polymerase chain reaction (qPCR) demonstrated this powerful killing effect was closely related to PBPs [71].

9.3.1.2 Delivery of Nucleic Acid Antibiotics Based on Tetrahedral Framework DNA

Zhang et al. [16] developed a biofilm-targeted drug delivery system using tetrahedral framework DNA as a vehicle for the first time. In this study, the specially designed antisense nucleotides (ASOs) targeting the VicK protein-related gene were attached to the tetrahedral framework DNA with an apical extension. Studies have shown that the VicK signaling system affected a variety of genes involved in *S. mutans* biofilm formation [76–78]. This study demonstrated that tetrahedral framework DNA-ASOs could successfully prevent bacterial biofilm formation, which was closely related to the inhibitory effect of tetrahedral framework DNA-ASOs on VicK protein-related genes. Furthermore, the ability of the nanomaterials to enter bacteria was examined by flow cytometry and confocal microscopy. The results indicated that tetrahedral framework DNA-ASOs were successfully internalized by *S. mutans* compared to tetrahedral framework DNA or ASOs alone. Similarly, Hu et al. [79] demonstrated that tetrahedral framework DNA loaded with ASOs effectively adhered to the surface of bacterial membranes and efficiently killed *S. aureus* and *E. coli*.

In addition to ASOs, another study by Zhang et al. [16] showed that antisense peptide nucleic acids (asPNAs), as synthetic DNA analogs of third-generation antisense nucleic acids [80], could also be successfully delivered by tetrahedral framework DNA. AsPNAs have been shown to inhibit bacterial gene expression and are potential antibacterial therapeutic [10, 11]. However, due to the lack of effective vehicles, asPNAs had a low rate of bacterial uptake, and their therapeutic effect was greatly limited [15, 81]. Tetrahedral framework DNA as a novel vehicle facilitated asPNAs to penetrate the cell wall of MRSA, which inhibited the expression of specific genes and ultimately effectively inhibited the growth of MRSA. In addition to G$^+$ bacteria such as MRSA, the penetrating ability of tetrahedral framework DNA loaded with nucleic acid antibiotics was also applicable to G$^-$ bacteria. A study showed that tetrahedral framework DNA successfully delivered anti-blaCTX-M-group 1 PNA (PNA4) to a typical G$^-$ bacteria *E. coli*, to eliminate *E. coli* effectively [15, 82].

9.3.1.3 Delivery of Polypeptide Antibiotics Based on Tetrahedral Framework DNA

AMPs, as cations and short peptides, usually rely on electrostatic interactions to interact with negatively charged microbial membranes, ultimately leading to bacterial death [9, 83, 84]. Although AMPs were not plagued by drug-resistant bacteria. However, AMPs were easily degraded by proteases produced by bacteria [12, 13]. For example, *Porphyromonas gingivalis* could degrade and overcome the AMP GL13K, rendering it ineffective. Therefore, how to protect AMPs from degradation is critical to address the resistance of drug-resistant bacteria. Tetrahedral framework DNA was successfully used for the delivery of AMPs for the first time. At appropriate ratios, tetrahedral framework DNA-GL13K (t-GL13K) eliminated *E. coli* due to increased bacterial uptake and enhanced interaction with microbial membranes [17].

Setyawati et al. [85] prepared a multifunctional platform based on tetrahedral framework DNA, which possessed both diagnostic and therapeutic functions. In this nanostructure, actinomycin D (AMD), as a model antimicrobial drug, was bound to the tetrahedral framework DNA in a dsDNA-intercalating manner, and when red-emitting glutathione-protected gold nanoclusters (GSH-Au NCs) were combined on the apex of tetrahedral framework DNA, the nanocomplex DPAu/AMD was finally formed. In this nanoplatform, tetrahedral framework DNA successfully delivered AMD through bacterial membranes into bacteria, stabilizing and protecting AMD, which resulted in antibacterial therapeutic effect of this nanoplatform. As for the diagnostic function, it depended on the function of Au NCs. The favorable diagnostic and therapeutic performance of DPAu/AMD was demonstrated in *E. coli* and *S. aureus*. The growth of both bacteria was significantly inhibited by DPAu/AMD.

9.3.2 DNA Six-Helix Bundle

Long et al. [50] utilized multiple chains to self-assemble into Y-shaped nanostructures. The two Y-shaped nanostructures were combined with each other through sticky ends and finally dimerized into a DNA six-helix bundle (6HB) structure (Figure 9.3b). The ASOs and Ag were integrated by 6HB and delivered to *E. coli* and *S. aureus*. Surprisingly, by observing the growth curve of bacteria, Ag-containing 6HB could inhibit the growth of *E. coli* and *S. aureus*, while ASO-containing 6HB was only selectively resistant to *S. aureus*. Furthermore, this powerful antibacterial effect persists for a longer period of time compared to previous means of delivery. The mechanism for this phenomenon might be that 6HB helped to increase local Ag and ASO concentrations, increasing the chance of contact with bacteria. In addition, this study demonstrated no adverse effects on normal mammalian cells by detecting the cell viability of L02 (human normal hepatocytes) after eight hours incubation with the nanomaterials.

9.3.3 DNA Nanoribbon

It was well known that β-lactamases were effective enzymes against the most common antimicrobial β-lactam antibiotics, such as penicillins [72, 86]. Under the catalysis of β-lactamase enzyme, β-lactam antibiotics were destroyed, resulting in bacterial resistance [87]. Class B enzymes, one of β-lactamases, were also known as metallo-β-lactamases (MβLs) [88]. Unfortunately, there were no MβL inhibitors in the clinic [89]. Therefore, the development of MβL inhibitors was critical to address drug resistance. Ouyang et al. [90] found that the DNA nanoribbons possessed antibacterial functions. Using DNA origami technology, three DNA nanoribbons with different widths (DNR1, DNR2, and DNR3) were created. Among them, DNR1 B with a width of 16.2 nm had a strong inhibitory effect on MβLs, while DNR2 and DNR3 did not. In addition, this study showed that this antibacterial effect was associated with specific nanostructures and specific sequences. That is to say, ssDNA, dsDNA, and tetrahedral DNA nanostructures had no inhibitory effect on MβLs. After changing the sequence, the inhibitory effect of DNR1 was also weakened. Finally, spectroscopic and atomic force microscopy results suggested that DNR1combined to *E. coli* through a minor groove binding mode and exerted an antibacterial effect. This study was also the first to discover DNA nanostructures that specifically inhibited β-lactamase.

9.3.4 DNA Pom-Pom Nanostructure

Zeng et al. [52] synthesized a multifunctional platform based on DNA Pom-Pom nanostructure (DNA PP-N). The platform integrated the detection and inactivation of pathogenic bacteria. Briefly, a specially designed initial strand (T1) opened metastable DNA hairpin structures (H1, H2, H3), enabling the self-assembly of DNA PP-N. Then, DNA PP-N was modified to successfully modify the aptamer to accurately recognize bacteria. On the one hand, PP was modified with the fluorescent DNA signaling probes, which were then magnetically separated by microbeads. Through this reaction, the DNA nanostructures could be used for target bacterial detection and sensing. This utility was validated by multiplex analysis of common *S. aureus* and *E. coli*. On the other hand, DNA PP-N could effectively carry DOX against *E. coli* bacterial infection. In this multifunctional platform, DNA PP-N could serve as a linker for detection elements and complete targeted delivery of antibiotics. This nanoplatform was expected to play a major role in detecting and reducing pathogenic bacteria.

9.4 Application of DNA Hydrogel Nanostructures in the Antibacterial Field

Obuobi et al. [57] constructed a DNA hydrogel delivering AMP L12 (Figure 9.4). First, the study utilized the cross-linking of Y-scaffold and L-linker to fabricate DNA hydrogel.

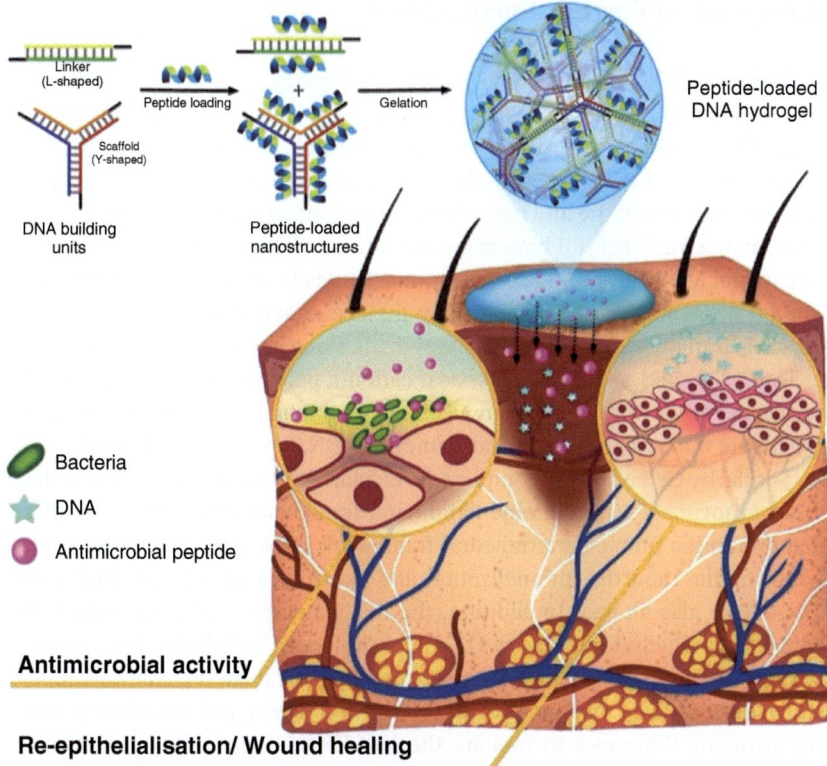

Figure 9.4 L12-loaded DNA hydrogel accelerated deep and chronic wounds through the antibacterial effects of L12 and the anti-inflammatory effects of DNA hydrogel *in vivo*. Source: Obuobi et al. [57]/Elsevier.

By sticky end hybridization, the Y-scaffold and L-linker accomplished gelatinization rapidly within one minute. L12 was a positively charged short peptide. Therefore, DNA hydrogel, as a negatively charged polymer, could bind to AMPs through electrostatic attraction, and the binding force was relatively strong.

Through the cell viability test, the DNA hydrogel delivery system showed good biosafety. More interestingly, this DNA hydrogel could achieve environmental nuclease release to L12. Specifically, when the DNA hydrogel was exposed to DNase, the bulky nucleic acid cross-linking system disintegrated (a hydrogel-to-sol state occurred during this period). Eventually, L12 was controllably released. This result indicated that the DNA hydrogel could intelligently control the release of L12, which had enlightening significance for the killing of pathogenic bacteria by releasing nuclease. In *in vitro* experiment, the results indicated that the L12-loaded DNA hydrogel was significantly resistant to *S. aureus* infection. And L12-loaded DNA hydrogel accelerated deep and chronic wounds through the antibacterial effects of L12 and the anti-inflammatory effects of DNA hydrogel *in vivo*. The above results suggested the great potential of DNA hydrogel as a vehicle for intelligent loading and release of AMPs.

9.5 Challenges and Further Perspectives

With the development of DNA nanotechnology in recent decades, a variety of DNA nanostructures have been created. In detail, 2D DNA nanostructures contained rectangular DNA, smiley DNA, etc. 3D DNA nanostructures contained tetrahedral DNA, octahedral DNA, etc. In addition, DNA hydrogels have also received extensive attention. Based on DNA origami technology, there will even be more and more fancy DNA nanostructures [27, 91].

However, DNA nanostructures still have some limitations. For example, although DNA nanostructures as nucleic acid substances can protect nucleic acids, polypeptides, and other antibacterial drugs to a certain extent, they may be attacked by nucleases *in vivo* [27]. To address this limitation, some scholars have made some enlightening explorations. For example, Tian et al. [92] applied the classical cationic polymer ethyleneimine (PEI) to protect tetrahedral framework DNA from DNase degradation. Moreover, the addition of PEI could facilitate the entry of tetrahedral framework DNA into cells and enable lysosomal escape. Similarly, Ge et al. [93] complexed PEGylated protamine with tetrahedral framework DNA to form a PEG-protamine-tetrahedral framework DNA (PPT) complex. The PEGylated protamine promoted tetrahedral framework DNA internalization in cells of three different tissues, and the internalization and excretion pathways of tetrahedral framework DNA were also altered. In addition, the PEGylated protamine promoted the lysosomal escape of tetrahedral framework DNA and prolonged circulation time *in vivo*. In the recent study of Ma et al. [94], the aptamer-modified tetrahedral framework DNA was successfully encapsulated by functionalizing the red blood cell membrane with pH-responsive synthetic liposomes to prolong the half-life of DNA nanostructures. At the same time, the protective effect of the cell membrane reduced the clearance of the mononuclear phagocytic cell system and increased its targeting of tumors.

Although various types of DNA nanostructures are currently involved in antibacterial field, more intelligent DNA nanorobots are necessary for more precise antibacterial needs. For example, the targeting performance of most DNA nanostructures to specific bacteria needs to be improved, especially the identification and targeted killing of pathogenic bacteria and normal flora. Previous studies on DNA nanostructures targeting tumors through aptamers and dynamically controlling the switch to release drugs may provide new ideas and strategies for this problem [95]. In a recent study, a novel DNA nanostructured DNA fiber enabled the specific capture of *E. coli*, which could be used for future studies on antimicrobial resistance [96]. With the help of easily incorporated smart molecules, including DNA aptamers and DNA origami-based robotics, the targeting problem of pathogenic bacteria may be solved.

In conclusion, plenty of DNA nanostructures showed non-negligible potential in the antibacterial field due to their multiple biological properties, including good biocompatibility, small size, unique editability, certain stability, and excellent delivery ability. According to the spatial conformation of DNA nanostructures, DNA nanostructures involved in antibacterial field can be divided into 2D DNA nanostructures, 3D DNA nanostructures, and DNA hydrogels (Table 9.1). Although the NPs based on various DNA nanostructures have demonstrated promising antibacterial effects *in vitro*, the final validation *in vivo* needs to be supplemented. In particular, how to improve the maintenance of DNA nanostructures *in vivo* and how to target and kill pathogenic bacteria are the current challenges. It is believed that more satisfactory answers will be given in the study in the near future.

Table 9.1 DNA nanostructures in the antibacterial field.

Number	DNA nanostructures	Spatial dimension	Payloads	Pathogens	References
1	Five "holes" DNA nanostructure	2D	Targeting aptamers, lysozyme, and fluorescent groups	E. coli, and Bacillus subtilis	[45]
2	Branched DNA	2D	AgNC	E. coli	[46]
3	Strand-shaped DNA	2D	Melamine AgNC	S. aureus and E. coli	[60]
4	DNA networks	2D	ZnO NPs	E. coli	[61]
5	Strand-shaped DNA	2D	PL	E. coli, S. aureus, Bacillus subtilis, Baumannii, and gas Monospores.	[63]
6	Tetrahedral framework DNA	3D	Erythromycin	E. coli	[5]
7	Tetrahedral framework DNA	3D	Ampicillin	MRSA	[71]
8	Tetrahedral framework DNA	3D	AsPNAs	MRSA	[16]
9	Tetrahedral framework DNA	3D	ASO	S. mutans	[16]
10	Tetrahedral framework DNA	3D	ASO	S. aureus and E. coli	[79]
11	Tetrahedral framework DNA	3D	PNA4	E. coli	[82]
12	Tetrahedral framework DNA	3D	AMD, Au NCs	S. aureus and E. coli	[85]
13	Tetrahedral framework DNA	3D	GL13K	Porphyromonas gingivalis	[17]
14	DNA six-helix bundle	3D	Ag and ASO	S. aureus and E. coli	[50]
15	DNA nanoribbon	3D	N/A	E. coli	[90]
16	DNA PP-N structure	3D	DOX	S. aureus and E. coli	[52]
17	Y-scaffold and L-linker	Hydrogel	L12	S. aureus	[57]

References

1 Boolchandani, M., D'Souza, A.W., and Dantas, G. (2019). Sequencing-based methods and resources to study antimicrobial resistance. *Nat. Rev. Genet.* 20 (6): 356–370.
2 Balaban, N.Q., Helaine, S., Lewis, K. et al. (2019). Definitions and guidelines for research on antibiotic persistence. *Nat. Rev. Microbiol.* 17 (7): 441–448.
3 Aminov, R.I. (2010). A brief history of the antibiotic era: lessons learned and challenges for the future. *Front. Microbiol.* 1: 134.
4 Veerapandian, M. and Yun, K. (2011). Functionalization of biomolecules on nanoparticles: specialized for antibacterial applications. *Appl. Microbiol. Biotechnol.* 90 (5): 1655–1667.
5 Sun, Y., Liu, Y., Zhang, B. et al. (2021). Erythromycin loaded by tetrahedral framework nucleic acids are more antimicrobial sensitive against *Escherichia coli* (*E. coli*). *Bioact. Mater.* 6 (8): 2281–2290.
6 Poirel, L., Jayol, A., and Nordmann, P. (2017). Polymyxins: antibacterial activity, susceptibility testing, and resistance mechanisms encoded by plasmids or chromosomes. *Clin. Microbiol. Rev.* 30 (2): 557–596.
7 Mohanam, L., Priya, L., Selvam, E.M. et al. (2016). Molecular mechanisms of efflux pump mediated resistance in clinical isolates of multidrug resistant pseudomonas aeruginosa. *Int. J. Infect. Diseases* 45: 104.
8 Dinos, G.P. (2017). The macrolide antibiotic renaissance. *Br. J. Pharmacol.* 174 (18): 2967–2983.
9 Lam, S.J., O'Brien-Simpson, N.M., Pantarat, N. et al. (2016). Combating multidrug-resistant gram-negative bacteria with structurally nanoengineered antimicrobial peptide polymers. *Nat. Microbiol.* 1 (11): 16162.
10 Grijalvo, S., Alagia, A., Jorge, A.F., and Eritja, R. (2018). Covalent strategies for targeting messenger and non-coding RNAs: an updated review on siRNA, miRNA and antimiR conjugates. *Genes* 9 (2): 74.
11 Hegarty, J.P. and Stewart, D.B.S. (2018). Advances in therapeutic bacterial antisense biotechnology. *Appl. Microbiol. Biotechnol.* 102 (3): 1055–1065.
12 Anaya-López, J.L., López-Meza, J.E., and Ochoa-Zarzosa, A. (2013). Bacterial resistance to cationic antimicrobial peptides. *Crit. Rev. Microbiol.* 39 (2): 180–195.
13 Andersson, D.I., Hughes, D., and Kubicek-Sutherland, J.Z. (2016). Mechanisms and consequences of bacterial resistance to antimicrobial peptides. *Drug Resist. Updates* 26: 43–57.
14 Wolfe, J.M., Fadzen, C.M., Holden, R.L. et al. (2018). Perfluoroaryl bicyclic cell-penetrating peptides for delivery of antisense oligonucleotides. *Angew. Chem. Int. Ed.* 57 (17): 4756–4759.
15 Bessa, L.J., Ferreira, M., and Gameiro, P. (2018). Evaluation of membrane fluidity of multidrug-resistant isolates of *Escherichia coli* and *Staphylococcus aureus* in presence and absence of antibiotics. *J. Photochem. Photobiol., B* 181: 150–156.
16 Zhang, Y., Ma, W., Zhu, Y. et al. (2018). Inhibiting methicillin-resistant *Staphylococcus aureus* by tetrahedral DNA nanostructure-enabled antisense peptide nucleic acid delivery. *Nano Lett.* 18 (9): 5652–5659.

17 Liu, Y., Sun, Y., Li, S. et al. (2020). Tetrahedral framework nucleic acids deliver antimicrobial peptides with improved effects and less susceptibility to bacterial degradation. *Nano Lett.* 20 (5): 3602–3610.

18 Zhang, Y., Xie, X., Ma, W. et al. (2020). Multi-targeted antisense oligonucleotide delivery by a framework nucleic acid for inhibiting biofilm formation and virulence. *Nano Micro Lett.* 12 (1): 74.

19 Stephenson, M.L. and Zamecnik, P.C. (1978). Inhibition of Rous sarcoma viral RNA translation by a specific oligodeoxyribonucleotide. *Proc. Natl. Acad. Sci U. S. A.* 75 (1): 285–288.

20 Kallenbach, N.R., Ma, R.-I., and Seeman, N.C. (1983). An immobile nucleic acid junction constructed from oligonucleotides. *Nature* 305 (5937): 829–831.

21 Rothemund, P.W. (2006). Folding DNA to create nanoscale shapes and patterns. *Nature* 440 (7082): 297–302.

22 Kong, Y., Du, Q., Li, J., and Xing, H. (2022). Engineering bacterial surface interactions using DNA as a programmable material. *Chem. Commun.* 58 (19): 3086–3100.

23 Douglas, S.M., Dietz, H., Liedl, T. et al. (2009). Self-assembly of DNA into nanoscale three-dimensional shapes. *Nature* 459 (7245): 414–418.

24 Zhang, Y., Tu, J., Wang, D. et al. (2018). Programmable and multifunctional DNA-based materials for biomedical applications. *Adv. Mater.* 30 (24): 1703658.

25 Dietz, H., Douglas, S.M., and Shih, W.M. (2009). Folding DNA into twisted and curved nanoscale shapes. *Science* 325 (5941): 725–730.

26 Han, D., Pal, S., Nangreave, J. et al. (2011). DNA origami with complex curvatures in three-dimensional space. *Science* 332 (6027): 342–346.

27 Ma, W., Zhan, Y., Zhang, Y. et al. (2021). The biological applications of DNA nanomaterials: current challenges and future directions. *Signal Transduction Targeted Ther.* 6 (1): 351.

28 Zhou, M., Zhang, T., Zhang, B. et al. (2021). A DNA nanostructure-based neuroprotectant against neuronal apoptosis via inhibiting toll-like receptor 2 signaling pathway in acute ischemic stroke. *ACS Nano*.

29 Zhu, J., Yang, Y., Ma, W. et al. (2022). Antiepilepticus effects of tetrahedral framework nucleic acid via inhibition of gliosis-induced downregulation of glutamine synthetase and increased AMPAR internalization in the postsynaptic membrane. *Nano Lett.* 22 (6): 2381–2390.

30 Jiang, Q., Liu, S., Liu, J. et al. (2019). Rationally designed DNA-origami nanomaterials for drug delivery in vivo. *Adv. Mater.* 31 (45): e1804785.

31 Kearney, C.J., Lucas, C.R., O'Brien, F.J., and Castro, C.E. (2016). DNA origami: folded DNA-nanodevices that can direct and interpret cell behavior. *Adv. Mater.* 28 (27): 5509–5524.

32 Wang, Y., Li, Y., Gao, S. et al. (2022). Tetrahedral framework nucleic acids can alleviate taurocholate-induced severe acute pancreatitis and its subsequent multiorgan injury in mice. *Nano Lett.* 22 (4): 1759–1768.

33 Kuzuya, A., Wang, R., Sha, R., and Seeman, N.C. (2007). Six-helix and eight-helix DNA nanotubes assembled from half-tubes. *Nano Lett.* 7 (6): 1757–1763.

34 Winfree, E., Liu, F., Wenzler, L.A., and Seeman, N.C. (1998). Design and self-assembly of two-dimensional DNA crystals. *Nature* 394 (6693): 539–544.

35 Park, S.H., Yin, P., Liu, Y. et al. (2005). Programmable DNA self-assemblies for nanoscale organization of ligands and proteins. *Nano Lett.* 5 (4): 729–733.

36 Yan, H., Park, S.H., Finkelstein, G. et al. (2003). DNA-templated self-assembly of protein arrays and highly conductive nanowires. *Science* 301 (5641): 1882–1884.

37 He, Y., Su, M., Fang, P.A. et al. (2010). On the chirality of self-assembled DNA octahedra. *Angew. Chem. Int. Ed.* 49 (4): 748–751.

38 Zinchenko, A., Miwa, Y., Lopatina, L.I. et al. (2014). DNA hydrogel as a template for synthesis of ultrasmall gold nanoparticles for catalytic applications. *ACS Appl. Mater. Interfaces* 6 (5): 3226–3232.

39 Li, J., Zheng, C., Cansiz, S. et al. (2015). Self-assembly of DNA nanohydrogels with controllable size and stimuli-responsive property for targeted gene regulation therapy. *JACS* 137 (4): 1412–1415.

40 Li, J., Mo, L., Lu, C.H. et al. (2016). Functional nucleic acid-based hydrogels for bioanalytical and biomedical applications. *Chem. Soc. Rev.* 45 (5): 1410–1431.

41 Hong, F., Zhang, F., Liu, Y., and Yan, H. (2017). DNA origami: scaffolds for creating higher order structures. *Chem. Rev.* 117 (20): 12584–12640.

42 Sun, Y., Meng, L., Zhang, Y. et al. (2021). The application of nucleic acids and nucleic acid materials in antimicrobial research. *Curr. Stem Cell Res. Ther.* 16 (1): 66–73.

43 Hu, Y. and Niemeyer, C.M. (2019). From DNA nanotechnology to material systems engineering. *Adv. Mater.* 31 (26): e1806294.

44 Zhang, T., Tian, T., and Lin, Y. (2021). Functionalizing framework nucleic-acid-based nanostructures for biomedical application. *Adv. Mater.* e2107820.

45 Mela, I., Vallejo-Ramirez, P.P., Makarchuk, S. et al. (2020). DNA nanostructures for targeted antimicrobial delivery. *Angew. Chem. Int. Ed.* 59 (31): 12698–12702.

46 Yang, L., Yao, C., Li, F. et al. (2018). Synthesis of branched DNA Scaffolded super-nanoclusters with enhanced antibacterial performance. *Small* 14 (16): e1800185.

47 Mei, Q., Wei, X., Su, F. et al. (2011). Stability of DNA origami Nanoarrays in cell lysate. *Nano Lett.* 11 (4): 1477–1482.

48 Lacroix, A. and Sleiman, H.F. (2021). DNA nanostructures: current challenges and opportunities for cellular delivery. *ACS Nano* 15 (3): 3631–3645.

49 Liu, Y., Liu, Z., Cui, W. et al. (2020). Tetrahedral framework nucleic acids as an advanced drug delivery system for oligonucleotide drugs. *APL Mater.* 8 (10): 100701.

50 Long, Q., Jia, B., Shi, Y. et al. (2021). DNA nanodevice as a co-delivery vehicle of antisense oligonucleotide and silver ions for selective inhibition of bacteria growth. *ACS Appl. Mater. Interfaces* 13 (40): 47987–47995.

51 Kim, M.G., Park, J.Y., Shim, G. et al. (2015). Biomimetic DNA nanoballs for oligonucleotide delivery. *Biomaterials* 62: 155–163.

52 Zeng, Y., Qi, P., Wang, Y. et al. (2021). DNA pom-pom nanostructure as a multifunctional platform for pathogenic bacteria determination and inactivation. *Biosens. Bioelectron.* 177: 112982.

53 Zhang, P., Ouyang, Y., Sohn, Y.S. et al. (2021). pH- and miRNA-responsive DNA-tetrahedra/metal-organic framework conjugates: functional sense-and-treat carriers. *ACS Nano* 15 (4): 6645–6657.

54 Wang, Z., Song, L., Liu, Q. et al. (2021). A tubular DNA nanodevice as a siRNA/chemo-drug co-delivery vehicle for combined cancer therapy. *Angew. Chem. Int. Ed.* 60 (5): 2594–2598.

55 Li, J., Fan, C., Pei, H. et al. (2013). Smart drug delivery nanocarriers with self-assembled DNA nanostructures. *Adv. Mater.* 25 (32): 4386–4396.

56 Liang, L., Li, J., Li, Q. et al. (2014). Single-particle tracking and modulation of cell entry pathways of a tetrahedral DNA nanostructure in live cells. *Angew. Chem. Int. Ed.* 53 (30): 7745–7750.

57 Obuobi, S., Tay, H.K., Tram, N.D.T. et al. (2019). Facile and efficient encapsulation of antimicrobial peptides via crosslinked DNA nanostructures and their application in wound therapy. *J. Controlled Release* 313: 120–130.

58 Whitfield, C.J., Zhang, M., Winterwerber, P. et al. (2021). Functional DNA-polymer conjugates. *Chem. Rev.* 121 (18): 11030–11084.

59 Trads, J.B., Tørring, T., and Gothelf, K.V. (2017). Site-selective conjugation of native proteins with DNA. *Acc. Chem. Res.* 50 (6): 1367–1374.

60 Eun, H., Kwon, W.Y., Kalimuthu, K. et al. (2019). Melamine-promoted formation of bright and stable DNA-silver nanoclusters and their antimicrobial properties. *J. Mater. Chem. B* 7 (15): 2512–2517.

61 Chen, Y.F., Chiou, Y.H., Chen, Y.C. et al. (2021). ZnO-loaded DNA nanogels as neutrophil extracellular trap-like structures in the treatment of mouse peritonitis. *Mater. Sci. Eng., C* 131: 112484.

62 Chang, S.-S., Lu, W.-Y.W., Park, S.-H., and Kang, D.-H. (2010). Control of foodborne pathogens on ready-to-eat roast beef slurry by ε-polylysine. *Int. J. Food Microbiol.* 141 (3): 236–241.

63 Jiang, S., Zeng, M., Zhao, Y. et al. (2019). Nano-complexation of ε-poly-l-lysine with DNA: improvement of antimicrobial activity under high phosphate conditions. *Int. J. Biol. Macromol.* 127: 349–356.

64 Zhang, T., Cui, W., Tian, T. et al. (2020). Progress in biomedical applications of tetrahedral framework nucleic acid-based functional systems. *ACS Appl. Mater. Interfaces* 12 (42): 47115–47126.

65 Chen, X., Xie, Y., Liu, Z., and Lin, Y. (2021). Application of programmable tetrahedral framework nucleic acid-based nanomaterials in neurological disorders: progress and prospects. *Front. Bioeng. Biotechnol.* 9: 782237.

66 Zhang, B., Tian, T., Xiao, D. et al. (2022). Facilitating in situ tumor imaging with a tetrahedral DNA framework-enhanced hybridization chain reaction probe. *Adv. Funct. Mater.* 32 (16): 2109728.

67 Qin, X., Xiao, L., Li, N. et al. (2022). Tetrahedral framework nucleic acids-based delivery of microRNA-155 inhibits choroidal neovascularization by regulating the polarization of macrophages. *Bioact. Mater.* 14: 134–144.

68 Eichenberger, E.M. and Thaden, J.T. (2019). Epidemiology and mechanisms of resistance of extensively drug resistant gram-negative bacteria. *Antibiotics* 8 (2).

69 Ghai, I. and Ghai, S. (2018). Understanding antibiotic resistance via outer membrane permeability. *Infect. Drug Resist.* 11: 523–530.

70 Laws, M., Shaaban, A., and Rahman, K.M. (2019). Antibiotic resistance breakers: current approaches and future directions. *FEMS Microbiol. Rev.* 43 (5): 490–516.

71 Sun, Y., Li, S., Zhang, Y. et al. (2020). Tetrahedral framework nucleic acids loading ampicillin improve the drug susceptibility against methicillin-resistant Staphylococcus aureus. *ACS Appl. Mater. Interfaces* 12 (33): 36957–36966.

72 Sauvage, E., Kerff, F., Terrak, M. et al. (2008). The penicillin-binding proteins: structure and role in peptidoglycan biosynthesis. *FEMS Microbiol. Rev.* 32 (2): 234–258.

73 Macheboeuf, P., Contreras-Martel, C., Job, V. et al. (2006). Penicillin binding proteins: key players in bacterial cell cycle and drug resistance processes. *FEMS Microbiol. Rev.* 30 (5): 673–691.

74 Hartman, B.J. and Tomasz, A. (1984). Low-affinity penicillin-binding protein associated with beta-lactam resistance in *Staphylococcus aureus*. *J. Bacteriol.* 158 (2): 513–516.

75 Sharma, V.K., Hackbarth, C.J., Dickinson, T.M., and Archer, G.L. (1998). Interaction of native and mutant MecI repressors with sequences that regulate mecA, the gene encoding penicillin binding protein 2a in methicillin-resistant staphylococci. *J. Bacteriol.* 180 (8): 2160–2166.

76 Senadheera, D.B., Cordova, M., Ayala, E.A. et al. (2012). Regulation of bacteriocin production and cell death by the VicRK signaling system in *Streptococcus mutans*. *J. Bacteriol.* 194 (6): 1307–1316.

77 Alves, L.A., Harth-Chu, E.N., Palma, T.H. et al. (2017). The two-component system VicRK regulates functions associated with *Streptococcus mutans* resistance to complement immunity. *Mol. Oral Microbiol.* 32 (5): 419–431.

78 Viszwapriya, D., Subramenium, G.A., Radhika, S., and Pandian, S.K. (2017). Betulin inhibits cariogenic properties of *Streptococcus mutans* by targeting vicRK and gtf genes. *Antonie van Leeuwenhoek* 110 (1): 153–165.

79 Hu, Y., Chen, Z., Mao, X. et al. (2020). Loop-armed DNA tetrahedron nanoparticles for delivering antisense oligos into bacteria. *J. Nanobiotechnol.* 18 (1): 109–109.

80 Nikravesh, A., Dryselius, R., Faridani, O.R. et al. (2007). Antisense PNA accumulates in *Escherichia coli* and mediates a long post-antibiotic effect. *Mol. Ther.* 15 (8): 1537–1542.

81 Weiss, J., Beckerdite-Quagliata, S., and Elsbach, P. (1980). Resistance of gram-negative Bacteria to purified bactericidal leukocyte proteins: relation to binding and bacterial lipopolysaccharide structure. *J. Clin. Invest.* 65 (3): 619–628.

82 Readman, J.B., Dickson, G., and Coldham, N.G. (2017). Tetrahedral DNA nanoparticle vector for intracellular delivery of targeted peptide nucleic acid antisense agents to restore antibiotic sensitivity in cefotaxime-resistant *Escherichia coli*. *Nucleic Acid Ther.* 27 (3): 176–181.

83 Nordström, R. and Malmsten, M. (2017). Delivery systems for antimicrobial peptides. *Adv. Colloid Interface Sci.* 242: 17–34.

84 Lee, H., Hwang, J.-S., Lee, J. et al. (2015). Scolopendin 2, a cationic antimicrobial peptide from centipede, and its membrane-active mechanism. *Biochim. Biophys. Acta, Biomembr.* 1848 (2): 634–642.

85 Setyawati, M.I., Kutty, R.V., Tay, C.Y. et al. (2014). Novel theranostic DNA nanoscaffolds for the simultaneous detection and killing of *Escherichia coli* and *Staphylococcus aureus*. *ACS Appl. Mater. Interfaces* 6 (24): 21822–21831.

86 King, D.T., Worrall, L.J., Gruninger, R., and Strynadka, N.C. (2012). New Delhi metallo-β-lactamase: structural insights into β-lactam recognition and inhibition. *JACS* 134 (28): 11362–11365.

87 Yang, S.K., Kang, J.S., Oelschlaeger, P., and Yang, K.W. (2015). Azolylthioacetamide: a highly promising scaffold for the development of Metallo-β-lactamase inhibitors. *ACS Med. Chem. Lett.* 6 (4): 455–460.

88 Hu, Z., Periyannan, G., Bennett, B., and Crowder, M.W. (2008). Role of the Zn_1 and Zn_2 sites in metallo-beta-lactamase L_1. *JACS* 130 (43): 14207–14216.

89 Bebrone, C., Lassaux, P., Vercheval, L. et al. (2010). Current challenges in antimicrobial chemotherapy: focus on ß-lactamase inhibition. *Drugs* 70 (6): 651–679.

90 Ouyang, X., Chang, Y.N., Yang, K.W. et al. (2017). A DNA nanoribbon as a potent inhibitor of metallo-β-lactamases. *Chem. Commun.* 53 (63): 8878–8881.

91 Li, S., Liu, Y., Tian, T. et al. (2021). Bioswitchable delivery of microRNA by framework nucleic acids: application to bone regeneration. *Small* 17 (47): e2104359.

92 Tian, T., Zhang, T., Zhou, T. et al. (2017). Synthesis of an ethyleneimine/tetrahedral DNA nanostructure complex and its potential application as a multi-functional delivery vehicle. *Nanoscale* 9 (46): 18402–18412.

93 Ge, Y., Tian, T., Shao, X. et al. (2019). PEGylated protamine-based adsorbing improves the biological properties and stability of tetrahedral framework nucleic acids. *ACS Appl. Mater. Interfaces* 11 (31): 27588–27597.

94 Ma, W., Yang, Y., Zhu, J. et al. (2022). Biomimetic nanoerythrosome-coated aptamer-DNA tetrahedron/maytansine conjugates: pH-responsive and targeted cytotoxicity for HER2-positive breast cancer. *Adv. Mater.* e2109609.

95 Tian, T., Xiao, D., Zhang, T. et al. (2021). A framework nucleic acid based robotic nanobee for active targeting therapy. 31 (5): 2007342.

96 Burns, J.R. (2021). Introducing bacteria and synthetic biomolecules along engineered DNA fibers. *Small* 17 (25): 2100136.

10

Framework Nucleic Acid Nanomaterial-Based Therapy for Osteoarthritis: Progress and Prospects

Yangxue Yao, Hongxiao Huang, and Sirong Shi

Sichuan University, State Key Laboratory of Oral Diseases, West China Hospital of Stomatology, No. 14, 3rd Sec, Ren Min Nan Road, Chengdu 610041, PR China

10.1 Introduction

Osteoarthritis (OA) is still one of the diseases that plague the world. It is a progressive disease in which the entire joint structure is altered. Patients with OA suffer from severe joint discomfort, with pain being the main symptom. In the early stages of the disease, pain is triggered by intense physical activity and can be predicted. Over time, it becomes more difficult to predict the pain and other joint symptoms that appear. Gradually, the pain becomes more stable and begins to interfere with daily activities. In the late stages, the pain becomes more severe, persistent, and unpredictable, leading to complete limitation of some activities [1]. Associated symptoms include joint cracking, joint effusion, bone swelling, deformity, and muscle weakness, which severely affect patients' normal lives [2]. Currently, OA is a major cause of disability burden. It is estimated that 300 million people worldwide suffer from OA [3]. In developed countries, medical expenditures for OA amount to 1 to 2.5% of GDP, with joint replacement surgery accounting for the largest share of these medical expenditures [4]. The individual costs of OA patients, such as lost income and the resulting national burden, far exceed the cost of medical care [5, 6].

OA occurs most frequently in the knee joint, followed by the hip and wrist joints [7]. It is also closely related to systemic diseases. Existing studies have shown an association between OA and atherosclerosis-related diseases [8]. Meta-analyses showed that people with OA have a slightly increased risk of cardiovascular disease and stroke than people without OA [9, 10]. In addition, more than half of older people with OA have hypertension, followed by cardiovascular disease, dyslipidemia, and diabetes [10, 11].

10.2 Pathology of OA

The weight-bearing surface of the bone is covered with articular cartilage, which forms an elastic, nearly frictionless layer that can spring back after deformation. However, the

---

Nucleic Acid-Based Nanomaterials: Stabilities and Applications, First Edition.
Edited by Yunfeng Lin and Shaojingya Gao.
© 2024 WILEY-VCH GmbH. Published 2024 by WILEY-VCH GmbH.

Diagram	OARSI grade	Pathological features
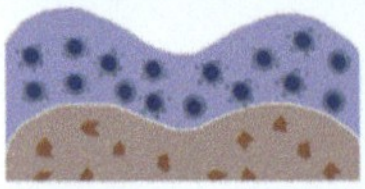	Grade 1 Surface intact	Some chondrocytes showed apoptosis and necrosis
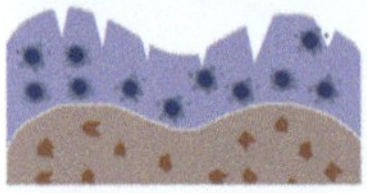	Grade 2 Surface discontinuity	Loss of superfical matrix, cell death
	Grade 3 Vertical fissures	Vertical clefts extends to mid-zone
	Grade 4 Erosion	Matrix loss superficial and mid-zone
	Grade 5 Denudation	Subchondral bone exposed
	Grade 6 Deformation	Subchondral bone is destroyed, accompanying restoration

Figure 10.1 OARSI grade and pathological features in each grade. Source: Drawn by Figdraw.

articular cartilage is avascular and nerve-free, so it cannot spontaneously repair itself when injured [12]. The pathological changes of OA affect the entire joint, including the articular cartilage, subchondral bone, ligaments, joint capsule, synovium, and muscles [13]. As shown in Figure 10.1, the degree of cartilage degeneration can be divided into six grades according to the destruction of the articular tissue, which is referred to as OARSI grade.

The pathological factors of OA are complex and include mechanical overload, inflammation, and altered metabolism, ultimately leading to structural destruction and loss of function [14, 15]. OA is an active dynamic change caused by an imbalance between repair and destruction of joint tissues. Most scientists believe that inflammation plays a crucial role in this process. To some extent, the deleterious effects of OA can be eliminated or compensated for by altering the metabolic activity of chondrocytes. However, matrix degradation occurs when these deleterious effects exceed the compensatory capacity of the system [16].

Another possible explanation for OA is that cartilage degeneration causes immune rejection within synovial cells, which may lead to the production of metalloproteinases and inflammatory cytokines, resulting in cartilage destruction [17]. Others theorize that the innate immune system and activated synovial macrophages are involved in OA development [17].

10.3 Risk Factors of OA

OA has a strong genetic background [18, 19]. There are more than 80 genes involved in the pathogenesis of OA. The contribution of genetics to OA has been estimated to be 40–80% [20].

Among the acquired factors, age is one of the major risk factors for OA. The increasing incidence of OA with age is due to cumulative exposure to various risk factors and age-related biological changes in joints, such as mitochondrial dysfunction and pathway alterations [21, 22]. Pathological loading is another important factor. Poor alignment of the knee joint, loss of the meniscus, repetitive stress injuries, and strenuous work activities can cause joint damage [23]. In addition, hormonal balance and estrogen are closely related to OA, making women more susceptible to developing it. Systemic factors such as obesity, aging, metabolic disorders, and atherosclerosis also play important roles in the occurrence of OA [17, 24].

It should not be overlooked that diet is also associated with OA. Patients who are deficient in vitamin C, D, and other trace elements have a higher risk of developing OA [25]. In addition, OA can result from any inflammatory events that lead to joint damage, such as joint fractures, cartilage, ligament, or meniscus injuries [26].

10.4 Challenges for OA Therapy

Cartilage is the main load-bearing part of the joint; thus, frequent loading will inevitably cause damage to it. Cartilage has no blood vessels or nerves. It has poor potential for spontaneous healing of lesions, resulting in clinical treatment challenges [27]. Many risk factors cause OA, and treatments that target one factor alone always do not work well. Age, gender, obesity, genetics, race, diet, and joint-related factors, including injury, joint alignment disorder, and abnormal joint loading, all might be the causative factors of OA. The interaction of these risk factors brings a great challenge to therapy. Moreover, the mechanism of pain in OA is complex, involving peripheral and central sensitization. This complexity may explain why traditional painkillers (paracetamol, tramadol, and opioids) are relatively ineffective in relieving pain or improving function in OA [28].

Current medical interventions for the treatment of OA only relieve pain and perhaps improve function, but have no effect on the onset or progression of the disease [29]. Nonsteroidal anti-inflammatory drugs (NSAIDs) and acetaminophen are the first-line agents for OA. However, when used systemically, they must pass through the synovial blood vessels to reach the local area. Therefore, extensive and prolonged usage is required, and other organs are exposed to high drug concentrations, which increases the possibility of undesirable side effects. In addition, current treatments based on direct intra-articular injection of anti-inflammatory drugs have short-term effects and unsatisfactory efficacy. Drugs that act directly on chondrocytes must penetrate the dense and nonvascular extracellular matrix, which limits their efficacy. Joint replacement surgery is the only option for end-stage OA, although the associated risks of venous thromboembolism, infection, neurovascular injury, and periprosthetic fracture are troubling. In addition, up to 25% of patients complain of pain and disability even after replacement surgery [30, 31]. Alleviating the progression of structural and symptomatic sequelae of OA has become the focus of treatment development.

10.5 Nucleic Acid Nanomaterial-Based Therapy for OA

Nucleic acids are the main material basis for the storage, duplication, and transmission of genetic information, which is a fundamental part of life. Nucleic acid nanomaterials are a general term for nucleic acids and their analogs, including oligonucleotides, plasmids, aptamers, and DNA origami, which have independent structures and perform specific biological functions [32–34].

Nucleic acids are widely used in the biomedical field, where they can regulate gene expression by providing specific nucleic acids. Some nucleic acids enter into cells independently. They remain in a specific location for a long time and are involved in transcription and translation processes that can rebalance the metabolism of the OA cartilage. These vector-independent nucleic acids include antisense oligonucleotides (ASOs), aptamers, and tetrahedral scaffold nucleic acids (tFNAs) [35, 36]. Some nucleic acids rely on vectors to be introduced into cells or damaged compounds, such as messenger RNA (mRNA), small interfering RNA (siRNA), and microRNA (miRNA) [37, 38]. In addition to gene coding, many other interesting functions of nucleic acids have been discovered. For example, it has been recognized that nucleic acid nanomaterials are well suited for drug delivery, especially in complex conditions such as blood and cells [39]. Plasmids are the traditional nucleic acid nanocarriers [40]. Aptamers have excellent targeting properties and are promising nucleic acid nanocarriers. Moreover, the nucleic acid polyhedron formed by DNA origami technology has intrinsic addressability, shape controllability, strong penetration ability, and good biocompatibility [41]. They are widely used in biosensing and biological therapy [42, 43].

Here, we will present a nucleic acid nanomaterial-based therapy for OA.

10.5.1 Vector-Independent Nucleic Acid Nanomaterials for OA Therapy

These materials have good penetration ability and stability, and they can enter cells without vectors. For the convenience of readers, we have listed the relevant research papers from the last five years in Table 10.1.

10.5.1.1 Tetrahedral Framework Nucleic Acids (tFNAs)
Therapeutic Effects of tFNAs In 2004, Turberfield and his colleagues first proposed a DNA polyhedron in which tFNAs would be the simplest polyhedron. Based on DNA base pairing

Table 10.1 Vector-dependent nucleic acid nanomaterials for OA therapy in recent five years.

Nucleic acid	Gene/target	Cell used	Animal used	Ref
tFNAs	Wnt/β-Catenin pathway	Primary rat chondrocytes	—	[44]
tFNAs	Nrf2/HO-1	Primary rat chondrocytes	—	[44]
tFNAs	RhoA, ROCK2, and vinculin	Primary rat chondrocytes	—	[45]
ASOs	miR-34a-5p	Human OA chondrocytes	OA mice	[46]
ASOs	COX-2	Human OA chondrocytes	—	[47]

and sticky-ended cohesion rules, tFNAs are synthesized from four single DNA strands [43, 48–52]. Apart from their stable structure, tFNAs have attracted much attention because of their good cell penetration and biocompatibility [53–58]. They can be readily taken up intracellularly without transfectants in a nest protein-dependent manner [59]. Like natural DNA, tFNAs can scavenge ROS, which has been shown to protect damaged cells and inflammatory diseases [60–67].

In OA research, tFNAs have been found to be taken up by chondrocytes and regulate cell proliferation, migration, and collagen secretion. The representative images are shown in Figure 10.2. Shi et al. discovered that tFNAs could regulate chondrocyte autophagy and maintain cellular metabolism and homeostasis. In their study, chondrocytes treated with 250 nM tFNAs showed accumulation of more autophagosomes. In addition, autophagy genes were clearly upregulated by activation of the PI3K/AKT/mTOR pathway [69]. Shi et al. also found that tFNAs promoted chondrocyte migration and increased the expression of RhoA, ROCK2, and vinculin [45]. Shao et al. reported that chondrocytes treated with 250 nM tFNAs for 24 hours maintained their original round and three-dimensional shape and exhibited significant cell proliferation [61]. Subsequently, Shi et al. stimulated chondrocytes with IL-1β to establish an OA model. They found that tFNAs could be taken up by IL-1β-induced chondrocytes, and the uptake rate of chondrocytes for 250 nM tFNAs was 72%. Later, they found that tFNAs could inhibit early and late apoptosis of cells and downregulate the expressions of apoptotic proteins Bcl2, Bax, and caspase-3. Under IL-1β stimulation, the production of ROS in chondrocytes increased and they were susceptible to damage, whereas after treatment with tFNAs, ROS significantly decreased and oxidative stress was reduced. This is because tFNAs significantly increased the expression of Nrf2 and HO-1, a critical signaling pathway for the maintenance of intracellular homeostasis [44]. In addition, tFNAs increased the autophagy body and protein expression of the autophagy-related marker LC3-II in IL-1β-induced chondrocytes. tFNAs inhibited the Wnt/β-catenin signaling pathway and decreased the occurrence and development of OA, as well as the associated proteins β-catenin, Lef1, and CyclinD1 [48]. The mechanism of regulation of chondrocytes by tFNAs is detailed in Figure 10.3.

Biodistribution of tFNAs In recent works, tFNAs have passed through the cell membrane via endocytosis [61]. Studies by Fan showed that the endocytosis process of tFNAs is mediated by caveolae and micropinocytosis [70]. The onset time was different in the different cells and ranged from 3 to 12 hours [71, 72]. The fate of tFNAs ends in lysosomes. Tian et al. discovered that tFNAs entered L929 and A549 cells after 6 hours and were digested in lysosomes after 24 hours [73]. Exocytosis of lysosomes is a promising process for the pathway of tFNA out of cells.

Regarding systemic biodistribution in animals, tFNAs accumulated preferentially in the kidney, liver, and gallbladder after injection via the tail vein of mice [65]. In addition, there were also fluorescently labeled tFNAs in the brain, indicating that they can cross the blood-brain barrier. tFNAs that entered the brain were stable for at least one hour [74]. The signal intensities in kidney and liver started to decrease after one hour. More tFNAs accumulated in the kidney than in the liver one hour later, indicating preferential renal clearance over hepatic metabolism. tFNAs subsequently appeared in the bladder for excretion from the body [54].

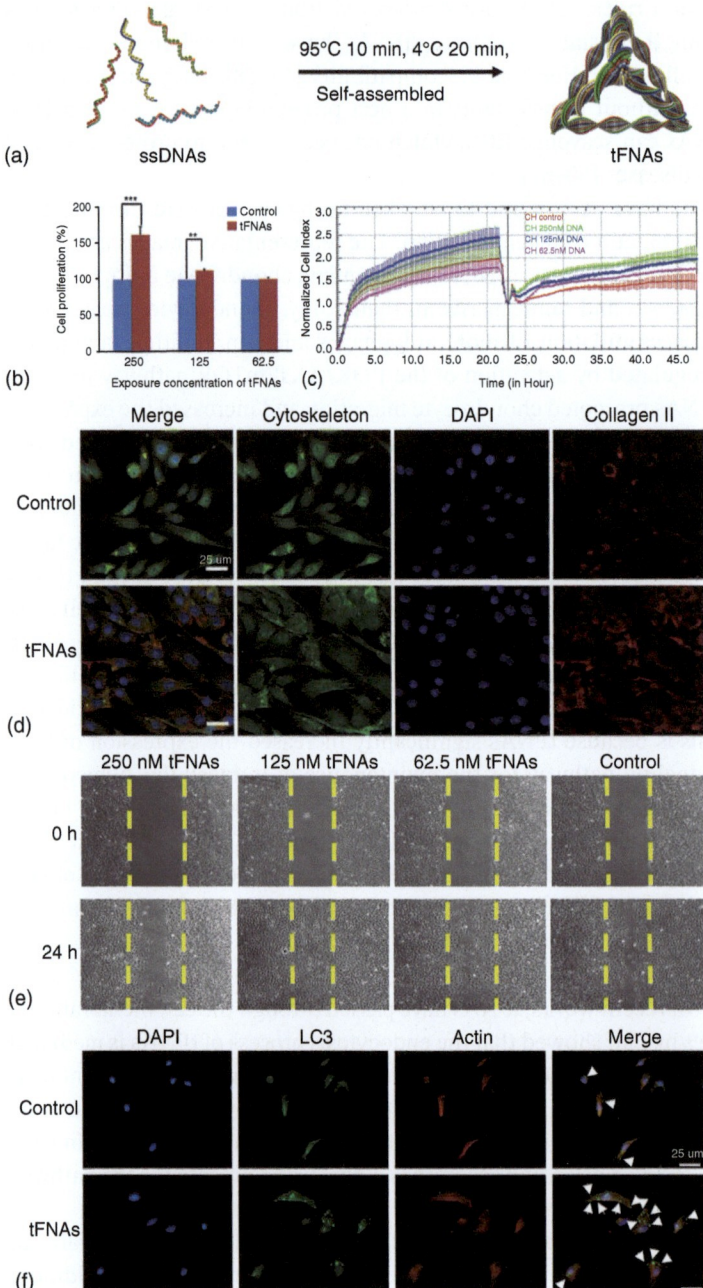

Figure 10.2 Effect of tFNAs on chondrocytes. (a) Schematic graph for the synthesis of tFNAs. (b) Cell proliferation detected by CCK-8 assay after treating chondrocytes with tFNAs for 24 h. (c) Cell proliferation detected by real-time cell analysis system (RTCA). (d) Immunofluorescent images of tFNAs-treated chondrocytes (cytoskeleton: green, nucleus: blue, collagen II: red). (e) Effect of tFNAs concentration on chondrocyte migration at 24 h, as determined by the wound healing assay. (f) Immunofluorescent images of tFNAs-treated chondrocytes (LC3: green, nucleus: blue, actin: red). Source: Sirong et al. [68]/Springer Nature/CC by 4.0 (a); Reproduced from Shao et al. [61], © 2017/John Wiley & Sons (b–d); Shi et al. [45], © 2017/John Wiley & Sons (e); Shi et al. [69], © 2018/Royal Society of Chemistry (f).

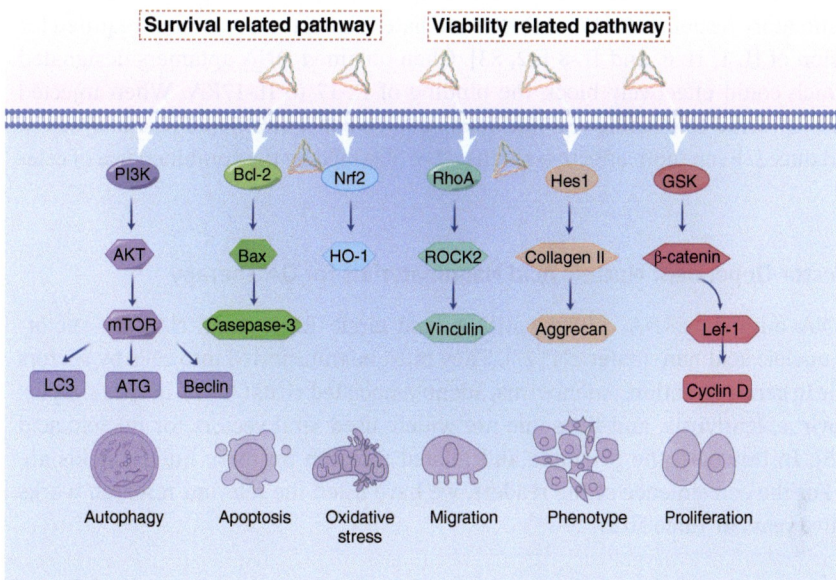

Figure 10.3 tFNAs regulate intracellular signaling pathways in chondrocytes. Source: Drawn by Figdraw.

10.5.1.2 Antisense Oligonucleotides (ASOs)

ASOs are composed of 15–30 nucleotides. They can bind to cell surface proteins and enter cells by endocytosis [75, 76]. After binding to the target mRNA, ASOs form an RNA-DNA hybridization and become a substrate of RNaseH, leading to RNA degradation and disruption of mRNA expression [77]. In exploring the pathogenesis of OA, it was found that after treating chondroblasts and chondrocytes with ASOs against miR-9, caspase-3 activity and the number of dead cells increased. The application of a miR-9 precursor was able to restore the inhibitory effect of protogenin (PRTG) on cartilage differentiation. These data suggest that miR-9 can inhibit sulfated proteoglycan accumulation and cartilage nodule formation by targeting PRTG and promoting cartilage differentiation [78]. Locked nucleic acid (LNA) technology is an improvement for ASOs. The ribose ring in the backbone is connected by a methylene bridge between the $2'$-O and $4'$ atoms, which significantly increases binding to the target RNA [79]. Local injection is a better method for OA delivery. It was discovered that miR-181a-5p could be blocked by intra-articular injection of LNA-miR-181a-5p, decreasing the expression of cartilage degradation genes and cell death markers [80]. Moreover, trauma-induced knee OA was alleviated in this manner. Baek et al. also discovered that LNA-miR-449a ASOs targeting LEF1 and SIRT1 could regenerate damaged cartilage and prevent the progression of OA [81].

10.5.1.3 Aptamers

Aptamers can not only be used as targeting vectors but have also been shown to regulate gene expression. Aptamers have certain application advantages. Due to the small size of DNA ligands, they are easily excreted from the blood. At the same time, they can be easily synthesized, chemically modified, and not immunogenic [81]. IL-17RA plays a central role in the inflammatory response. IL-17/ IL-17RA-mediated signal transduction is required for the expression of IL-1, IL-6, and IL-8 [82, 83]. Chen obtained DNA aptamers designated RA10-6, which could effectively block the binding of IL-17 to IL-17RA. When injected into OA mice, synovial thickening was inhibited, and the concentration of IL-6 in synovial tissue was reduced. Even more effective results were obtained by the combined use of celecoxib [84].

10.5.2 Vector-Dependent Nucleic Acid Nanomaterials for OA Therapy

miRNA, RNA mimics, siRNA, cDNA, mRNA, and circle RNA (CircRNA) are vector-dependent nucleic acid nanomaterials [77]. They must be transported into cells by vectors to play a role in gene regulation. Adenovirus, adeno-associated virus (AAV), herpes simplex virus, retrovirus, lentivirus, and liposome are widely used viral vectors for nucleic acid transfer [85]. In this part, the principle and related research on these nucleic acids are presented. For the convenience of the readers, we have listed the relevant research works in the last five years in Table 10.2.

10.5.2.1 MicroRNA (miRNA) Mimics

miRNA is a single-stranded RNA consisting of 20–24 nucleotides. It belongs to a class of non-coding RNAs. In most cases, miRNA pairs with the 3'-noncoding region of the target mRNA, blocking gene translation [126, 127]. It has been widely studied that miRNA regulates autophagy, apoptosis, inflammation, and cartilage degradation in OA [128–130]. Therefore, various miRNA mimics have been developed for the treatment of OA. Liang encapsulated miR-140 mimics with chondrocyte affinity peptide (CAP) exosomes and introduced them into chondrocytes, which inhibited the expression of MMP-13 and ADAMT-5. The increased miR-140 decreased the expression of MMP-13 and ADAMT-5 in rat OA model and alleviated the cartilage damage [107]. Liang et al. stimulated human synovial fibroblasts with IL-1β and then treated them with miR-26a-5p mimic-bearing exosomes. *In vitro*, apoptosis of synovial fibroblasts was observed. In addition, synovial hyperplasia was alleviated in OA rats. In addition, inflammatory cells and serum IL-1β were reduced [131].

10.5.2.2 Small Interfering RNA (siRNA)

siRNA is an artificially produced double-stranded RNA. It consists of about 20 base pairs and switches off various genes. During RNA interference, silencing protein complexes bind to siRNA and form an RNA-induced silencing complex that cleaves the complementary mRNA transcript, resulting in gene repression [132]. 15- LO -1 in primary chondrocytes was silenced by si-15-LO-1, which significantly reduced the secretion of cartilage extracellular matrix, cartilage degrading enzymes, and the production of reactive oxygen species induced by mechanical stress [133]. As reported by Xu et al., si-SOX11-1 was added to the human chondrocyte line CHON -001 using Lipofectamine 2000. Knockout of SOX11 reduced the expression of MMP-13 and caspase-3, increased the expression of collagen II, and inhibited cell apoptosis [134]. In addition, Li et al. encapsulated Lipofectamine 2000 with si-XIST and transfected it onto M1 macrophages, which inhibited the XIST/miR-376c-5p/OPN axis and the inflammatory markers secreted by M1 macrophages IL-1β, IL-6, and TNF-α.

Table 10.2 Vector-dependent nucleic acid nanomaterials for OA therapy in recent five years.

Nucleic acid	Gene/target	Vector	Cell used	Animal used	Ref
miRNA mimics	miR-103a-3p	Lipofectamine 2000	LPS-induced OA rat chondrocytes	—	[86]
miRNA mimics	miR-24	Lipofectamine 2000	OA rat chondrocytes	—	[87]
miRNA mimics	miR-146a	PmirGLO	OA patients' chondrocytes	—	[88]
miRNA mimics	miR-26a	—	—	OA rat	[89]
miRNA mimics	miR-30a	Lentivirus	Rat bone marrow cells	—	[90]
miRNA mimics	miR-93	PcDNA3.1 eukaryotic expression vector	Primary mouse articular chondrocytes	OA mice	[91]
miRNA mimics	miR-197	PcDNA3.1-EIF4G2	Human OA cartilage chondrocytes	—	[92]
miRNA mimics	miR-21-5p	Lipofectamine 2000	Primary mouse condylar chondrocytes	SPF-grade white egg	[93]
miRNA mimics	miR-145	Not mentioned	OA mice's primary chondrocytes	—	[94]
miRNA mimics	miR-146a-5p	PmirGLO vector	IL-1β-induced human and mouse chondrocytes	OA rat	[95]
miRNA mimics	miR-204-5p	Lipofectamine 2000	Human osteoarthritic synovial fibroblasts	—	[96]
miRNA mimics	miR-467b	Lipofectamine 2000	LPS-induced mouse chondrogenic ATDC5 cells	OA rat	[97]
miRNA mimics	miR-520d-5p	Lipofectamine 3000	Human mesenchymal stem cells/human OA chondrocytes	—	[98]
miRNA mimics	miR-590-5p	Lipofectamine 3000	IL-1β-induced human chondrocytes	—	[99]
miRNA mimics	miR-877-5p	Lipofectamine 2000	IL-1β-treated primary mouse chondrocytes	OA mice	[100]
miRNA mimics	miR-101	Lipofectamine RNAiMAX	Rat mesenchymal stem cells	—	[101]

(Continued)

Table 10.2 (Continued)

Nucleic acid	Gene/target	Vector	Cell used	Animal used	Ref
miRNA mimics	miR-411	Lipofectamine 3000	IL-1β-induced human chondrocyte C28/12 line	—	[102]
miRNA mimics	miR-140-5p	RiboFECTTM CP transfection kit	IL-1β-induced primary mice chondrocytes	—	[103]
miRNA mimics	miR-135b	Lipofectamine 2000	Synovial macrophages	—	[104]
miRNA mimics	miR-140-5p	Lipofectamine 2000	Synovial cells	OA rat	[105]
miRNA mimics	miR-27a	PcDNA3.1	—	OA rat	[106]
miRNA mimics	miR-140	Exosomes	IL-1β-induced human primary chondrocytes	OA rat	[107]
miRNA mimics	miR-122	Lipofectamine	IL-1β-induced rat primary chondrocytes	—	[108]
siRNA	Notch1 siRNA	PLGA-PEG encapsulate NO-HB@siRNA	Macrophages	OA mice	[109]
siRNA	MMP-13 siRNA	PLGA based microplates	Murine chondrogenic cells (ATDC5)	OA mice	[110]
siRNA	MMP-13& Adamts5 miRNA	not mentioned	—	OA mice	[111]
siRNA	YAP siRNA	Lipofectamine 2000	IL-1β-induced mice primary chondrocytes	OA mice	[112]
siRNA	MMP-13 siRNA	not used	—	OA mice	[113]
siRNA	IκBζ siRNA	Lipofectamine	IL-1β-induced mouse chondrocytes	—	[114]
siRNA	SGK1	Lipofectamine	IL-1β-induced human chondrocytes	—	[115]
siRNA	SGK1	Lipofectamine 2000	OA mouse chondrocytes	—	[116]
siRNA	SOX11	Lipofectamine 2000	IL-1β-induced human chondrocytes CHON-001	—	[117]

siRNA	XIST	Lipofectamine 2000	M1 macrophages, human primary cartilage chondrocytes	—	[118]
cDNA	CIRS-7 cDNA	PcDNA3.1	IL-1β-induced C28/I2 chondrocytes	—	[119]
cDNA	Hsox9	RAAV-FLAG	—	Sheep	[120]
cDNA	SOX-6, 9	Parental plasmid of minicircle	HASCs	Goat	[121]
mRNA	IGF-1	MessengerMAX transfected ADSCs	Mouse primary chondrocytes	C57BL/6 mice	[122]
mRNA	Int-1 (WNT) 16	Peptide-based nanoplatforms	Human cartilage chondrocytes	—	[46]
CircRNA	CircATRNL1	Lipofectamine 3000	IL-1β-treated human cartilage chondrocytes	—	[123]
CircRNA	Circ3503	Small extracellular vesicles derived from synovium mesenchymal stem cells (SMSCs)	Primary SMSCs and IL-1β-induced chondrocytes	OA rat	[124]
CircRNA	Circ0083429	AAVs	IL-1β-treated OA patient chondrocytes	OA mice	[125]

In addition, the migration ability of co-cultured chondrocytes was improved, apoptosis was inhibited, and the protein content of anabolic cartilage markers was increased [118].

10.5.2.3 cDNA

cDNA is a DNA strand complementary to RNA after reverse transcription *in vitro*. Expression of the corresponding genes was increased by exogenous transfer using vectors. IL-1 is a mediator of the inflammatory response. IL-1 receptor antagonist (IL-1ra) interferes with the binding of IL-1 with the IL-1 receptor, which is an anti-inflammatory mechanism *in vivo* [135]. IL-10 inhibits the synthesis of macrophage-derived inflammatory cytokines. In Zhang's study, IL-1ra and IL-10 cDNA were inserted into a plasmid and transduced into primary rabbit synovial cells. Then, intra-articular injection was performed, and the cargos were transplanted into the knee joint of rabbits with OA. Rabbits receiving IL-1ra and IL-10 gene therapy showed a significant reduction in cartilage pathology [136]. Fat-1 gene reduced IL-1 and TNF in patients with OA and upregulated IL-4, IL-10, TGF-β, and IGF-1. When Fat-1 cDNA was transfected into synovial cells and chondrocytes using lentivirus, degradation of inflammatory cartilage matrix was permanently inhibited, and articular cartilage was maintained in a healthy state [137]. Ko et al. loaded the parental plasmid of minicircle with SOX-6, 9, then the load was transfected into human adipose stem cells (ASCs) and injected into OA knee joints of rats. These rats obtained complete, smooth cartilage surfaces and good extracellular matrix condition [121].

10.5.2.4 mRNA

The therapeutic potential of mRNA-based nanoparticles is enormous. Because mRNA does not integrate into the host genome, there is no risk of mutagenesis, which has become a common method in gene therapy. Continuous transfection of recombinant human connective tissue growth factor (CTGF) mRNA into bone mesenchymal stem cells (BMSCs) and inoculation on poly(lactic-*co*-glycolic acid) (PLGA) scaffolds was used. CTGF treatment increased the expression of GAG and collagen, as well as the differentiation of BMSCs into chondrocytes and the production of cartilage matrix. In the OA model, the implantation of CTGF-modified BMSCs/NaOH-treated PLGA scaffold stimulates the biosurface development of hyaline cartilage [138]. Aini et al. developed a micelle vector based on a PEG-polyamino acid block copolymer. The vector delivered mRNA for cartilage synthesis-related transcription factor (Runx) 1 into the knee joint of OA mice. The mice were genetically modified to prevent cartilage breakdown and osteophyte production. Sox9 expression in articular chondrocytes and type II collagen increased, which was associated by enhanced articular chondrocyte differentiation and proliferation [139].

10.5.2.5 Circular RNA (CircRNA)

CircRNA is a noncoding RNA molecule with no 5′cap and 3′tail that forms a circular configuration by covalent binding. It is RNase-insensitive and more stable than linear RNA. CircRNA acts as a miRNA sponge, competitively binding to miRNAs to modulate downstream gene expression. Furthermore, CircRNA interacts with mRNA regulation-binding proteins and influences mRNA stability. CircRNA may also bind to ribosomes and disrupt translation [140–142]. In mice knee joints, AAVs-Circ0083429 was injected. MMP-3, MMP-13, ACOL2A1, and aggrecan mRNA and protein levels were reduced in cartilage, and

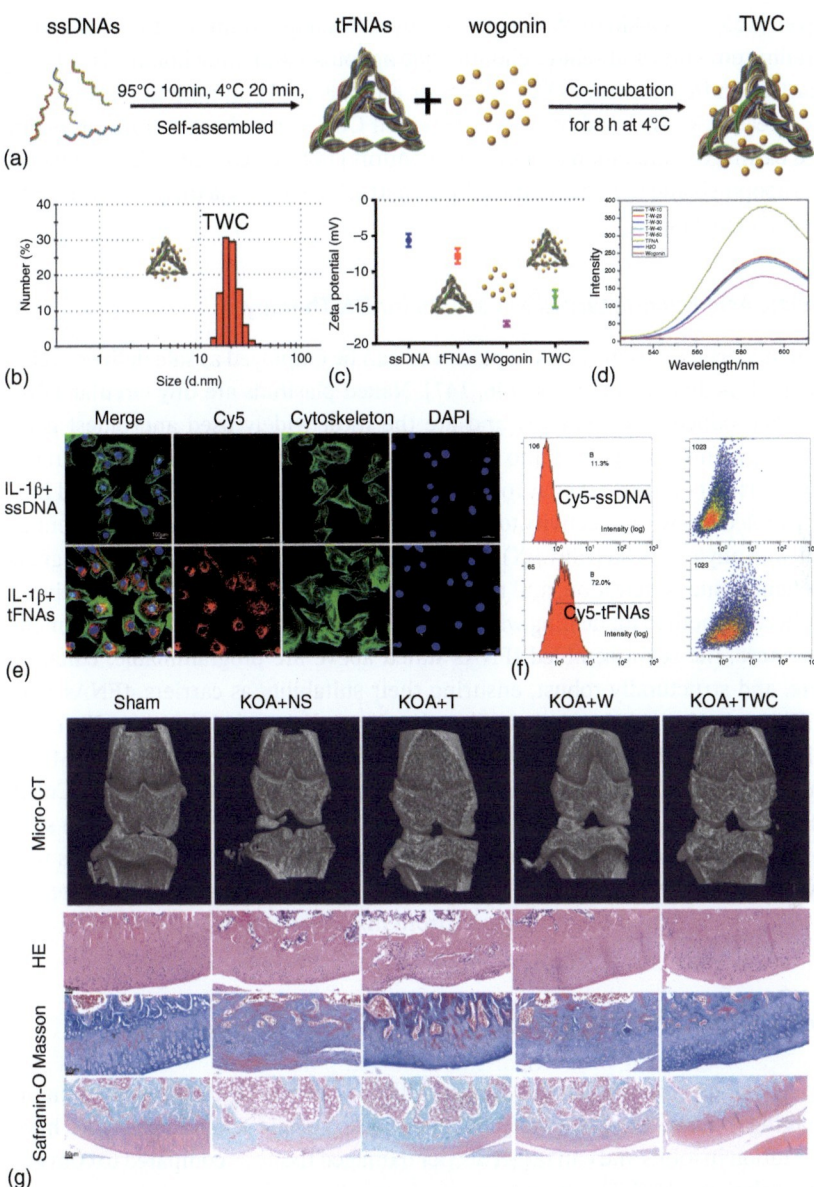

Figure 10.4 tFNAs were used to load wogonin as a vector to treat OA. (a) Schematic of the synthesis of tFNAs and the preparation of TWC. (b) Analysis of the hydrodynamic size of tFNAs and TWC via DLS. (c) Zeta potentials of ssDNA, tFNAs, wogonin, and TWC were determined. Fluorescence emission spectra of a mixture of Gel-Red, λ (Ex)-312 nm, and Gel-Red-tFNAs (250 nmol L^{-1}) in the presence of increasing concentrations of wogonin (10–50 μmol L^{-1}) in ddH$_2$O. The entrapment efficiency was calculated according to the fluorescence intensity at 600 nm (TWC, tFNAs/wogonin complexes). (e) Images illustrating CY5-ssDNA and CY5-tFNAs uptake into IL-1β-induced chondrocytes. Red, green, and blue indicate CY5, F-actin, and nuclei, respectively. (f) Quantitative analysis of IL-1β-induced chondrocyte uptake of CY5-ssDNA and CY5-tFNAs by flow cytometry. (g) Micro-CT, HE, Masson, Safranin-O staining analyses were used to evaluate the knee joints of rats after treatment with different materials at 2 months, respectively. Source: Sirong et al. [68], © 2020/Springer Nature/CC BY 4.0 (a–d, g); Reproduced from Shi et al. [44], © 2020/American Chemical Society (e, f).

OA was relieved [125]. CircFAM160A2 transfected by AAVs might diminish the production of ECM-degrading enzymes and relieve chondrocyte apoptosis and mitochondrial dysfunction in primary chondrocytes from OA patients or animals [143]. In a mouse OA model, CircPDE4B decreased MMP-3 and MMP-13, increased COL2A1 and aggrecan expression, and alleviated cartilage degradation and knee discomfort [144]. CircSERPINE2 was shown to prevent cell apoptosis and enhance extracellular matrix formation via the miR-1271/ERG pathway by Shen et al. [145].

10.5.3 Nucleic Acid Nanomaterials as Carriers for OA Therapy

Nucleic acid nanoparticles, such as naked plasmids, can be employed as safe delivery vehicles for medicinal medicines into cells [146, 147]. Naked plasmids are tiny circular DNA units that can reproduce independently and are the most widely used and oldest gene transfer technology. Inserting genes onto plasmids allows for gene transfer, albeit transfection efficiency is not optimal and has to be increased [148]. Aptamers have evolved into good targeting molecules with selectivity to certain cells as a result of cell-based exponential enrichment ligand phylogeny (cell SELEX). They are simple to produce and alter, have great stability, excellent tissue penetration, and minimal immunogenicity. The aptamer delivery system is an effective method for dealing with the specificity of nucleic acid-based treatment [149]. Furthermore, the self-assembled tFNAs stated above are programmable, editable, biocompatible, and structurally robust, ensuring their suitability as carriers. tFNAs may transport miRNA, siRNA, asPNA, and small-molecule medicines [150]. When Shi et al. co-cultured IL-1-induced chondrocytes with tFNAs, inflammatory factors MMP-1, MMP-3, MMP-13, and TNF-α were downregulated, while chondrogenic markers AGC and Col-II were elevated. tFNAs-wogonin lowered the expressions of IL-1, TNF-, and MMP-3 in synovial fluid in *in vivo* OA models. The organization of chondrocytes in the knee joints was more orderly, the loss of cartilage collagen was decreased, and bone formation was hastened [68]. Figure 10.4 depicts representative pictures from this investigation.

10.6 Conclusion and Prospects

OA is a complicated, diverse, whole-joint disease with various etiologies and unique manifestations, making it difficult to design effective therapies. Nucleic acid nanoparticles have superior penetration in joints and can target deeper damaged tissue as compared to conventional therapy techniques [68]. This paper reviews the nucleic acid nanomaterials currently used in OA treatment, which is summarized in Figure 10.5. The use of targeting carriers such as aptamers can help to drastically reduce the dose of drugs, alleviating their toxicity and side effects [151]. In the future, precise therapy for OA may become popular. ROS is well acknowledged to be important for the activation and activity of M1 macrophages in inflammatory responses [152]. Inducing the shift from the M1 to the M2 phenotype can prevent macrophages from lowering unwanted inflammatory cytokines and may be a promising technique for treating OA. Existing functional nucleic acid nanoparticles, such as tFNAs, have been shown to accomplish this, making them potential therapeutic materials [72]. Furthermore, autologous chondrocyte implantation is the gold standard of cell therapy for

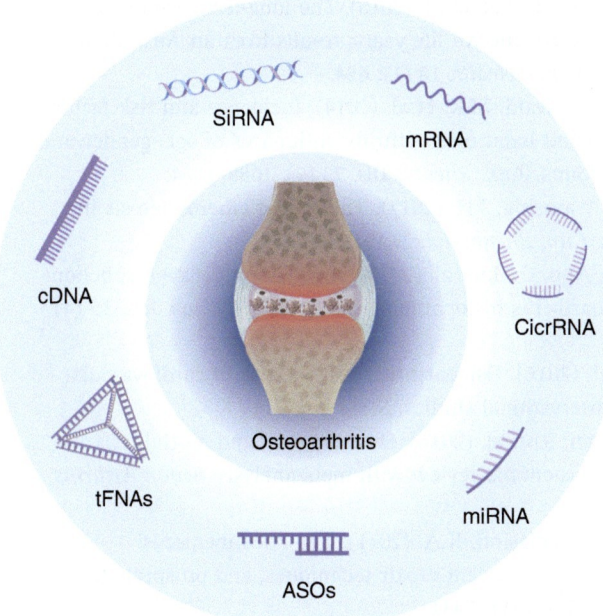

Figure 10.5 Nucleic acid nanomaterials in OA therapy. Source: Drawn by Figdraw.

healing articular cartilage abnormalities [153]. Nucleic acid nanomaterial-based treatments may overlap with cell therapies. To demonstrate this, further, high-quality clinical trials are required. Overall, improvements in nucleic acid nanotherapeutic therapy may transform the treatment of OA.

References

1 Hawker, G.A., Stewart, L., French, M.R. et al. (2008). Understanding the pain experience in hip and knee osteoarthritis--an OARSI/OMERACT initiative. *Osteoarthritis Cartilage* 16 (4): 415–422.
2 Abramoff, B. and Caldera, F.E. (2020). Osteoarthritis: pathology, diagnosis, and treatment options. *Med. Clin. North. Am.* 104 (2): 293–311.
3 Vos, T., Barber, R.M., Bell, B. et al. (2015). Global, regional, and national incidence, prevalence, and years lived with disability for 301 acute and chronic diseases and injuries in 188 countries, 1990–2013: a systematic analysis for the global burden of disease study 2013. *Lancet* 388 (10053): 1545–1602.
4 Hunter, D.J., Schofield, D., and Callander, E. (2014). The individual and socioeconomic impact of osteoarthritis. *Nat. Rev. Rheumatol.* 10 (7): 437–441.
5 Schofield, D.J., Shrestha, R.N., Percival, R. et al. (2013). The personal and national costs of lost labour force participation due to arthritis: an economic study. *BMC Public Health* 13: 188.

6 Schofield, D., Cunich, M., Shrestha, R.N. et al. (2018). The long-term economic impacts of arthritis through lost productive life years: results from an Australian microsimulation model. *BMC Public Health* 18 (1): 654.

7 Prieto-Alhambra, D., Judge, A., Javaid, M.K. et al. (2014). Incidence and risk factors for clinically diagnosed knee, hip and hand osteoarthritis: influences of age, gender and osteoarthritis affecting other joints. *Ann. Rheum. Dis.* 73 (9): 1659–1664.

8 Bierma-Zeinstra, S.M.A. and Waarsing, J.H. (2017). The role of atherosclerosis in osteoarthritis. *Best Pract. Res. Clin. Rheumatol.* 31 (5): 613–633.

9 Finnila, M.A.J., Thevenot, J., Aho, O.M. et al. (2017). Association between subchondral bone structure and osteoarthritis histopathological grade. *J. Orthop. Res.* 35 (4): 785–792.

10 Wang, H., Bai, J., He, B. et al. (2016). Osteoarthritis and the risk of cardiovascular disease: a meta-analysis of observational studies. *Sci. Rep.* 6 (1).

11 Veronese, N., Cereda, E., Maggi, S. et al. (2016). Osteoarthritis and mortality: a prospective cohort study and systematic review with meta-analysis. *Semin Arthritis Rheum.* 46 (2): 160–167.

12 Makris, E.A., Hadidi, P., and Athanasiou, K.A. (2011). The knee meniscus: structure-function, pathophysiology, current repair techniques, and prospects for regeneration. *Biomaterials* 32 (30): 7411–7431.

13 Brandt, K.D., Radin, E.L., Dieppe, P.A., and Putte, L.J.A. o. t. R. D(2006). Yet more evidence that osteoarthritis is not a cartilage disease. *Ann. Rheum. Dis.* 65 (10): 1261–1264.

14 Fu, K., Robbins, S.R., and McDougall, J.J. (2018). Osteoarthritis: the genesis of pain. *Rheumatology* 57 (suppl_4): iv43–iv50.

15 Dougados, M. (2006). Symptomatic slow-acting drugs for osteoarthritis: what are the facts? *Joint Bone Spine* 73 (6): 606–609.

16 Michael, J.W., Schluter-Brust, K.U., and Eysel, P. (2010). The epidemiology, etiology, diagnosis, and treatment of osteoarthritis of the knee. *Dtsch. Arztebl. Int.* 107 (9): 152–162.

17 Berenbaum, F. (2013). Osteoarthritis as an inflammatory disease (osteoarthritis is not osteoarthrosis!). *Osteoarthritis Cartilage* 21 (1): 16–21.

18 Goldring, M.B. (2012). Chondrogenesis, chondrocyte differentiation, and articular cartilage metabolism in health and osteoarthritis. *Ther. Adv. Musculoskelet Dis.* 4 (4): 269–285.

19 Sandell, L.J. (2012). Etiology of osteoarthritis: genetics and synovial joint development. *Nat. Rev. Rheumatol.* 8 (2): 77–89.

20 van Meurs, J.B. (2017). Osteoarthritis year in review 2016: genetics, genomics and epigenetics. *Osteoarthritis Cartilage* 25 (2): 181–189.

21 Neogi, T. and Zhang, Y. (2013). Epidemiology of osteoarthritis. *Rheum Dis. Clin. North Am.* 39 (1): 1–19.

22 Lotz, M. and Loeser, R.F. (2012). Effects of aging on articular cartilage homeostasis. *Bone* 51 (2): 241–248.

23 Guilak, F. (2011). Biomechanical factors in osteoarthritis. *Best Pract. Res. Clin. Rheumatol.* 25 (6): 815–823.

24 Sacitharan, P.K. (2019). Ageing and osteoarthritis. *Subcell Biochem.* 91: 123–159.

25 Felson, D.T., Zhang, Y., Hannan, M.T. et al. (2010). The incidence and natural history of knee osteoarthritis in the elderly. *Framingham Osteoarthritis Study* 38 (10): 1500–1505.

26 Brown, T.D., Johnston, R.C., Saltzman, C.L. et al. o. O. T(2006). Posttraumatic osteoarthritis: a first estimate of incidence, prevalence, and burden of disease. *J. Orthop. Trauma.* 20 (10): 739–744.

27 Szwedowski, D., Szczepanek, J., Paczesny, L. et al. (2021). The effect of platelet-rich plasma on the intra-articular microenvironment in knee osteoarthritis. *Int. J. Mol. Sci.* 22 (11): 5492.

28 Latourte, A., Kloppenburg, M., and Richette, P. (2020). Emerging pharmaceutical therapies for osteoarthritis. *Nat. Rev. Rheumatol.* 16 (12): 673–688.

29 Bennell, K.L., Hunter, D.J., and Hinman, R.S. (2012). Management of osteoarthritis of the knee. *BMJ* 345: e4934.

30 Raynauld, J.P., Buckland-Wright, C., Ward, R. et al. (2003). Safety and efficacy of long-term intraarticular steroid injections in osteoarthritis of the knee: a randomized, double-blind, placebo-controlled trial. *Arthritis Rheum* 48 (2): 370–377.

31 Dai, L., Zhang, X., Hu, X. et al. (2015). Silencing of miR-101 prevents cartilage degradation by regulating extracellular matrix-related genes in a rat model of osteoarthritis. *Mol. Ther.* 23 (8): 1331–1340.

32 Wang, L., Liu, X., Zhang, Q. et al. (2012). Selection of DNA aptamers that bind to four organophosphorus pesticides. *Biotechnol. Lett* 34 (5): 869–874.

33 Joyce, F.G. (2001). RNA cleavage by the 10-23 DNA enzyme. In: *Ribonucleases - Part A*, 503–517.

34 Rothemund, P.W. (2006). Folding DNA to create nanoscale shapes and patterns. *Nature* 440 (7082): 297–302.

35 Hu, Y., Ren, J., Lu, C.H., and Willner, I. (2016). Programmed pH-driven reversible association and dissociation of interconnected circular DNA dimer nanostructures. *Nano Lett.* 16 (7): 4590–4594.

36 Chen, R., Wen, D., Fu, W. et al. (2022). Treatment effect of DNA framework nucleic acids on diffuse microvascular endothelial cell injury after subarachnoid hemorrhage. *Cell Proliferation* 55 (4): e13206.

37 Xu, J. and Wei, C. (2017). The aptamer DNA-templated fluorescence silver nanoclusters: ATP detection and preliminary mechanism investigation. *Biosens. Bioelectron.* 87: 422–427.

38 Zhou, Z., Du, Y., and Dong, S. (2011). Double-strand DNA-templated formation of copper nanoparticles as fluorescent probe for label-free aptamer sensor. *Anal. Chem.* 83 (13): 5122–5127.

39 Wang, H., Yang, R., Yang, L., and Tan, W. (2009). Nucleic acid conjugated nanomaterials for enhanced molecular recognition. *ACS Nano* 3 (9): 2451–2460.

40 Scholz, C. and Wagner, E. (2012). Therapeutic plasmid DNA versus siRNA delivery: common and different tasks for synthetic carriers. *J. Controlled Release* 161 (2): 554–565.

41 Smith, D.M., Schuller, V., Forthmann, C. et al. (2011). A structurally variable hinged tetrahedron framework from DNA origami. *J. Nucleic Acids* 2011: 360954.

42 Zhang, T., Cui, W., Tian, T. et al. (2020). Progress in biomedical applications of tetrahedral framework nucleic acid-based functional systems. *ACS Appl. Mater. Interfaces* 12 (42): 47115–47126.

43 Zhang, B., Tian, T., Xiao, D. et al. (2022). Facilitating in situ tumor imaging with a tetrahedral DNA framework-enhanced hybridization chain reaction probe. *Adv. Funct. Mater.* 2109728.

44 Shi, S., Tian, T., Li, Y. et al. (2020). Tetrahedral framework nucleic acid inhibits chondrocyte apoptosis and oxidative stress through activation of autophagy. *ACS Appl. Mater. Interfaces* 12 (51): 56782–56791.

45 Shi, S., Lin, S., Shao, X. et al. (2017). Modulation of chondrocyte motility by tetrahedral DNA nanostructures. *Cell Proliferation* 50 (5).

46 Endisha, H., Datta, P., Sharma, A. et al. (2021). MicroRNA-34a-5p promotes joint destruction during osteoarthritis. *Arthritis Rheumatol.* 73 (3): 426–439.

47 Cai, Y., Lopez-Ruiz, E., Wengel, J. et al. (2017). A hyaluronic acid-based hydrogel enabling CD44-mediated chondrocyte binding and gapmer oligonucleotide release for modulation of gene expression in osteoarthritis. *J. Controlled Release* 253: 153–159.

48 Zhou, M., Gao, S., Zhang, X. et al. (2021). The protective effect of tetrahedral framework nucleic acids on periodontium under inflammatory conditions. *Bioact. Mater.* 6 (6): 1676–1688.

49 Wang, Y., Li, Y., Gao, S. et al. (2022). Tetrahedral framework nucleic acids can alleviate taurocholate-induced severe acute pancreatitis and its subsequent multiorgan injury in mice. *Nano Lett.* 22 (4): 1759–1768.

50 Zhu, J., Yang, Y., Ma, W. et al. (2022). Antiepilepticus effects of tetrahedral framework nucleic acid via inhibition of gliosis-induced downregulation of glutamine synthetase and increased AMPAR internalization in the postsynaptic membrane. *Nano Lett.* 22 (6): 2381–2390.

51 Zhou, M., Zhang, T., Zhang, B. et al. (2021). A DNA nanostructure-based neuroprotectant against neuronal apoptosis via inhibiting toll-like receptor 2 signaling pathway in acute ischemic stroke. *ACS Nano* 16 (1): 1456–1470.

52 Chen, X., Cui, W., Liu, Z. et al. (2022). Positive neuroplastic effect of DNA framework nucleic acids on neuropsychiatric diseases. *ACS Mater. Lett.* 4 (4): 665–674.

53 Ma, W., Yang, Y., Zhu, J. et al. (2022). Biomimetic Nanoerythrosome-coated aptamer-DNA tetrahedron/Maytansine conjugates: pH-responsive and targeted cytotoxicity for HER2-positive breast cancer. *Adv. Mater.* e2109609.

54 Zhang, T., Tian, T., and Lin, Y. (2021). Functionalizing framework nucleic-acid-based nanostructures for biomedical application. *Adv. Mater.* e2107820.

55 Jun Li, Y.L., Li, M., Chen, X. et al. (2022). Repair of infected bone defect with clindamycin-tetrahedral DNA nanostructure. *Chem. Eng. J.* 435 (1): 134855.

56 Zhao, D., Xiao, D., Liu, M. et al. (2022). Tetrahedral framework nucleic acid carrying angiogenic peptide prevents bisphosphonate-related osteonecrosis of the jaw by promoting angiogenesis. *Int. J. Oral Sci.* 14 (1): 23.

57 Jiang, Y., Li, S., Zhang, T. et al. (2022). Tetrahedral framework nucleic acids inhibit skin fibrosis via the pyroptosis pathway. *ACS Appl. Mater. Interfaces* 14 (13): 15069–15079.

58 Chen, Y., Shi, S., Li, B. et al. (2022). Therapeutic effects of self-assembled tetrahedral framework nucleic acids on liver regeneration in acute liver failure. *ACS Appl. Mater. Interfaces* 14 (11): 13136–13146.

59 Liang, L., Li, J., Li, Q. et al. (2014). Single-particle tracking and modulation of cell entry pathways of a tetrahedral DNA nanostructure in live cells. *Angew. Chem. Int. Ed. Engl.* 53 (30): 7745–7750.

60 Ma, W., Shao, X., Zhao, D. et al. (2018). Self-assembled tetrahedral DNA nanostructures promote neural stem cell proliferation and neuronal differentiation. *ACS Appl. Mater. Interfaces* 10 (9): 7892–7900.

61 Shao, X., Lin, S., Peng, Q. et al. (2017). Tetrahedral DNA nanostructure: a potential promoter for cartilage tissue regeneration via regulating chondrocyte phenotype and proliferation. *Small* 13 (12).

62 Zhou, M., Liu, N., Zhang, Q. et al. (2019). Effect of tetrahedral DNA nanostructures on proliferation and osteogenic differentiation of human periodontal ligament stem cells. *Cell Proliferation* 52 (3): e12566.

63 Liu, N., Zhang, X., Li, N. et al. (2019). Tetrahedral framework nucleic acids promote corneal epithelial wound healing in vitro and in vivo. *Small* 15 (31): e1901907.

64 Qin, X., Li, N., Zhang, M. et al. (2019). Tetrahedral framework nucleic acids prevent retina ischemia-reperfusion injury from oxidative stress via activating the Akt/Nrf2 pathway. *Nanoscale* 11 (43): 20667–20675.

65 Zhang, Q., Lin, S., Wang, L. et al. (2021). Tetrahedral framework nucleic acids act as antioxidants in acute kidney injury treatment. *Chem. Eng. J.* 413: 127426.

66 Zhao, D., Cui, W., Liu, M. et al. (2020). Tetrahedral framework nucleic acid promotes the treatment of bisphosphonate-related osteonecrosis of the jaws by promoting angiogenesis and M2 polarization. *ACS Appl. Mater. Interfaces* 12 (40): 44508–44522.

67 Zhang, M., Zhang, X., Tian, T. et al. (2021). Anti-inflammatory activity of curcumin-loaded tetrahedral framework nucleic acids on acute gouty arthritis. *Bioact. Mater.* 2022 (8): 368–380.

68 Sirong, S., Yang, C., Taoran, T. et al. (2020). Effects of tetrahedral framework nucleic acid/wogonin complexes on osteoarthritis. *Bone Res.* 8: 6.

69 Shi, S., Lin, S., Li, Y. et al. (2018). Effects of tetrahedral DNA nanostructures on autophagy in chondrocytes. *Chem. Commun.* 54 (11): 1327–1330.

70 Tian, T., Zhang, C., Li, J. et al. (2021). Proteomic exploration of endocytosis of framework nucleic acids. *Small* 17 (23): e2100837.

71 Yao, Y., Wen, Y., Li, Y. et al. (2021). Tetrahedral framework nucleic acids facilitate neurorestoration of facial nerves by activating the NGF/PI3K/AKT pathway. *Nanoscale* 13 (37): 15598–15610.

72 Zhang, Q., Lin, S., Shi, S. et al. (2018). Anti-inflammatory and anti-oxidative effects of tetrahedral DNA nanostructures via the modulation of macrophage responses. *ACS Appl. Mater. Interfaces* 10 (4): 3421–3430.

73 Tian, T., Zhang, T., Zhou, T. et al. (2017). Synthesis of an ethyleneimine/tetrahedral DNA nanostructure complex and its potential application as a multi-functional delivery vehicle. *Nanoscale* 9 (46): 18402–18412.

74 Shi, S., Li, Y., Zhang, T. et al. (2021). Biological effect of differently sized tetrahedral framework nucleic acids: endocytosis, proliferation, migration, and biodistribution. *ACS Appl. Mater. Interfaces* 13 (48): 57067–57074.
75 Koller, E., Vincent, T.M., Chappell, A. et al. (2011). Mechanisms of single-stranded phosphorothioate modified antisense oligonucleotide accumulation in hepatocytes. *Nucleic Acids Res.* 39 (11): 4795–4807.
76 Butler, M., Stecker, K., and Bennett, C.F. (1997). Cellular distribution of phosphorothioate oligodeoxynucleotides in normal rodent tissues. *Lab. Invest.* 77 (4): 379–388.
77 Wahane, A., Waghmode, A., Kapphahn, A. et al. (2020). Role of lipid-based and polymer-based non-viral vectors in nucleic acid delivery for next-generation gene therapy. *Molecules* 25 (12).
78 Song, J., Kim, D., Chun, C.H. et al. (2013). MicroRNA-9 regulates survival of chondroblasts and cartilage integrity by targeting protogenin. 11 (1): 66–66.
79 Vester, B. and Wengel, J.J.B. (2004). LNA (locked nucleic acid): high-affinity targeting of complementary RNA and DNA. *Biochemistry* 43: 13233–13241.
80 Nakamura, A., Rampersaud, Y.R., Nakamura, S. et al. (2018). MicroRNA-181a-5p antisense oligonucleotides attenuate osteoarthritis in facet and knee joints. *Ann. Rheumatic Dis.* 78.
81 Baek, D., Lee, K.M., Park, K.W. et al. (2018). Inhibition of miR-449a promotes cartilage regeneration and prevents progression of osteoarthritis in in vivo rat models. *Mol. Ther. Nucleic Acids* 13: 322–333.
82 Zrioual, S., Toh, M.L., Tournadre, A. et al. (2008). IL-17RA and IL-17RC Receptors are essential for IL-17A-induced ELR+ CXC chemokine expression in synoviocytes and are overexpressed in rheumatoid blood. *J. Immunol.* 180 (1): 655–663.
83 Maitra, A., Fang, S., Hanel, W. et al. (2007). Distinct functional motifs within the IL-17 receptor regulate signal transduction and target gene expression. *Biol. Sci.* 104 (18): 7506–7511.
84 Chen, L., Li, D.Q., Zhong, J. et al. (2011). IL-17RA aptamer-mediated repression of IL-6 inhibits synovium inflammation in a murine model of osteoarthritis. *Osteoarthritis Cartilage* 19 (6): 711–718.
85 McMahon, J.M., Conroy, S., Lyons, M. et al. (2006). Gene transfer into rat mesenchymal stem cells: a comparative study of viral and nonviral vectors. *Stem Cells Dev.* 15 (1): 87–96.
86 Cheng, M. and Wang, Y. (2020). Downregulation of HMGB1 by miR-103a-3p promotes cell proliferation, alleviates apoptosis and in Flammation in a cell model of osteoarthritis. *Iran. J. Biotechnol.* 18 (1): e2255.
87 Wu, Y.H., Liu, W., Zhang, L. et al. (2018). Effects of microRNA-24 targeting C-myc on apoptosis, proliferation, and cytokine expressions in chondrocytes of rats with osteoarthritis via MAPK signaling pathway. *J. Cell. Biochem.* 119 (10): 7944–7958.
88 Zhong, J.H., Li, J., Liu, C.F. et al. (2017). Effects of microRNA-146a on the proliferation and apoptosis of human osteoarthritis chondrocytes by targeting TRAF6 through the NF-kappaB signalling pathway. *Biosci. Rep.* 37 (2).
89 Zhao, Z., Dai, X.S., Wang, Z.Y. et al. (2019). MicroRNA-26a reduces synovial inflammation and cartilage injury in osteoarthritis of knee joints through impairing the NF-kappaB signaling pathway. *Biosci. Rep.* 39 (4).

90 Tian, Y., Guo, R., Shi, B. et al. (2016). MicroRNA-30a promotes chondrogenic differentiation of mesenchymal stem cells through inhibiting Delta-like 4 expression. *Life Sci.* 148: 220–228.

91 Ding, Y., Wang, L., Zhao, Q. et al. (2019). MicroRNA93 inhibits chondrocyte apoptosis and inflammation in osteoarthritis by targeting the TLR4/NFkappaB signaling pathway. *Int. J. Mol. Med.* 43 (2): 779–790.

92 Gao, S., Liu, L., Zhu, S. et al. (2020). MicroRNA-197 regulates chondrocyte proliferation, migration, and inflammation in pathogenesis of osteoarthritis by targeting EIF4G2. *Biosci. Rep.* 40 (9).

93 Ma, S., Zhang, A., Li, X. et al. (2020). MiR-21-5p regulates extracellular matrix degradation and angiogenesis in TMJOA by targeting Spry1. *Arthritis Res. Ther.* 22 (1): 99.

94 Wang, W.F., Liu, S.Y., Qi, Z.F. et al. (2020). MiR-145 targeting BNIP3 reduces apoptosis of chondrocytes in osteoarthritis through notch signaling pathway. *Eur. Rev. Med. Pharmacol. Sci.* 24 (16): 8263–8272.

95 Zhang, H., Zheng, W., Li, D., and Zheng, J. (2021). miR-146a-5p promotes chondrocyte apoptosis and inhibits autophagy of osteoarthritis by targeting NUMB. *Cartilage* 13 (2_suppl): 1467S–1477S.

96 He, X. and Deng, L. (2021). miR-204-5p inhibits inflammation of synovial fibroblasts in osteoarthritis by suppressing FOXC1. *J. Orthop. Sci.*

97 Jin, F., Liao, L., and Zhu, Y. (2021). MiR-467b alleviates lipopolysaccharide-induced inflammation through targeting STAT1 in chondrogenic ATDC5 cells. *Int. J. Immunogenet.* 48 (5): 435–442.

98 Lu, J., Zhou, Z., Sun, B. et al. (2020). MiR-520d-5p modulates chondrogenesis and chondrocyte metabolism through targeting HDAC1. *Aging* 12 (18): 18545–18560.

99 Jiang, P., Dou, X., Li, S. et al. (2021). miR-590-5p affects chondrocyte proliferation, apoptosis, and inflammation by targeting FGF18 in osteoarthritis. *Am. J. Transl. Res.* 13 (8): 8728–8741.

100 Zhu, S., Deng, Y., Gao, H. et al. (2020). miR-877-5p alleviates chondrocyte dysfunction in osteoarthritis models via repressing FOXM1. *J. Gene. Med.* 22 (11): e3246.

101 Gao, F., Peng, C., Zheng, C. et al. (2019). miRNA-101 promotes chondrogenic differentiation in rat bone marrow mesenchymal stem cells. *Exp. Ther. Med.* 17 (1): 175–180.

102 Yang, F., Huang, R., Ma, H. et al. (2020). miRNA-411 regulates chondrocyte autophagy in osteoarthritis by targeting hypoxia-inducible factor 1 alpha (HIF-1alpha). *Med. Sci. Monit.* 26: e921155.

103 Li, W., Zhao, S., Yang, H. et al. (2019). Potential novel prediction of TMJ-OA: MiR-140-5p regulates inflammation through Smad/TGF-beta Signaling. *Front. Pharmacol.* 10: 15.

104 Wang, R. and Xu, B. (2021). TGF-beta1-modified MSC-derived exosomal miR-135b attenuates cartilage injury via promoting M2 synovial macrophage polarization by targeting MAPK6. *Cell Tissue Res.* 384 (1): 113–127.

105 Huang, X., Qiao, F., and Xue, P. (2019). The protective role of microRNA-140-5p in synovial injury of rats with knee osteoarthritis via inactivating the TLR4/Myd88/NF-kappaB signaling pathway. *Cell Cycle* 18 (18): 2344–2358.

106 Liu, W., Zha, Z., and Wang, H. (2019). Upregulation of microRNA-27a inhibits synovial angiogenesis and chondrocyte apoptosis in knee osteoarthritis rats through the inhibition of PLK2. *J. Cell. Physiol.* 234 (12): 22972–22984.

107 Liang, Y., Xu, X., Li, X. et al. (2020). Chondrocyte-targeted microRNA delivery by engineered exosomes toward a cell-free osteoarthritis therapy. *ACS Appl. Mater. Interfaces* 12 (33): 36938–36947.

108 Scott, K.M., Cohen, D.J., Hays, M. et al. (2021). Regulation of inflammatory and catabolic responses to IL-1β in rat articular chondrocytes by microRNAs miR-122 and miR-451. *Osteoarthritis Cartilage* 29 (1): 113–123.

109 Chen, X., Liu, Y., Wen, Y. et al. (2019). A photothermal-triggered nitric oxide nanogenerator combined with siRNA for precise therapy of osteoarthritis by suppressing macrophage inflammation. *Nanoscale* 11 (14): 6693–6709.

110 Bedingfield, S.K., Colazo, J.M., Di Francesco, M. et al. (2021). Top-down fabricated microPlates for prolonged, intra-articular matrix metalloproteinase 13 siRNA nanocarrier delivery to reduce post-traumatic osteoarthritis. *ACS Nano* 15 (9): 14475–14491.

111 Hoshi, H., Akagi, R., Yamaguchi, S. et al. (2017). Effect of inhibiting MMP13 and ADAMTS5 by intra-articular injection of small interfering RNA in a surgically induced osteoarthritis model of mice. *Cell Tissue Res.* 368 (2): 379–387.

112 Gong, Y., Li, S.J., Liu, R. et al. (2019). Inhibition of YAP with siRNA prevents cartilage degradation and ameliorates osteoarthritis development. *J. Mol. Med.* 97 (1): 103–114.

113 Nakagawa, R., Akagi, R., Yamaguchi, S. et al. (2019). Single vs. repeated matrix metalloproteinase-13 knockdown with intra-articular short interfering RNA administration in a murine osteoarthritis model. *Connect. Tissue Res.* 60 (4): 335–343.

114 Choi, M.C., MaruYama, T., Chun, C.H., and Park, Y. (2018). Alleviation of murine osteoarthritis by cartilage-specific deletion of IkappaBzeta. *Arthritis Rheumatol.* 70 (9): 1440–1449.

115 Huang, W., Cheng, C., Shan, W.S. et al. (2020). Knockdown of SGK1 alleviates the IL-1beta-induced chondrocyte anabolic and catabolic imbalance by activating FoxO1-mediated autophagy in human chondrocytes. *FEBS J.* 287 (1): 94–107.

116 Wang, Z., Ni, S., Zhang, H. et al. (2021). Silencing SGK1 alleviates osteoarthritis through epigenetic regulation of CREB1 and ABCA1 expression. *Life Sci.* 268: 118733.

117 Xu, S., Yu, J., Wang, Z. et al. (2019). SOX11 promotes osteoarthritis through induction of TNF-alpha. *Pathol. Res. Pract.* 215 (7): 152442.

118 Li, L., Lv, G., Wang, B., and Kuang, L. (2020). XIST/miR-376c-5p/OPN axis modulates the influence of proinflammatory M1 macrophages on osteoarthritis chondrocyte apoptosis. *J. Cell. Physiol.* 235 (1): 281–293.

119 Zhou, X., Jiang, L., Fan, G. et al. (2019). Role of the ciRS-7/miR-7 axis in the regulation of proliferation, apoptosis and inflammation of chondrocytes induced by IL-1beta. *Int. Immunopharmacol.* 71: 233–240.

120 Lange, C., Madry, H., Venkatesan, J.K. et al. (2021). rAAV-mediated sox9 overexpression improves the repair of osteochondral defects in a clinically relevant large animal model over time in vivo and reduces perifocal osteoarthritic changes. *Am. J. Sports Med.* 49 (13): 3696–3707.

121 Ko, J.Y., Lee, J., Lee, J. et al. (2019). SOX-6, 9-transfected adipose stem cells to treat surgically-induced osteoarthritis in goats. *Tissue Eng. Part A.* 25 (13–14): 990–1000.

122 Wu, H., Peng, Z., Xu, Y. et al. (2022). Engineered adipose-derived stem cells with IGF-1-modified mRNA ameliorates osteoarthritis development. *Stem Cell Res. Ther.* 13 (1): 19.

123 Wang, K.F., Shi, Z.W., and Dong, D.M. (2021). CircATRNL1 protects against osteoarthritis by targeting miR-153-3p and KLF5. *Int. Immunopharmacol.* 96: 107704.

124 Tao, S.C., Huang, J.Y., Gao, Y. et al. (2021). Small extracellular vesicles in combination with sleep-related circRNA3503: a targeted therapeutic agent with injectable thermosensitive hydrogel to prevent osteoarthritis. *Bioact. Mater.* 6 (12): 4455–4469.

125 Yao, T., Yang, Y., Xie, Z. et al. (2020). Circ0083429 regulates osteoarthritis progression via the Mir-346/SMAD3 axis. *Front. Cell Dev. Biol.* 8: 579945.

126 Friedman, R.C., Farh, K.K., Burge, C.B., and Bartel, D.P. (2009). Most mammalian mRNAs are conserved targets of microRNAs. *Genome Res.* 19 (1): 92–105.

127 Catalanotto, C., Cogoni, C., and Zardo, G. (2016). MicroRNA in control of gene expression: an overview of nuclear functions. *Int. J. Mol. Sci.* 17 (10).

128 Yu, Q., Zhao, B., He, Q. et al. (2019). microRNA-206 is required for osteoarthritis development through its effect on apoptosis and autophagy of articular chondrocytes via modulating the phosphoinositide 3-kinase/protein kinase B-mTOR pathway by targeting insulin-like growth factor-1. *J. Cell. Biochem.* 120 (4): 5287–5303.

129 Sui, C., Zhang, L., and Hu, Y. (2019). MicroRNAlet7a inhibition inhibits LPSinduced inflammatory injury of chondrocytes by targeting IL6R. *Mol. Med. Rep.* 20 (3): 2633–2640.

130 Guo, Y., Tian, L., Du, X., and Deng, Z. (2020). MiR-203 regulates estrogen receptor alpha and cartilage degradation in IL-1beta-stimulated chondrocytes. *J. Bone Miner. Metab.* 38 (3): 346–356.

131 Jin, Z., Ren, J., and Qi, S. (2020). Human bone mesenchymal stem cells-derived exosomes overexpressing microRNA-26a-5p alleviate osteoarthritis via down-regulation of PTGS2. *Int. Immunopharmacol.* 78: 105946.

132 Tomari, Y. and Zamore, P.D. (2005). Perspective: machines for RNAi. *Genes Dev.* 19 (5): 517–529.

133 Chen, K., Yan, Y., Li, C. et al. (2017). Increased 15-lipoxygenase-1 expression in chondrocytes contributes to the pathogenesis of osteoarthritis. *Cell Death Dis.* 8 (10): e3109.

134 Xu, L., Sun, C., Zhang, S. et al. (2015). Sam68 promotes NF-kappaB activation and apoptosis Signaling in articular chondrocytes during osteoarthritis. *Inflam. Res.* 64 (11): 895–902.

135 Weber, C., Armbruster, N., Scheller, C. et al. (2013). Foamy virus-adenovirus hybrid vectors for gene therapy of the arthritides. *J. Gene. Med.* 15 (3–4): 155–167.

136 Chen, L.X., Lin, L., Wang, H.J. et al. (2008). Suppression of early experimental osteoarthritis by in vivo delivery of the adenoviral vector-mediated NF-kappaBp65-specific siRNA. *Osteoarthritis Cartilage* 16 (2): 174–184.

137 Huang, M.J., Wang, L., Zheng, X.C. et al. (2012). Intra-articular lentivirus-mediated insertion of the fat-1 gene ameliorates osteoarthritis. *Med. Hypotheses* 79 (5): 614–616.

138 Zhu, S., Zhang, B., Man, C. et al. (2014). Combined effects of connective tissue growth factor-modified bone marrow-derived mesenchymal stem cells and NaOH-treated PLGA scaffolds on the repair of articular cartilage defect in rabbits. *Cell Transplant.* 23 (6): 715–727.

139 Aini, H., Itaka, K., Fujisawa, A. et al. (2016). Messenger RNA delivery of a cartilage-anabolic transcription factor as a disease-modifying strategy for osteoarthritis treatment. *Sci. Rep.* 6 (1).

140 Salzman, J. (2016). Circular RNA expression: its potential regulation and function. *Trends Genet.* 32 (5): 309–316.

141 Han, D., Li, J., Wang, H. et al. (2017). Circular RNA circMTO1 acts as the sponge of microRNA-9 to suppress hepatocellular carcinoma progression. *Hepatology* 66 (4): 1151–1164.

142 Pamudurti, N.R., Bartok, O., Jens, M. et al. (2017). Translation of CircRNAs. *Mol. Cell* 66 (1): 9–21. e7.

143 Bao, J., Lin, C., Zhou, X. et al. (2021). circFAM160A2 promotes mitochondrial stabilization and apoptosis reduction in osteoarthritis chondrocytes by targeting miR-505-3p and SIRT3. *Oxid. Med. Cell. Longevity* 2021: 5712280.

144 Shen, S., Yang, Y., Shen, P. et al. (2021). circPDE4B prevents articular cartilage degeneration and promotes repair by acting as a scaffold for RIC8A and MID1. *Ann. Rheum. Dis.* 80 (9): 1209–1219.

145 Shen, S., Wu, Y., Chen, J. et al. (2019). CircSERPINE2 protects against osteoarthritis by targeting miR-1271 and ETS-related gene. *Ann. Rheum. Dis.* 78 (6): 826–836.

146 Uzieliene, I., Kalvaityte, U., Bernotiene, E., and Mobasheri, A. (2020). Non-viral gene therapy for osteoarthritis. *Front. Bioeng. Biotechnol.* 8: 618399.

147 Lu, H., Dai, Y., Lv, L., and Zhao, H. (2014). Chitosan-graft-polyethylenimine/DNA nanoparticles as novel non-viral gene delivery vectors targeting osteoarthritis. *PLoS One* 9 (1): e84703.

148 Clanchy, F.I. and Williams, R.O. (2008). Plasmid DNA as a safe gene delivery vehicle for treatment of chronic inflammatory disease. *Expert Opin. Biol. Ther.* 8 (10): 1507–1519.

149 Zhang, Y., Lai, B.S., and Juhas, M. (2019). Recent advances in aptamer discovery and applications. *Molecules* 24 (5).

150 Qin, X., Xiao, L., Li, N. et al. (2022). Tetrahedral framework nucleic acids-based delivery of microRNA-155 inhibits choroidal neovascularization by regulating the polarization of macrophages. *Bioact. Mater.* 14: 134–144.

151 Kim, Y., Liu, C., and Tan, W. (2009). Aptamers generated by cell SELEX for biomarker discovery. *Biomarker Res.* 3 (2): 193–202.

152 Covarrubias, A., Byles, V., and Horng, T. (2013). ROS sets the stage for macrophage differentiation. *Cell Res.* 23 (8): 984–985.

153 Xiao, S. and Chen, L. (2020). The emerging landscape of nanotheranostic-based diagnosis and therapy for osteoarthritis. *J. Controlled Release* 328: 817–833.

Index

a

acetaminophen   183
actinomycin D (AMD)   169
acyclic L-threoninol nucleic acid (L-aTNA)
    17, 18
adenosine triphosphate (ATP)   70, 110, 116
adipo-8   148
adipose tissue   147, 148
allele mutation detection method   107
allicin   148
ampicillin-loaded tFNAs   44
angiogenic peptide loaded tFNAs   15
antibiotic resistance   45, 51, 161, 162, 168
antibodies vs. aptamers   5
antibody-based HCR system   107, 108
antimicrobial peptides (AMPs)   161, 162,
    166, 169–171
antisense oligonucleotides (ASOs)   64–65,
    70, 116, 129, 141, 149, 162, 163, 168,
    169, 184, 187
antisense peptide nucleic acids (asPNAs)
    168, 169
apolipoprotein B (ApoB)   142, 149
aptamers   5, 129, 142, 150, 184, 187
    antibodies vs.   5
    AS1411-modified tFNAs   41
    based HCR system   108–109
    functionalized DNA origami delivery
        nanostructures   51
    modified DNA origami   45
    modified tetrahedral framework DNA   172
    targeted drug delivery   49
    tFNAs   126

arm-acceptor hairpins   107
arm-donating hairpins   107
artificial DNA nanostructures   141
artificial nucleic acids (XNAs)   4, 17–18
ATP-responsive HCR-based logic gate   110
autocatalytic HCR biocircuit   103–105

b

bacterial magnetosomes   86, 87
β-lactamases   170
bioimaging of bone tissue regeneration
    93–94
biological polymer DNA   37
bionanomaterials   7, 86
bioscaffold materials   81
biosensing   2, 3, 7, 37, 82, 84–86, 139, 141,
    184
bone defect   15, 81, 90, 92–94
bone-related diseases   81–94
bone tissue engineering   81, 93, 94
"bottom-up" approach   1
BSA-coated DNA origami   21

c

carbazole derivative BMEPC   44
carbon-based nanomaterials   84, 85, 87
caveolin-mediated endocytosis   41, 150
cDNA   106, 188, 192
cell targeting   47, 49, 53, 65, 66, 68, 184
chemical antibodies   49
chemotherapy   46, 51, 68, 115, 116, 123, 124,
    127, 129, 130

chitosan-based injectable thermosensitive hydrogel scaffold structure   93
chondrocyte affinity peptide (CAP) exosomes   188
circular RNA (CircRNA)   192, 194
clearing mutant mtDNA   62–64
clindamycin-tFNAs complex   15
composite nanomaterials   7, 84, 87
curcumin-loaded tFNAs   15
cytosine-phosphate-guanine oligodeoxynucleotides (CpG ODNs)   127

d

dendritic oligonucleotide-coated DNA brick nanostructures   21
deoxyribonucleic acid (DNA)   3
  based nanocomposite hydrogels   146
  DNA-based nanomaterials   2
  history of   3–9
  hydrogel nanostructures, in antibacterial field   170
  hydrogels   162
  logic gate-based HCR   112
  nanomesh   26
  nanoribbons   170
  nanostructures   16, 162
    advantages in antibacterial field compatibility   163
    challenges and perspectives   172
    compatibility   163
    drug-loading performance   164
    editability   163
    ε-poly-L-lysine-DNA nanocomplex   166
    five "holes" origami framework   164
    melamine-DNA-AgNC complex   165–166
    NET-like nanogel based on 2D DNA networks   166
    stability   163
    super silver nanoclusters based   164, 165
    3D application   166, 170
  nanotechnology   161
  origami   6, 7, 22, 127–129, 161
  origami techniques   38, 144
  origami technology   124
  six-helix bundle (6HB)   169
  tetrahedron   39, 144
  tiles   6
diabetes mellitus (DM)   144–146
diacylglycerol acyltransferase-2 (DGAT2)   142, 146, 149
disulfide crosslinking   23
DNA Pom-Pom nanostructure (DNA PP-N)   170
DNA scaffold strand   38
DNAzymes   4, 5, 103
double-crossed (DX) molecules   6
DOX-loaded DNA origami nanosystem   52
doxorubicin (DOX) origami delivery system   128
dual miRNA detection system   111
dynamic DNA nanostructures   123, 124, 130–133

e

Emapticap pegol (NOX-E36)   145
enhanced permeability and retention (EPR) effects   128
enzymatic ligation   22–23
ε-poly-L-lysine-DNA nanocomplex   166
erythromycin-loaded tFNA   44
exosomes   86, 188

f

FD-RNA   68
fluorescence-based detection   101
2′-fluoroarabinonucleic acid (FANA)   18
FNA-MSN nanomaterial   85
folate   46, 47
framework nucleic acids (FNAs)   37, 124
  advantages   37, 38
  biological properties   40–41
  classification and construction   38–40
  delivery systems in biomedical application
    controlled drug release   49–51
    efficient drug delivery   46
    overcoming drug resistance   49–51
    targeted drug delivery   46–49
  DNA tetrahedron   39

fabrication and properties 39
physical and chemical properties 40
phytochemicals 45
small-molecule drugs 41
  antibiotic agents 44–45
  antitumor agents 42–45
  phytochemicals 45
functional nucleic acids (FNAs) 4
aptamers 5
DNA microchips 84
DNA nanobots 84
DNA origami 6, 83
DNA tile 83
DNA tiles 6
DNAzymes 4
three-dimensional DNA self-assembly 83–84
triplex DNA 5–6

g

G4 17, 18
gamma-modified PNA(γPNA) 17
glioma-specific peptide-modified tFNAs 49
glucagon 145, 146
glutaraldehyde-mediated chemical crosslinking, of oligolysine-PEG5K coated DNA nanostructures 21
graphene-encapsulated DNA nanostructures 22

h

HApt-tFNA 126
hybridization chain reaction (HCR) 101
based assembly nanoplatforms 113, 114
based in situ fluorescence imaging and biotherapy 102
based oligonucleotide drug delivery system 116
based tumor biotherapy 115
  chemotherapy 115
  photodynamic therapy 115–116
  RNA interfering therapy 116
components 102
hairpin design principles 102
miRNA detection 103

autocatalytic HCR biocircuit 103–104
nonlinear HCR system 104–105
multiple target detection 109
  combined HCR-based probe 109–110
  HCR-based logic gate 110–113
vs. polymerase chain reaction 102
protein detection
  antibody-based HCR system 107, 108
  aptamer-based HCR system 108–109
single-nucloetide variants detection 105–107
hybrid protein-DNA hydrogel 93

i

immunotherapy 2, 53, 123, 126, 127

l

label-free HCR-based PTD method 115
L-DNA 17
ligand-modified FNAs 46
L12-loaded DNA hydrogel 171
LncRNAs 65
locked nucleic acid (LNA) 20, 88, 187

m

magnetic bead nanoparticles 86
magnetic nanomaterials 86–87
magnetotaxis 87
melamine-DNA-AgNC complex 165–166
melittin-loaded DNA nanostructure 124
mesoporous silica nanoparticles (MSN) 85
messenger RNA (mRNA) 64, 65, 91, 92, 102, 103, 113, 126, 142, 146, 147, 149, 184, 187, 188, 192
metabolic diseases 61, 139–150
metal-based nanomaterials 7, 84–85, 87
metallo-β-lactamases (MβLs) 170
metal nanoclusters 85
methionine adenosyltransferase (MAT) 147
8-methoxypsoralen (8-MOP) 23
microfluidic technology 110, 117
micropinocytosis-related proteins 41
microRNA-155 loaded tFNAs 15

microRNAs (miRNAs)   126
  detection
    autocatalytic HCR biocircuit   103–104
    nonlinear HCR system   104–105
  mimics   188
mitochondrial DNA (mtDNA)   61
  clearing mutant   62–64
  inhibiting replication   64
mitochondria targeting   66–68
mitochondria treatment, nucleic acid-based delivery system in   68–71
molecular data storage   8
molecular nanotechnology   1
mtRNA
  increase normal RNA   64
  non-coding RNA treating   65
  silencing abnormal RNA   64–65
multifunctional DNA origami nanostructure   129
multiple target detection
  combined HCR-based probe   109–110
  HCR-based logic gate   110–113

n

naked plasmids   194
nanotechnology   1, 3, 4, 6, 37, 52, 81–83, 88, 103, 124, 139, 161, 164, 172
natural nucleic acids   3
NET-like DNA-HCl-ZnO nanogel   166
NET-like nanogel based on 2D DNA networks   166
nonalcoholic fatty liver disease (NAFLD)   140, 148, 149
nongenetic nucleic acids   4, 141
nonlinear HCR system   104–105, 116
nonsteroidal anti-inflammatory drugs (NSAIDs)   183
Notch receptors   147
NOX-G15   145
nuclease degradation   16, 20, 21, 26
nucleic acid-based amplification reaction   101
nucleic acid-based antitumor treatment   101
nucleic acid-based functional nanomaterials   82
  bionanomaterials   86
    in bone tissue repair and regeneration   89
    bioimaging   93
    bone targeting   91–92
    scaffold material   92–93
    sustained-release effect   89–91
  carbon-based   85
  characteristics   84
  composite nanomaterials   87
  magnetic nanomaterials   86–87
  metal-based   84–85
  quantum dots   86
nucleic acid-based gene therapy   141
nucleic acid-based nanomaterials (NAN)
  improve stability methods   15
    artificial nucleic acids   17–18
    coating with protective structures   20–22
    construction   26
    covalent crosslinking   22–24
    improve stability   15
    nucleobase/ribose modification   19–20
    phosphate group modification   18–19
    tuning buffer conditions   23, 25
  two-or three-dimensional nanostructures   15
nucleic acid, description   3
nucleic acid nanomaterials   123, 139, 141, 150
  advances in endocrine and metabolic diseases   143
  application directions   82
  development of   124, 125
  for diabetes mellitus   144–146
  for nonalcoholic fatty liver disease   148–149
  for obesity   147–148
  for osteoporosis   146–147
  properties and applications
    DNA origami   127–129
    dynamic DNA nanostructure   130–133
    tFNAs   125–127
nucleic acid nanoparticles, as carriers for OA therapy   194

nucleic acid nanotechnology   82, 83, 124, 139, 141
nucleic acid sequences   6, 15, 62

o

obesity   139, 140, 147–149, 183
oral hypoglycemic agents   144
osteoarthritis (OA)   181
    nucleic acid nanomaterials-based therapy   184
    pathology of   181–182
    risk factors   183
    therapy challenges   183
    vector-dependent   188–194
    vector-independent   184–188
osteoporosis   81, 91, 93, 146–147
overcoming drug resistance   43, 51, 52

p

PEG-DNA structure   20
PEG-protamine-tetrahedral framework DNA (PPT) complex   172
penicillin-binding proteins (PBPs)   168
peptide-decorated FNAs   49
peptide nucleic acids (PNA)   17, 64, 69, 88
pharmacological therapy   140
photochemical crosslinking   23
photodynamic therapy (PDT)   43, 44, 68, 70, 102, 115–116, 123, 127
photothermal therapy (PTT)   7, 43, 46, 89, 130
pH-sensitive FNA design   51
phytochemicals   45, 49, 148
plasmids   93, 184, 192, 194
point of care (POC) detection   102
polymerase chain reaction (PCR)   102, 142
polymeric DNA-based nanomaterials   2
proinflammatory chemokine C-C motif-ligand 2   145
protein-coated DNA origami   20
protein detection
    antibody-based HCR system   107, 108
    aptamer-based HCR system   108–109
protein kinase A (PKA) phosphorylates sub-peptide   113

protein tyrosine phosphatase-1B (PTP1B)   146
proteomic identification method   41

q

quantitative polymerase chain reaction (qPCR)   168
quantum dots   84, 86, 87, 93, 94, 108

r

recombinase polymerase amplification (RPA)   106, 107
resveratrol   45
ribonucleic acid (RNA)   3
    interfering therapy   116

s

scaffold material, for bone regeneration   92–93
SELEX *in vitro* screening technique   4
SELEX system   148
self-assembled 3D FNA nanostructures   7
self-assembled DNA nanostructure   38
self-assembled DNA nanotechnology   88
self-assembled FNAs   40
self-assembled nucleic acid nanostructures   88
serine/threonine protein kinase (STK)25   149
short tube DNA origami (STDO)   21
single-nucleotide variants (SNV) detection   105–107
siRNA-templated 3D spherical FNAs   50
small interfering RNAs (siRNAs)   126
    nucleic acid drugs   141
    vector-dependent nucleic acid nanomaterials for OA therapy   188, 192
small-molecule drugs   41
    antibiotic agents   44–45
    antitumor agents   42–45
        chemotherapeutic drugs   42–43
        phototherapeutic agents   43–44
    phytochemicals   45
sodium glucose cotransporter-2 (SGLT2)   146

spherical nucleic acids (SNAs)   91, 148, 149
stimuli-sensitive FNAs   51
sub-peptide modified silica nanoparticles   113
super silver nanoclusters based on branched DNA   164–165
systematic evolution of ligands by the exponential enrichment (SELEX) process   49
systemic drug delivery, of gene-related therapeutics   142

t

tensegrity triangle   124
tetrahedral DNA-based nanomaterials (TDNs)   2, 9
tetrahedral framework DNA   166
   nucleic acid antibiotics delivery   168–169
   polypeptide antibiotics delivery   169
   traditional antibiotics delivery   168
tetrahedral framework nucleic acids (tFNAs)   15, 124, 144, 145
   ampicillin-loaded   44
   angiogenic peptide loaded   15
   aptamer AS1411-modified   41
   aptamers   126
   clindamycin-tFNAs complex   15
   curcumin-loaded   15
   erythromycin-loaded   44
   glioma-specific peptide-modified   49
   HApt   126
   microRNA-155 loaded   15
   nanostructures   88
   nucleic acid nanomaterials   125–127

tetrahedron-loaded hairpins   113
TE20-type kinase MST3   149
3D DNA-based nanomaterials   8, 9
toehold-mediated strand displacement (TMSD)   103
"top-down" approach   1
traditional bone tissue regeneration treatment methods   81
transferrin   129
triplex DNA   5–6
triplex-forming oligonucleotides (TFOs)   5
tumor cell membrane antigen detection   107

v

vector-dependent nucleic acid nanomaterials, for OA therapy   189–191
   cDNA   192
   CircRNA   192, 194
   miRNA mimics   188
   mRNA   192
   small interfering RNA   188, 192
vector-independent nucleic acid nanomaterials, for OA therapy   184
   antisense oligonucleotides   187
   aptamers   187–188
   tFNA
      biodistribution of   185
      therapeutic effect of   184–185, 187
VicK protein-related gene   168
vildagliptin   145
vitamin B12 (VB12)   46

w

wild-type mtRNA (WT-mRNA)   64